THE PHILOSOPHY AND HISTORY OF MOLECULAR BIOLOGY:
NEW PERSPECTIVES

BOSTON STUDIES IN THE PHILOSOPHY OF SCIENCE

Editor

ROBERT S. COHEN, *Boston University*

Editorial Advisory Board

THOMAS F. GLICK, *Boston University*
ADOLF GRÜNBAUM, *University of Pittsburgh*
SYLVAN S. SCHWEBER, *Brandeis University*
JOHN J. STACHEL, *Boston University*
MARX W. WARTOFSKY, *Baruch College of the City University of New York*

VOLUME 183

THE PHILOSOPHY AND HISTORY OF MOLECULAR BIOLOGY: NEW PERSPECTIVES

Edited by

SAHOTRA SARKAR
McGill University, Montréal, Québec, Canada

KLUWER ACADEMIC PUBLISHERS
DORDRECHT / BOSTON / LONDON

Library of Congress Cataloging-in-Publication Data

```
The philosophy and history of molecular biology : new perspectives /
  edited by Sahotra Sarkar.
      p.   cm. -- (Boston studies in the philosophy of science ; v.
183)
   Based on papers from two conferences, the first held in Boston,
April 14-15, 1991, and the second held in New York, N.Y., Jan.
16-1993; with new papers added.
   Includes bibliographical references and index.
   ISBN 0-7923-3947-9 (hb : alk. paper)
   1. Molecular biology--Philosophy--Congresses.  2. Molecular
biology--History--Congresses.   I. Sarkar, Sahotra.  II. Series.
Q174.B67  vol. 183
[QH506]
001'.01 s--dc20
[574.8'9'01]                                                  96-153
```

ISBN 0-7923-3947-9

Published by Kluwer Academic Publishers,
P.O. Box 17, 3300 AA Dordrecht, The Netherlands.

Kluwer Academic Publishers incorporates
the publishing programmes of
D. Reidel, Martinus Nijhoff, Dr W. Junk and MTP Press.

Sold and distributed in the U.S.A. and Canada
by Kluwer Academic Publishers,
101 Philip Drive, Norwell, MA 02061, U.S.A.

In all other countries, sold and distributed
by Kluwer Academic Publishers Group,
P.O. Box 322, 3300 AH Dordrecht, The Netherlands.

Printed on acid-free paper

All Rights Reserved
© 1996 Kluwer Academic Publishers
No part of the material protected by this copyright notice may be reproduced or
utilized in any form or by any means, electronic or mechanical,
including photocopying, recording or by any information storage and
retrieval system, without written permission from the copyright owner.

Printed in the Netherlands

TABLE OF CONTENTS

SAHOTRA SARKAR / Philosophy, History, and Molecular Biology – Introduction	1
JOSHUA LEDERBERG / What the Double Helix (1953) Has Meant for Basic Biomedical Science. *A Personal Commentary*	15
KENNETH F. SCHAFFNER / Theory Structure and Knowledge Representation in Molecular Biology	27
DORIS T. ZALLEN / Redrawing the Boundaries of Molecular Biology: The Case of Photosynthesis	47
RICHARD M. BURIAN / Underappreciated Pathways Toward Molecular Genetics as Illustrated by Jean Brachet's Cytochemical Embryology	67
LILY E. KAY / Life as Technology: Representing, Intervening, and Molecularizing	87
SCOTT F. GILBERT / Enzymatic Adaptation and the Entrance of Molecular Biology into Embryology	101
ALFRED I. TAUBER / The Molecularization of Immunology	125
JON BECKWITH / The Hegemony of the Gene: Reductionism in Molecular Biology	171
SAHOTRA SARKAR and DAVID S. THALER / Introductory Note to the Contributions by Sarkar and Thaler	185
SAHOTRA SARKAR / Biological Information: A Skeptical Look at Some Central Dogmas of Molecular Biology	187
DAVID S. THALER / Paradox as Path: Pattern as Map. *Classical Genetics as a Source of Non-Reductionism in Molecular Biology*	233
NAME INDEX	249

SAHOTRA SARKAR

PHILOSOPHY, HISTORY, AND MOLECULAR BIOLOGY – INTRODUCTION

1. When the logical empiricists reset the direction of philosophy of science in the 1920s and 1930s, the loci of their attention were mathematics (almost entirely mathematical logic) and physics (initially relativity, later quantum mechanics). It is doubtful – to say the least – whether this philosophical scrutiny and analysis resulted in any substantial contribution to either mathematics or physics. However, it not only set the agenda, but also the tone for the philosophy of science. The relatively simple axiomatic structure of relativity theory and quantum mechanics – at least how the professional philosophers viewed those disciplines – became the yardstick for the comparison of other disciplines. If they were found to be less general in their intended scope, to be using different criteria of rigor (that is, different from the type of mathematics used in relativity theory and quantum mechanics), or simply different, they were presumed to be wanting. This applied not only to biology or chemistry (or, for that matter, the social sciences) but even to other areas of physics, particularly what came to be called condensed matter physics which, arguably, went through a period of conceptual ferment (scaling, renormalization, etc.) that is as profound as the changes initiated by relativity and quantum mechanics.[1]

If the extent of attention, or competence of discussion, is used as a standard, biology fared better during the early decades of the logical empiricist regime, that is, from 1930 to 1950, than during the next fifteen years.[2] This is not only because many biologists – H. Driesch, J.B.S. Haldane, J.S. Haldane, L. Hogben, J. Loeb, and S. Wright (to mention some better-known examples) – explicitly espoused philosophical positions during those early decades. That certainly helped, but there were also some significant attempts by the philosophers themselves to come to terms with the exciting developments that had taken place in biology, particularly in (classical) genetics and evolution, during the first three decades of the 20th century. J.H. Woodger – perhaps best-known to the logical empiricists as the translator of Tarski's papers – attempted to axiomatize parts of genetics as early as 1937 though his axiomatization suffered from the worst excesses of early logical empiricism. Out of devotion of operationalism, he refused to introduce the "gene" in his axioms – it was far too theoretical a term for admission! Carnap (1958) developed some of Woodger's themes. By 1952, Woodger had clearly articulated what, after elaboration by Nagel (1949, 1951, 1961), came to be regarded as the standard model of theory reduction. Nagel used this model in an attempted explication of "mechanistic" explanation in biology that was not completely unreasonable. Less successfully, he even attempted to use it to provide a deflationary account of teleological explanation in biology (see, especially, Nagel (1961)).[3]

In the case of mechanistic explanation, Ernest Nagel achieved little more than Hogben (1930), as far as attention to substantive biological problems is concerned. All he did was translate the simplest of questions into the logical empiricists' framework. Following in an unfortunately well-established philosophical tradition, Nagel's writings on biology contributed virtually nothing that biologists – even philosophically-oriented theoretical biologists – found valuable. Worse than that, Nagel demonstrated a quite remarkable refusal to follow developments in modern biology.[4] Between 1949 and 1961 he saw no reason to temper his bleak assessment of the state of mechanistic/reductionist explanation in genetics – the events of 1953 either completely slipped by him, or had failed to impress him for some unknown reason. *The Structure of Science* (1961) has several sections devote to reductionism in biology, but makes no mention of the double helix or, for that matter, any other development in molecular biology that had brought the potential for reduction in biology to an entirely new level of sophistication. Nagel was a step backwards from Woodger, let alone Hogben. Meanwhile the philosophy of physics continued to be explored systematically. By the late 1960s and early 1970s, many philosophers such as Abner Shimony and Howard Stein had succeeded in raising philosophical discussions of physics to an entirely new level of sophistication. In comparison, the philosophy of biology lagged behind.

2. The situation only began to change in the late 1960s and 1970s. David Hull (1965, 1967, 1968) began to explore the conceptual structure of evolutionary biology. W.C. Wimsatt (1971, 1972) provided a detailed analysis of teleological explanation (and biological "feedback"), drawing extensively on recent work in theoretical biology – the analyses of the conceptual relations between fitness, evolutionary purpose, teleology and feedback that he provided have not needed any basic revision. In a series of papers, Kenneth Schaffner (1967a,b, 1969) began to argue the case for reductionism in molecular genetics while Hull (1972, 1974), questioned Schaffner's assessment. Ruse (e.g., 1971, 1973) and Wimsatt (1976) were among those who joined the debate. The general consensus was against reduction, provided that reduction was construed in the fashion that had been inherited from logical empiricism (as, for instance, it was construed by Schaffner). Using Salmon's (1971) account of statistical explanation, Wimsatt (1976) attempted an altogether novel approach to reduction. He also endorsed a sort of realism and freely talked of causes – philosophy of biology played its part, though only rather late, in the liberation from logical empiricism.

Since the late 1960s, philosophy of biology has had a continuous and increasingly prominent presence in the philosophy of science, though occasional abuse of biology by philosophers continued – as late as 1974, Popper would claim that Darwinism was not a scientific enterprise.[5] Over the years, the philosophy of biology has contributed to the development of the various alternatives to logical empiricism that have sprung up, including scientific realism, the semantic (loosely "model-based") conception of theories and, in particular, naturalized epistemology. The last of these developments has been particularly natural and fecund – to judge by the volume of work that has been produced – perhaps for the obvious reason that philosophers of biology are most likely to see how humans are evolutionarily produced, constrained and challenged as biological organisms.[6] In fact, barring a very few exceptions (including the present Editor), there is a general consensus among philosophers of biology of the great value of a naturalized perspective in philosophy, where "naturalism" is interpreted purely in evolutionary terms.[7] Moreover, philosophers of biology have routinely attempted to "do biology", that is to enter into the biological arena, without demanding any special status of permissible biological incompetence because they are, by profession, philosophers. If philosophy is to be done in continuity with science – as Quine would have it – no area within philosophy of science has followed that dictum more systematically (though with no discernible direct influence of Quine) than the philosophy of biology.

However, in the late 1970s, in an unfortunately constrictive development, the philosophy of biology became almost entirely concerned only with evolutionary theory. Even within evolutionary theory almost all philosophers of biology gravitated to a single problem – the units of selection. Philosophers provided some useful analyses of this (and related) problems (see, especially, Wimsatt (1980) and Lloyd (1988)), while some prominent biologists, including Richard Lewontin, Ernst Mayr and John Maynard Smith, also made important philosophical contributions.[8] However, the narrowness of the focus hurt the development of the discipline. In particular, much of the philosophical writing about biology from this period remained blissfully inattentive to molecular biology where, for better or for worse, most biological work had become concentrated, and where most of the significant developments in biology since 1950 took place. Perhaps the most blatant example of this kind of narrow philosophical focus is Sober's (1993) *Philosophy of Biology*, a book that is written as if nothing except evolutionary theory, not even molecular biology, existed. Sober even ignores the transformation wrought by molecular biology, once again for better or for worse, on the practice of evolutionary biology. Given this state of the field, it is hard not to sympathize with the molecular biologists' lack of concern for philosophical critiques of their enterprise, a lack that has often been noticeable during the debates surrounding the initiation of the Human Genome Project, a debate on which philosophers have exerted no perceptible influence.

3. Unlike the rather straightforward situation in the philosophy of biology, it is impossible – and beyond the professional competence of this Editor – to summarize all past trends in the history of biology. Restricting attention to molecular biology, however, it is easy to point to one very important beginning: Olby's (1974) *The Path to the Double Helix* was the first scholarly attempt by a historian to treat the origins of molecular biology.[9] Though further work has led to the inevitable reassessment of many of Olby's positions, this book laid out a chronology, and a consistent interpretation of that chronology, which served as a point of departure for many, though not all, subsequent histories. It provided a detailed analysis of the structural studies that led to the search for the structure of proteins and nucleic acids. At times it seemed to suggest – though never explicitly – that these structural studies alone led to the double helix model of DNA, the event that this history was written to celebrate. Perhaps even more troublesome is the book's lack of relative attention to classical genetics, how it was part of the transformation that led to the establishment of molecular biology. This seems to have been due to the undue influence of the British figures involved, and to their historiography (see e.g., Kendrew (1967), which will be mentioned again in this Introduction).

The 1970s also saw the publication of another useful book, Judson's (1979) *The Eighth Day of Creation*, which provided a breathtaking account of the emergence of molecular biology. It was a journalistic tour-de-dorce, even more triumphalist in tone than Olby's book, and broke new ground by carrying the story beyond 1953 (where Olby had stopped) to the mid 1960s. It relied heavily on recorded interviews – this is the books' greatest strength. The interviews provided the widely differing perspectives of most of the prominent scientists involved in the development of molecular biology. Even if it does nothing else, Judson's book, and his recorded interviews (which were deposited with the American Philosophical Society), will continue to be important primary sources for other historians of molecular biology. Beyond that, the book supported – though, unfortunately, not with much detailed argument – a provocative thesis, that the modern development of biology was basically an attempt to explicate the concept of biological specificity, a concept that went back to the work of Ehrlich in the 1890s.

It is surprising, given the format of his book, that Judson did not pursue his history of molecular biology any further, beyond the mid-1960s, into the 1970s when, starting with the invention of recombinant DNA techniques, molecular biology enjoyed a whole new spate of discoveries and technical achievements. What is perhaps even more surprising is that these stories continued to be largely shunned by professional historians of biology. Krimsky's (1982) *Genetic Alchemy*, an account of the recombinant DNA controversy, is a remarkable exception, at least as far as the political developments are concerned (see, also, Yoxen, 1983). The recent industry of books devoted to the Human Genome Project has taken up some of the slack but these are all popularizations and cannot be adequate substitutes for careful histories. In fact, as will be emphasized below, the history of molecular biology since the mid-1960s remains generally unexplored territory. Moreover, there was a general lack of attention to molecular biology from any period by professional historians of science prior to the mid-1980s (with Olby being, of course, the notable exception).

4. The late 1980s and 1990s, however, have seen a renewed attention on the part of philosophers, and a new and increasing interest, on the part of historians and the other practitioners of the new discipline of "science studies", in molecular biology. Along with this new interest has come a greater sophistication, mainly technical (that is, scientific) on the part of the philosophers, and methodological on the part of the historians. In philosophy, this sophistication is already apparent in Rosenberg's (1985) *The structure of Biological Science*, the first book-length explicitly philosophical work that took at least some of the advances of molecular biology into account. Schaffner's recent (1993) *Discovery and Explanation in Biology and Medicine* continues this important trend, and does so with remarkable breadth. Interest in the units of selection certainly has not disappeared but it (and evolutionary biology in general) is no longer the only philosophical obsession. Philosophy of biology today is broader than it has been at any point during the last two decades. This increase in breadth is also a result of increasing – and welcome – philosophical attention to ecology (see, e.g., Shrader-Frechette and McCoy, 1993), which probably reflects the growing social concern with the rampant destruction of biological diversity.

The new historical work on molecular biology has led to many important collections, though comprehensive book-length single-author works are yet to appear.[10] At the cost of simplifying a complex web of developments, at least five broad approaches can be distinguished:

(i) an attention to intellectual detail and context. This is the most conventional of the approaches used in recent years – both Olby (1974) and Judson (1979) can be regarded as using it – but it should be emphasized that this approach, when used exclusively, is not consonant with the "externalist" historiographies that are currently the height of fashion. Nevertheless, especially when used along with other approaches, it has many insights to offer – see Keller (1992, 1995), as well as any of the historical papers in this volume;

(ii) there has been serious concern with experimental issues, including the choice of laboratory systems. For instance, considerable attention has been paid to the process by which certain organisms and molecules such as phage, *Escherichia coli*, and hemoglobin became established in molecular biology laboratories – see, for example, Kevles (1993),

Lederman and Tolin (1993), Sarkar (1991), Summers (1993), and Zallen (1993);

(iii) there has been an increasing attention to the social/cultural matrix of laboratory and research group interactions (see, for example, Rheinberger (1993) and, especially, the forthcoming works of Creager and Gaudillière (e.g., 1995)). Broadly, this might be called an anthropological approach to the history of science;

(iv) there has been attention to national (and other) styles of research. While Anglo-American traditions have received the most attention, the French (see, e.g., Burian et al., 1988; Gaudillière, 1992, 1993) and Japanese (Uchida, 1993) programs in molecular biology have also been analyzed. It should be emphasized that, in general, "national" in this context is a (philosophically) pragmatic geographical/political designation rather than a (philosophically) categorical one;

(v) there has been some focus on the ideological constraints and forces on the development of biology. The approaches to ideology that these efforts use go beyond traditional marxism, whose direct influence seems to have decreased as other modes of analysis took over the discussion of "externalist" factors in science. Feminist analyses have been the most prominent of these, but there are others (see Abir-Am, 1993; Kay, 1992). However, it is probably uncontroversial to say that these analyses have so far been less prevalent in the history of molecular biology than, for instance, in the history of evolutionary theory.

These new approaches have not, of course, been confined to the history of molecular biology. They represent general trends in the history of science. They also show the extent of philosophical engagement on the part of the historians – the different historiographical choices reflect strong philosophical commitments. In particular, each of the approaches except the first represents at least a partial rejection of that scientific realism which the philosophers of biology helped usher in! Meanwhile, of course, the history of science is routinely invoked by the same philosophers to argue for scientific realism. On philosophy, there is little agreement between the philosophers and historians. In the Editor's opinion, so far the historians have had the better of this exchange over realism.

The historians who have contributed to this volume have used most of these approaches, sometimes several at the same time, both here and in other works: for instance, Burian and Zallen have used (iv); Burian et al. have used (ii); Gilbert and Kay have used (v); and all have used (i) at some point or the other. Among the other works that have already been mentioned, those of Angela Creager and Jean-Paul Gaudillière (1995), who have provided a scrupulous reconstruction of Monod's work in the 1950s, and Evelyn Keller (1995), who has explored the role of various metaphors in the development of twentieth-century biology, are particularly important.

INTRODUCTION 7

5. This volume emerged primarily from two conferences. The first was a two-day event, with the same title as this volume, held under the auspices of the Boston Colloquium for the Philosophy of Science on April 14–15, 1991. This conference brought together philosophers, historians, and four scientists with historical or philosophical interests (broadly construed). Both philosophers and historians were deliberately included with a hope that this would help clarify their differences. Not all the contributions at the symposium emerged in publishable format. Contributions by Pnina Abir-Am, Walter Gilbert, Matthew Meselson and Robert Olby are not represented here, while those by Joshua Lederberg and Sahotra Sarkar are not the papers delivered at the conference. Lederberg's paper is included because of its obvious relevance to the theme of this volume. Sarkar's paper, and that by David Thaler, emerged from a session of the Conference on Methods in Philosophy and the Sciences, New York (January 16, 1993). Finally, Tauber's paper was written independent of both events. The other papers were delivered in (roughly) this form at the Boston conference.

6. The Boston conference was quite successful not only because it generated lively discussion among the participants but also because, at least in the opinion of this Editor, many points of consensus about the history of molecular biology emerged. This probably indicates that the field has achieved a certain maturity. Unfortunately, with the exception of reductionism, most philosophical issues, including scientific realism, did not receive much attention – in this respect, the conference was a failure. It is impossible to summarize the papers or the discussions, which were much more varied than in most conferences. Suffice it here, merely to note ten points of consensus that emerged at the Boston conference, and one additional point about which there probably is consensus, that emerged from the New York meeting. An attempt has been made to indicate how these points are related to the papers included in this volume. There is also a separate introductory note to the last two papers:

(i) What constitutes molecular biology should be construed *much* more broadly than what has conventionally been done by both philosophers and historians. perhaps the only guideline for demarcating the boundaries of molecular biology is that research is guided by an exploration of interactions at the molecular or sub-molecular level. However, if this characterization is pushed to its extreme, there is a problem. Since all of biology seems to be using molecular techniques, is there any "non-molecular biology" left?;

(ii) In particular, it should be emphasized that there is more to molecular biology than molecular genetics. Among philosophers, at least, the increasing dominance of genetics over the rest of biology has tended to obscure this otherwise obvious point. The early discussions of reductionism in molecular biology (e.g., Hull, 1972, 1974; Schaffner, 1967b, 1969, 1974), and even some of the more recent ones (Kitcher, 1984; Waters, 1990), have been particularly guilty of treating molecular biology as if it consisted solely of molecular genetics. (There is, however, an ironic counterpoint to this – classical genetics has not been given the kind of philosophical attention it deserves (see (xi) below).);

(iii) Similarly, there were many independent pathways to molecular biology. In particular, the traditional view going back to Kendrew (1967), and implicitly endorsed by Olby (1974) and Judson (1979), that there were two schools – the functional/informational school associated with Delbrück and the Phage Group (and its intellectual descendants), and the structural school of Perutz and the MRC – needs radical revision. The papers by Burian and Zallen in this volume underscore all of these points;

(iv) In fact, what remains unclear is whether there is any part of classical biochemistry that does not fall within the domain of molecular biology.[11] The boundaries of molecular biology appear to be blurred but this might simply be a reflection of the fact that research in molecular biology has always favoured research programs that attempted to push forward its frontiers. (Should the locution "molecular biology" reflect only methodology, and not subject?) But then, the problem "What *is* molecular biology?" returns with renewed force;

(v) In the same vein, to a large extent, research in molecular biology has involved the "molecularization" of some existing subdiscipline of biology. Genetics came first. Gilbert explores the case of embryology in this volume; Tauber takes up immunology. It is at least arguable that the recent attention to DNA sequence evolution represents a similar process in evolutionary biology;

(vi) Whether such a "molecularization" constitutes a scientific advance remains an open question. Both Gilbert and Tauber implicitly suggest that molecularization led to a myopic redefinition of important biological questions which, in fact, prevented the development of some promising and important lines of inquiry. The conventional view, of course, is that molecularization constitutes not only a technical, but even a conceptual, advance. Which view is more apposite remains an open and, in some ways, unanswerable question at present simply because it is still far too early for the full impact of molecular biology to have been felt on these areas, let alone be assessed. Meanwhile, it is hard not to suspect that vehement unequivocal answers to this question (from either side) reflect little more than prejudice induced by professional self-interest;

(vii) There was a general feeling – though certainly not a consensus – that understanding molecular biology requires an acknowledgment that it is an *intensely* theoretical science, but with few significant well-articu-

INTRODUCTION 9

lated theories. This point – at least the part that molecular biology has few, if any, theories – seems to have been first articulated by Hull (1972), and developed by Wimsatt (1976) and Sarkar (1989). It is explicitly developed in this volume – in different ways – by Burian and Sarkar. Meanwhile, Schaffner's contribution shows that there is no consensus on this point;

(viii) It is far from clear that classical (Nagel-Woodger type) accounts of reduction are even interesting in the context of molecular biology. To some extent, because these classical accounts view *reduction* as a type of inter-theoretic explanation, this point follows from point (vii). However, it is possible that reduction could take place from biology to the theories of physics and chemistry (as mechanists/reductionists from the pre-molecular era would have proposed). Nevertheless, most participants at the meeting, though not all, were less than interested in pursing this line of analysis;

(ix) There should be more exploration of the theoretical concepts and principles used in molecular biology, that is, those theories and concepts that make it an intensely theoretical science. Where these explorations will go is hard to say. Sarkar's contribution to this volume expresses skepticism about the value of "information", at least in the forms in which that concept has so far been interpreted. Keller (1995) develops what is, at least implicitly, an even more skeptical attitude, not only about "information", but about many of the other organizing concepts of molecular biology. Obviously, these arguments have important deflationary relevance for the types of scientific realism that are popular among philosophers of biology;

(x) To mention a very obvious point, the history of molecular biology since the late 1960s remains virtually unexplored. There was some discussion at the conference about whether the character of molecular biology has changed during that period. It was also clear that attempts to reconstruct that history will require significantly greater technical competence in molecular biology than that which was necessary for previous periods;

(xi) Finally, Thaler's contribution to this volume makes the important point, that the relation of genetics to molecular biology is not quite what philosophers and historians often assume. Briefly, two independent themes emerge from Thaler's discussion: (a) that classical genetics as an important source from which molecular biology emerged in the 1950s; and (b) in an interesting sense, classical genetics was not reductionist. Though these were not themes addressed at the Boston meeting, they are consistent with, and help explain, several of the points mentioned above. In particular, they also demonstrate a need for breadth in exploring the origins of molecular biology, and point out how the classical analyses of reductionism avoided important issues. But, perhaps, what is most surprising about Thaler's point is that it needed to be made at all. Presumably no one would deny the significance of genetics to biology in the 1920s and 1930s. Similarly, it is hard to doubt that molecular genetics has begun to establish a hegemony over molecular biology, if not biology as a whole. As has been noted before, philosophers entered into this game early and treated molecular biology as if it consisted entirely of molecular genetics. Yet, classical genetics has not received the kind of attention it deserves, and its role in the genesis of molecular biology has been surprisingly ignored – Thaler's reminder is important.

ACKNOWLEDGMENTS

The Boston conference was funded in part by a grant from the Dibner Institute for the History of Science and Technology. Thanks are due to Robert S. Cohen, Guido Sandri and Robert Tamarin for chairing sessions, and to Tracy Lubas and Deborah Wilkes for help with organization and publicity. Thanks are due to Marx Wartofsky for organizing the New York conference from which some of the contributions to this volume emerged.

McGill University,
Montréal, Québec, Canada

NOTES

[1] See Leggett (1987) and Shimony (1987) for developments of these themes. Note that these developments in condensed matter physics only occurred in the 1950s and 1960s – the first generation of logical empiricists cannot fully be blamed for ignoring them.
[2] There are two notable exceptions to this assessment – Beckner (1959) and Goudge (1961) – both of whom deserve more attention than they have received from the current crop of philosophers of biology. Other exceptions include Grene (1959) and Lehman (1965).
[3] For a relatively sympathetic account of Nagel's achievements, see Sarkar (1989).
[4] At least Nagel, unlike Popper (1974) did not denigrate Darwinism (see below).
[5] See Popper (1974). This claim is retracted later, in Popper (1979).
[6] See Callebaut (1993) for a detailed examination of this issue.
[7] To the present Editor, this has been a continual source of mystery. If one takes biology, especially evolutionary theory, seriously, what biology can explain about human culture (including science) is minuscule compared to what it cannot. One would suppose that philosophers of biology would be more attentive to this point than either biologists or other types of philosophers. They are not. What is even more odd is that the same philosophers who expound evolutionary naturalism in one context – in science or mathematics (e.g., Wimsatt or Kitcher) – go on to reject sociobiology which, arguably, requires less adaptationist faith than evolutionary epistemology.
[8] The most important early papers were collected – and sometimes excerpted – by Brandon and Burian (1984). For later work, see Lloyd (1988), the most important book-length philosophical account of evolutionary theory written so far. For a response to Lloyd's point of view, see Sarkar (1994).

[9] In fact, its only significant predecessor was Carlson's (1976) *The Gene: A Critical History* which, in the limited context of molecular genetics, traced parts of molecular biology into the mid-1960's.

[10] A particularly valuable collection is the Fall 1993 issue of the *Journal of the History of Biology*, which grew out of the second Mellon Workshop at MIT (April 3–4, 1992). Keller (1995) will probably be the first book to exemplify the point being developed here.

[11] There was some discussion at the conference of the claim that biochemistry was more "dynamic" and more "system-oriented" than what came to be called "molecular biology". It can also be argued that biochemistry has always been more coherent with standard physics and chemistry than molecular biology, with the latter's reliance on rather novel concepts such as that of "information" (see Sarkar's contribution to this volume). However, there was no explicit consensus on these points. The point needs much more philosophical exploration.

REFERENCES

Abir-Am, P. (1993). 'The Politics of Macromolecules: Molecular Biology, Biochemists and the Biomolecular Revolution'. *Osiris* **7**: 164–99.
Beckner, M. (1959). *The Biological Way of Thought*. New York: Columbia University Press.
Brandon, R. and Burian, R. (eds.) (1984). *Genes, Organisms, Populations: Controversies Over the Units of Selection*. Cambridge, MA: MIT Press.
Burian, R., Gayon, J. and Zallen, D. (1988). 'The Singular Fate of Genetics in the History of French Biology'. *Journal of the History of Biology* **21**: 357–402.
Callebaut, W. (1993). *Taking the Naturalistic Turn*. Chicago: University of Chicago Press.
Carlson, E.A. (1966). *The Gene: A Critical History*. Philadelphia: W.B. Sauders.
Carnap, R. (1958). *Introduction to Symbolic Logic and Its Applications*. New York: Dover.
Clarke, A.E. and Fujimura, J. (eds.) (1992). *The Right Tools for the Job*. Princeton: Princeton University Press.
Creager, A. and Gaudillière, J.-P. (1995). 'Meaning in Search of Experiments or *Vice-Versa*: The Invention of *Allosteric Regulation* in Paris and Berkeley'. Draft.
Gaudillière, J.-P. (1992). 'J. Monod, S. Spiegelman et l'adaptation enzymatique: programmes de recherche, cultures locales et traditions disciplinaires'. *History and Philosophy of the Life Sciences* **14**: 29–98.
Gaudillière, J.-P. (1993). 'Molecular Biology in the French Tradition? Redefining Local Traditions and Disciplinary Patterns'. *Journal of the History of Biology* **26**: 473–98.
Goudge, T.A. (1961). *The Ascent of Life*. Toronto: University of Toronto Press.
Grene, M. (1959). 'Two Evolutionary Theories'. *British Journal for the Philosophy of Science* **9**: 110–27, 185–93.
Hogben, L. (1930). *The Nature of Living Matter*. London: Kegan Paul.
Hull, D. (1965). 'The Effects of Essentialism on Taxonomy: Two Thousand Years of Stasis'. *British Journal for the Philosophy of Science* **15**: 314–26.
Hull, D. (1967). 'Certainty and Circularity in Evolutionary Taxonomy'. *Evolution* **21**: 174–89.
Hull, D. (1968). 'The Operational Imperative: Sense and Nonsense in Operationism'. *Systematic Zoology* **17**: 438–57.
Hull, D. (1972). 'Reductionism in Genetics – Biology or Philosophy?'. *Philosophy of Science* **39**: 491–9.
Hull, D. (1974). *Philosophy of Biological Science*. Englewood Cliffs: Prentice-Hall.
Judson, J.F. (1979). *The Eighth Day of Creation*. New York: Simon and Schuster.
Kay, L. (1992). *The Molecular Vision of Life: CalTech, the Rockefeller Foundation, and the Rise of New Biology*. New York: Oxford University Press.
Keller, E.F. (1992). *Secrets of Life/Secrets of Death*. London: Routledge.

Keller, E.F. (1995). *Refiguring Life: Metaphors of Twentieth Century Biology*. New York: Columbia University Press.
Kendrew, J.C. (1967). 'How Molecular Biology Was Started'. *Scientific American* **216**: 141–4.
Kevles, D.J. (1993). 'Renato Dulbecco and the New Animal Virology: Medicine, Methods, and Molecules'. *Journal of the History of Biology* **26**: 409–42.
Kitcher, P. (1984). '1953 and All That. A Tale of Two Sciences'. *Philosophical Review* **93**: 335–73.
Krimsky, S. (1982). *Genetic Alchemy: The Social History of the Recombinant DNA Controversy*. Cambridge: MIT Press.
Lederman, M. and Tolin, S.A. (1993). 'OVATOOMB: Other Viruses and the Origins of Molecular Biology'. *Journal of the History of Biology* **26**: 239–54.
Leggett, A.J. (1987). *The Problems of Physics*. Oxford: Oxford University Press.
Lehman, H. (1965). 'Functional Explanation in Biology'. *Philosophy of Science* **32**: 1–20.
Lloyd, E.A. (1988). *The Structure and Confirmation of Evolutionary Theory*. Westport: Greenwood Press.
Nagel, E. (1949). 'The Meaning of Reduction in the Natural Sciences'. In: Stauffer, R.C. (ed.), *Science and Civilization*. Madison: University of Wisconsin Press, pp. 99–135.
Nagel, E. (1951). 'Mechanistic Explanation and Organismic Biology'. *Philosophy and Phenomenological Research* **11**: 327–38.
Nagel, E. (1961). *The Structure of Science*. New York: Harcourt, Brace and World.
Olby, R.C. (1974). *The Path to the Double Helix*. Seattle: University of Washington Press.
Popper, K. (1974). 'Intellectual Autobiography'. In: Schilpp, P.A. (ed.), *The Philosophy of Karl Popper*. La Salle: Open Court, pp. 3–181.
Popper, K. (1979). *Objective Knowledge*. Oxford: Clarendon.
Rheinberger, H.-J. (1993). 'Experiment and Orientation: Early Systems of in Vitro Protein Synthesis'. *Journal of the History of Biology* **26**: 443–71.
Rosenberg. (1985). *The Structure of Biological Science*. Cambridge: Cambridge University Press.
Ruse, M. (1971). 'Reduction, Replacement, and Molecular Biology'. *Dialectica* **25**: 39–72.
Ruse, M. (1973). *The Philosophy of Biology*. London: Hutchinson.
Salmon, W. (1971). *Statistical Explanation and Statistical Relevance*. Pittsburgh: University of Pittsburgh Press.
Sarkar, S. (1989). 'Reductionism and Molecular Biology: A Reappraisal'. Ph.D. Dissertation, Department of Philosophy, University of Chicago.
Sarkar, S. (1991). 'Lamarck *contre* Darwin, Reducation versus Statistics: Conceptual Issues in the Controversy over Directed Mutagenesis in Bacteria'. *Boston Studies in the Philosophy of Science* **129**: 235–71.
Sarkar, S. (1992). 'Models of Reduction and Categories of Reductionism'. *Synthese* **91**: 167–94.
Sarkar, S. (1994). 'The Selection of Alleles and the Additivity of Variance'. In: Hull, D., Forbes, M. and Burian, R.M. (eds.), *PSA 1994: Proceedings of the 1990 Biennial Meeting of the Philosophy of Science Association*. East Lansing: Philosophy of Science Association, vol. 1, pp. 3–12.
Schaffner, K.F. (1967a). 'Approaches to Reduction'. *Philosophy of Science* **34**: 137–47.
Schaffner, K.F. (1967b). 'Antireductionism and Molecular Biology'. *Science* **157**: 644–7.
Schaffner, K.F. (1969). 'The Watson-Crick Model and Reductionism'. *British Journal for the Philosophy of Science* **20**: 235–48.
Schaffner, K.F. (1993). *Discovery and Explanation in Biology and Medicine*. Chicago: University of Chicago Press.
Shimony, A. (1987). 'The Methodology of Synthesis: Parts and Wholes in Low-Energy Physics'. In: Kargon, R. and Achinstein, P. (eds.), *Kelvin's Baltimore Lectures and Modern Theoretical Physics*. Cambridge, MA: MIT Press, pp. 399–423.
Shrader-Frechette, K.S. and McCoy, E.D. (1993). *Method in Ecology: Strategies for Conservation*. New York: Cambridge University Press.

Sober, E. (1993). *Philosophy of Biology*. Boulder: Westview Press.
Summers, W.C. (1993). 'How Bacteriophage Came to Be Used by the Phage Group'. *Journal of the History of Biology* **26**: 255–67.
Uchida, H. (1993). 'Building a Science in Japan: The Formative Decades of Molecular Biology'. *Journal of the History of Biology* **26**: 499–517.
Waters, C.K. (1990). 'Why the Anti-Reductionist Consensus Won't Survive the Case of Classical Mendelian Genetics'. In: Fine, A., Forbes, M. and Wessels, L. (eds.), *PSA 1990: Proceedings of the 1990 Biennial Meeting of the Philosophy of Science Association*. East Lansing: Philosophy of Science Association, vol. 1, pp. 125–39.
Wimsatt, W.C. (1971). 'Some Problems with the Concept of "Feedback"' [Proceedings of the 1970 Meeting of the US Philosophy of Science Association]. *Boston Studies in the Philosophy of Science* **8**: 241–56.
Wimsatt, W.C. (1972). 'Teleology and the Logical Structure of Function Statements'. *Studies in the History and Philosophy of Science* **3**: 1–80.
Wimsatt, W.C. (1976). 'Reductive Explanation: A Functional Account' [Proceedings of the 1974 Meeting of the US Philosophy of Science Association]. *Boston Studies in the Philosophy of Science* **32**: 671–710.
Wimsatt, W.C. (1980). 'Reductionist Research Strategies and Their Biases in the Units of Selection Controversy'. *Boston Studies in the Philosophy of Science* **60**: 213–59.
Woodger, J.H. (1937). *The Axiomatic Method in Biology*. Cambridge: Cambridge University Press.
Woodger, J.H. (1952). *Biology and Language*. Cambridge: Cambridge University Press.
Yoxen, E. (1983). *The Gene Business, Who Should Control Biotechnology?* New York: Harper and Row.
Zallen, D.T. (1993). 'The "Light" Organism for the Job: Green Algae and Photosynthesis Research'. *Journal of the History of Biology* **26**: 269–79.

JOSHUA LEDERBERG

WHAT THE DOUBLE HELIX (1953) HAS MEANT FOR BASIC BIOMEDICAL SCIENCE*

A Personal Commentary

The article published by Watson and Crick in 1953 (Watson and Crick, 1953a) was the landmark pointer to our contemporary model of DNA as a macromolecular structure. This lay on a well-worn path of biophysical analysis, reducing microscopic anatomy to the molecular level. It also helped inspire an enormous body of biochemical research that has defined DNA as *the* informational molecule, a discontinuity that has been labeled the Biological Revolution of the 20th Century. As a piece of structural analysis, the idea of the double helix includes the concepts (1) that DNA is a duplex structure, comprising two paired complementary strands, associated by secondary, non-covalent bonds; (2) that the strand pairs are coiled, forming a double helix; and (3) that these are antiparallel – the orientation of one strand being in the opposite polarity from the other.

The most novel features of DNA are associated with its duplicity, rather than its helicity. Linear polymers rarely form stiff straight rods; folding into coils is the norm. The genetic functions of DNA are inextricably associated with its duplex structure, and hardly at all with its helical shape; this is reflected in the preoccupation of DNA research with its role as an informational molecule. However, we shall see a recent concentration of interest in supercoiling. Inevitably, the biochemical interactions of DNA with other molecules, be they regulatory proteins or chemotherapeutic inhibitors, will often be intimately wound up with the precise three-dimensional conformation of the helix. This is also proxy for higher orders of coiling, interactions with histones and other DNA-binding proteins, and the organization of DNA into chromosomes.

DNA can be built in either an antiparallel or a parallel format, although the former adds a note of symmetry that may account for the prevalence of the antiparallel in nature. For parallel DNA a different enzyme would be needed to recognize and replicate the left-compared with the right-hand end. Recognizing this asymmetry, Watson and Crick (1953a) speculated that DNA was antiparallel prior to concrete observational evidence for this conformation.

Rarely has a structural determination been coupled so promptly with functional implications. Watson and Crick (1953a) immediately inferred that DNA duplexes were formed automatically when each strand was replicated, and that this involved the assembly of nucleotides, one by one, complementary

---

* Reprinted with kind permission from *JAMA* **269**: 1981–1985. © 1993 American Medical Association.

to the existing structure (Watson and Crick, 1953b). They overreached the mark by suggesting that this might be possible even without the intervention of specific anabolic enzymes, the discovery of which we owe to the prodigious labors of Arthur Kornberg and his school in the 1960s. But in imputing autocatalytic powers to the DNA double helix, Watson and Crick (1953a) might lay claim to having anticipated the enzymatic functions of RNA (if not DNA), an iconoclasm that earned the Nobel Prize in 1989 for Sidney Altman and Thomas Cech.

Despite the intellectual revolution initiated by Watson and Crick (1953a), we might still ask the question, at what point was the welfare of any patient altered by specific knowledge of the double helix? This is a question I agonized over during the 1970s, and its first answer was perhaps the work of Y.W. Kan on the prenatal diagnosis of hemoglobin disorders, using DNA hybridization (1978). How rapidly we have moved in the interval is recounted by Caskey (1993). Why did that take 25 years? One may simply point to the enormous edifice of contributory knowledge that now bridges the most reductionist aspects of DNA structure to pathological manifestations.

HISTORICAL BACKGROUND OF WATSON AND CRICK

The biological role of DNA was still enmeshed in controversy in 1953. Nucleic acids had been extracted from pus cells by Miescher in 1869, and from the beginning were associated with cell nuclei. These substances are now known to be macromolecules composed of a linear array of nucleotides joined by phosphodiester bonds. Cytologists writing in the early 1900s remarked on the association of nucleic acids with chromosomes and speculated that this basophilic material in chromatin might be the substance of genetic continuity. This brilliant anticipation was, however, submerged by a misleading observation, namely, the apparent loss of basophilia in the chromosomes of oocytes, leading E.B. Wilson (1925) to remark "That the continued presence of 'chromatin' [i.e., basic-chromatin] is essential to the genetic continuity of the chromosome has, however, become an antiquated notion." We now know that these chromosomes become remarkably unraveled in keeping with their massive involvement in transcription, associated proteins then overshadowing the continuity of the DNA.

This skepticism was reinforced by the apparent monotony of DNA structure embedded in Phoebus Levene's first analyses of DNA. They contained only four constituent nucleotides – each comprising a phosphate group, a sugar, and one of the four bases: cytosine (C), thymine (T), adenine (A), or guanine (G). Within the limited analytical precision available in the 1920s, these appeared to be present in the exact stoichiometric equivalence. Hence the provisional hypothesis of DNA as a tetranucleotide, although it was well recognized that its molecular weight and other key parameters had yet to be ascertained. Nor was there any biological system or array of sources to tell that one DNA preparation was in any way different from any other. Such a simple molecule seemed a poor candidate for the miraculous capabilities of the gene. On the other hand, proteins contained an abundant variety of constituent amino acids (eventually 20). More important, dozens, even hundreds of proteins were isolated with vastly different biological, physical, and chemical properties, including wide disparities in composition. The 1920s saw the most exciting developments in protein chemistry, even the crystallization of urease and of pepsin and the demonstration that enzymes were pure proteins (Sumner, 1926; Northrop, 1930). The cap seemed to be a similar characterization of the tobacco mosaic virus, claimed to be pure protein by Wendell Stanley in 1935. This was, however, soon to be corrected by Bawden and Pirie in 1937, who found phosphorus and carbohydrate in infectious concentrates of tobacco mosaic virus and inferred the presence of RNA. Stanley, nevertheless, received the Nobel Prize in chemistry in 1946, together with Sumner and Northrop. By that time, Stanley acknowledged "that the nucleic acid could not be removed without causing loss of virus activity and there was general agreement that the virus was a nucleoprotein". Thus, this prize was a noble reinforcement of the primacy of proteins as the seat of biological specificity.

The breakthrough challenge to that dogma was thrust forth in 1944 by Oswald T. Avery, Colin MacLeod, and Maclyn McCarty. They had studied the diverse serological types of the pneumococcus and followed up Griffith's report (1928) that these could be altered or transformed by extracts of other strains. The gist of the 1944 study was that the transforming substance was DNA! This was contrary to expectations that the carbohydrate antigen or some associated protein would be the transforming substance. Avery, a member of the same Rockefeller Institute as Wendell Stanley, was intimately familiar and impressed with the difficulties of characterizing biopolymers. Though fully cognizant of the biological implications of the discovery, he was even more hesitant to dwell on them – but did include a remark that "The inducing substance has been likened to a gene. . . ."

Their claims, of course, aroused intense critical controversy, largely around the obvious question whether their DNA preparations were still contaminated with traces of biologically active protein. Avogadro's number, 6×10^{23} per mole, would allow a residuum of 10^7 protein molecules per microgram of a preparation that was 99.99% protein free, at the limit of analytical detectability. The sensitivity of the active materials to deoxyribonuclease might be ascribed to a protective rather than informational function of the DNA. Likewise, the insensitivity to proteases might be an attribute of a nucleoprotein complex.

My own role in the debate was a willingness, even desire, to believe – but a sense of responsibility that the issue was too important to be regarded as closed until there was no escape. It was not clear what feasible experiments (short of *ab initio* synthesis of DNA) could ultimately seal all these infinitesimal loopholes. One might go along with "DNA" as a working hypothesis, and some did. Most biologists blurred their judgments by talking about nucleoproteins – not necessarily informed by the distinction they were implying. Some might have meant something like "protein" or "nucleic acid" or a combination thereof, but please do not ask the role of the constituents. A rare few gambled on the DNA – as in some sense did Watson and Crick (1953a), although they would have enjoyed working out its structure regardless of its biological implications. In the event, the final elucidation of DNA structure was a horse race. By Watson's own account, only a few weeks would have separated their priority from the looming insights of Maurice Wilkins and Rosalind Franklin (who had provided the critical experimental data) or of Linus Pauling.

The biological significance of the pneumococcus transformation was also problematical. It looked like a transfer of genetic information; but until 1951, the only markers tested were the serotype antigens. Could one extrapolate from those to genes in general, particularly given that the very idea of a bacterial genetics was in its infancy?

After the 1944 bombshell, more chemical attention was given to the tetranucleotide model, and signs of greater chemical complexity emerged. Of particular import were the deviations of the four bases from the simplistic 1:1:1:1 ratio, found by Erwin Chargaff. Furthermore, DNA from different sources exhibited different base composition. So perhaps DNA could be more complex, more diversified than previously thought – could be rehabilitated as a candidate for the gene. During the 1940s, the Feulgen cytochemical test for DNA and analyses indicating constancy of DNA per genome in somatic cells and a halving in germ cells also added to DNA's respectability. But these findings did not necessarily prove more than a structural or scaffolding role for the DNA. The pneumococcus transformation remained the only biological assay for a genetic role for DNA – in contrast to the innumerable enzyme and immunological assays available for candidate proteins.

This impasse was alleviated by the broadening of phage research, sternly governed by Max Delbrück's genius, to embrace a wider range of chemical studies of phage infection. A critical one was the 1952 double-labeling experiment of Hershey and Chase. Most of the S-35 label (capsid protein) was excluded from infected cells; most of the P-32 (DNA) entered and was transmitted to the phage progeny. This experiment is often cited as the crowning blow on behalf of the "DNA-only" model. But Hershey himself did not go so far – well aware that "most" is not "all", he was still referring to "nucleoprotein" in 1953 – and this at the same Cold Spring Harbor Symposium that sponsored a critical discussion of the paper by Watson and Crick (1953a).

The article by Watson and Crick (1953a) did not, of course, bear directly on the loopholes in Avery's claims. It did add a further note of plausibility to a DNA-only concept of the gene. In the absence of any serious contradiction, this gradually hardened from working hypothesis to central dogma. The most serious challenge today is the prion hypothesis: that some "infectious" agents may be devoid of nucleic acid. This is still contentious at an experimental level: the hypothesis least in conflict with nucleic doctrine is that the infectious prion is a sort of epitaxial primer of aggregation of a host-determined protein. This still leaves obscure how and whether different prions could maintain and propagate their identity in a genetically defined host.

Long after many other lines of evidence converged to support an informational role of DNA – e.g., the colinearity of DNA sequences with protein products (Yanofsky), genetically active DNA was eventually synthesized in the chemical laboratory (Khorana) and replicated enzymologically (Kornberg), fully vindicating Avery et al. and those who gave their faith to these propositions.

THE FLOWERING OF MOLECULAR GENETICS

Since the rediscovery in 1900 of Mendel's 1865 work, genetics has had an extraordinary development, even without the benefit of tangible physical and chemical models of the genetic material. The biological phenomena of mutation and of sexual crossing (genetic recombination) opened the door to experiments in which existing organisms were the reagents. Genomes could be mixed by crossing,and new combinations of factors segregated into the offspring. Likewise, fruit flies could be subjected to radiation, and variant or mutant forms discovered. Genetic information is organized into linear chromosomes, and the processes of meiosis in gametogenesis: precise synapsis of homologues and crossing-over or segmental exchange of chromosome parts allowed powerful dissection of fine structure on a scale that rivals that of microchemical analysis. These methods continue to play an indispensable role in the denomination and mapping of mutant genes. By 1941, through the work of Beadle and Tatum, the groundwork of biochemical genetics had been laid – the role of genes in the prescription of protein products, and the use of mutations in the dissection of metabolic pathways. Indeed, many of these ideas had been anticipated by Archibald Garrod's studies of human biochemical defects at the very dawn of genetics.

Since 1953, we have had a new language for the description of genes: they are now segments of DNA that can be defined and manipulated as chemical entities. The linguistic transition has been conceptually smooth, though marked by occasional generational quarrels. Understandably, very few individuals can combine erudite knowledge of the life histories of a wide range of organisms in their natural habitats with focused and specialized knowledge of biochemical manipulations in the laboratory. Nor have many radical revisions of genetic doctrine issued from the molecular perspective. We have had to acknowledge that genes, as bits of DNA, are subject to a wider range of chemical and biological interactions than was previously thought – especially with other DNA. The icon of stability of genomes has been shaken by the discovery of transposable elements, first noted in maize by McClintock in 1951; these remained inexplicable until they could be studied as DNA molecules. And concentrating on DNA now allows us to inject genes with viruses, needles, even "shotguns", into a range of cellular targets including the germ line, providing a technical revolution in the construction of new genotypes in all kinds of organisms – bacteria, plants, and mammals.

Meanwhile, other advances, notably the extension of recombination analysis to somatic cells in culture by cell fusion, have extended the technical power of genetic analysis in ways compatible with, but not dependent on, the double helix. It is paradoxical that the human chromosome number, $2n = 46$, was not correctly understood until 1956 (Tjio and Levan), and that for about 20 years thereafter this was at least as important in the development of human genetics as was the structure of DNA.

The adumbration of DNA-based research, molecular genetics, since 1953 would embrace a substantial fraction of world science. Many encyclopedic monographs struggle to record the details and promptly become obsolete. We can hardly do more herein than summarize the major headings, following an imprecise dichotomy distinguishing topological DNA – an information duplex – from mechanical DNA – a three-dimensional geometric object.

DNA AS AN INFORMATIONAL DUPLEX

Denaturation and Hybridization

The most elementary aspect of the duplex is the separability of its strands, using temperature or chemical denaturants. A-T base pairs melt (separate from one another) at a lower temperature than G-C pairs, so melting curves can distinguish DNA of different base composition. Single strands once separated can also be reannealed, allowed to rejoin, the kinetics allowing the discovery that much DNA (in eukaryotes) has a repetitive or a redundant sequence. Radioactively labeled probes can be used to ferret out target homologous DNA with high precision.

Homology and Evolution; Polymorphism within the Species

These and related methods can be used as quantitative indices of the genetic relatedness of diverse species, supplanting the subjectively evaluated morphological criteria used in systematics heretofore. Within the species, genetic polymorphism can now be described at the DNA level – one astonishing finding is that humans are typically heterozygous with a prevalence of two or three per 1000, i.e., almost once in every gene. As most of these base substitutions have no perceptible phenotypic effect, random drift (rather than selectible or adaptive change) may predominate in evolutionary change (Kimura).

Mutagenesis and DNA Repair

The vulnerability of genes to mutational change in response to X-rays was known empirically since 1927 (Muller), and to chemicals since 1944 (Auerbach). Early hopes that chemical mutagenesis would be a direct path to the chemistry of the gene were not substantiated. Most chemical mutagens react with amino acids as well as DNA bases. The exceptions are nuclein base analogues, which may be misincorporated into DNA; but these were discovered much later. Above all, we now understand that the initial lesions in DNA would usually be lethal, and that eventual mutations are the result of intricate repair metabolism that occasionally misfires.

Transcription; Genetic Code

The "central dogma" of information flow has emerged, that DNA → (transcription) RNA → (translation) protein. The base sequence of DNA is transcribed faithfully into a messenger RNA copy. This in turn governs the assembly of a polypeptide sequence, each three-base frame of RNA encoding one particular amino acid. The polypeptide then folds (perhaps with the guidance of a chaperone) into a preordained protein three-dimensional shape, which can then function as an enzyme, antibody, hormone, structural unit, and so forth. This folding process is not yet fully computable. There may even be circumstances where a given polypeptide might have alternative foldings – but this is not accepted dogma.

The details of messenger RNA synthesis have become much more intricate. Primary transcripts are usually processed, only some of the RNA tracts being spliced together to form the final message. The other "intervening sequences", or introns, may be the major part of the RNA – their functions remain obscure. As with repeated sequences, they may reflect "selfish DNA", whose presence in the genome has little to do with their adaptive value to the overall organism. In other examples, RNA may be edited in other ways before translation is completed.

Enzymology: Nucleases, Ligase, Replication; Reverse Transcriptase

For a legion of brilliant and tireless investigators, the DNA structural model has been the platform for isolating a host of enzymes involved in every aspect of DNA metabolism. Besides giving us that metabolic map, explaining how

DNA is replicated, sliced, stitched, spliced, and repaired, these enzymes are the vital technical tools for further study of DNA and for the engineering of new constructs.

Some viruses, notoriously the retroviruses (including human immunodeficiency virus), exhibit a reverse transcriptase, whereby RNA → DNA. This knowledge is indispensable to the virologist. it has also given some of the most valuable tools for studying RNA, e.g., messenger, by allowing the production of DNA copies for input into other technology.

Tools for Engineering: DNA Splicing; PCR

These sempstering tools have founded the multibillion-dollar biotechnology industry. DNA tailored *in vitro*, with inserts from human or a variety of other sources, can be patched into convenient host garments (from bacteria to cows) for the easier exhibition of a variety of products – growth factors, enzymes, immunizing antigens, replacement therapeutics (like clotting factors) – in unlimited variety. Related technology is used to target specific host genes, to elucidate their functions in physiology and development.

The PCR (polymerase chain reaction) has been the instrument of the "democratization of molecular biology", With it a single DNA molecule in a messy mixture can be fished out and amplified ad libitum, most importantly at low cost and with simple instruments. High school students do experiments today that would have been doctoral dissertations 15 years ago. The applications range widely, from forensics and diagnosis of genetic disease to the hunt for new viruses and the revival of fossil DNA. At its heart, a synthetic DNA probe is a rational, linear, digital signature to locate any counterpart in the analysand. Its core of combinatorial specificity can be contrasted with that of antibodies, which is founded on three-dimensional shapes of the immunoglobulin and its targets.

Drug Discovery

DNA combinatorics has reached a new peak in a paradigm for drug discovery that mimics natural evolution (Pei et al., 1991). Randomized DNA sequences are expressed on host cells (or phages), and these are then selectively screened for specificities of binding to specific reagents – usually receptors for which agonists or antagonists are sought. The cell expressing the desired epitope can then be grown out for larger scale production and testing. In one application, the mammalian antibody-forming mechanism can be emulated, and mutant immunoglobulin polypeptides selected for the desired specificity. RNA can fold into stereospecific objects; hence, randomized RNA molecules can be directly selected and replicated with reverse transcriptase.

Human Genome Project

With the availability of all of these tools, the image has firmed of establishing the complete DNA sequence of the human genome. As a scientific objective, this is uncontroversial. The controversy pertains to the primacy given to the staging of the effort. Should it be a once and for all technological production, mindless of the ancillary interest in some genes or DNA tracts compared with others? Does it need to be a centralized project, administered top-down with the trappings (and political appeal) of other Big Science? Or can it be left to the cumulative efforts of hundreds or thousands of laboratories, each digging more deeply at some features of the terrain, and intent on going much further than establishing a sequence of bases? In fact, we are seeing the emergence of constructive compromise among these visions; and at the same time the technologies of mapping and sequencing are advancing to where the costs of a unified project need no longer prejudice more individualized efforts.

In any case, sequence information is but the beginning of more intensive inquiry into the polymorphisms, regulatory factors, and gene functions associated with any DNA segment.

DNA AS A HELIX

Higher Orders of Organization

The visible chromosome is a packaging of DNA, histones, and accessory proteins three or four orders of coiling beyond the double helix. Cytological observation leaves no doubt that the morphological expression of the chromosome reflects functional allocation of different genes; but we are at the mere beginning of understanding.

Gene Regulation and Morphogenesis

The basic outlines of the central dogma now consensually agreed, the core challenge of molecular biology has been the path from the gene to the organism. Given that, to some approximation, each somatic cell has the identical genotype, (1) how is gene expression differentially modulated, and (2) how is this transmitted in cell lineages?

A multitude of DNA-binding proteins have been found that do modulate gene expression: transcriptional regulators. As a three-dimensional interaction, protein binding is fully sensitive to three-dimensional shape and the major and minor grooves of the double helix, as well as the base sequences contained therein. In addition, if not in consequence of bound proteins, some tracts of

DNA are methylated shortly after DNA replication, in ways correlated with gene activation.

How these properties are locally transmitted remains a matter of speculation, but may well be bound up with local methylation.

DNA Supercoiling; Topoisomerases; Other Conformations

The standard double helix exhibits a pitch of about 10 base pairs per complete turn. If nothing else, the processes of replication and transcription would entail the unraveling and rewinding of the helices: this is the task of enzymes generically called topoisomerases. These can transiently cut single strands to permit the relief of torsional stress, then rejoin them. In its natural habitat, DNA is often found in states of positive or negative supercoiling, often correlated with maintained gene expression. In addition, many cytotoxic and cancer chemotherapeutic agents seem to be topoisomerase inhibitors, and most owe some of their specificity to the momentary DNA-supercoil status of a given cell. It is particularly intriguing that environmental signals can modulate that status, often by regulating the production of the various topoisomerases.

At least *in vitro*, DNA can undergo a spontaneous transition to a totally different, kinked and left-handed conformation called Z-DNA. Tracts rich in G-C pairs are especially prone to this shift. The importance of Z-DNA *in vivo* is hotly contested.

DNA conformations plainly confer different chemical reactivity on the bases, a principle exploited by the footprinting methods used to study conformation. This must have some implications for localized chemical mutagenesis – a matter not yet systematically studied.

TRIUMPH OF MECHANISM

The dominion of the DNA paradigm has been the triumph of mechanistic interpretation in 20th-century biology. It is sometimes remarked that human personality is nothing but the individual's 3 billion base pairs – an assertion that fascinates some, terrifies others, and has much to do with the debate about the Human Genome Project. If we could believe that existing genotypes had achieved more than a tiny fraction of the human potential – in culture, in intellect, in compassion, in a sane ordering of affairs – we could elevate the genome to that pedestal of nemesis. On the other hand, we do know that many, probably most, individuals labor under some potentially remediable burden of hereditary origin. As much to understand the better nurturing of human development, a euphenics, as to intervene in genetic constitution, eugenics, it does behoove us to learn all we can about genetic polymorphisms and their impact on human health and capability. It is particularly important to distinguish interventions in germ cells from those in the somatic cells, and to communicate that it is only the latter that are intended to be the targets of the new gene therapies.

Rockefeller University,
New York, USA

SELECTED READINGS

It would be a precious exercise to provide specific documentation for every detail of this commentary; it would be both arduous and redundant – many single points would deserve a library. The up-to-date detail can be found in standard texts of molecular biology (a few are listed) and in the volumes of *Annual Review of Biochemistry*. The leading historical monographs on DNA are also listed.

Molecular Biology

Alberts, B., Bray, D., Lewis, J., Raff, M., Roberts, K. and Watson, J.D. (1989). *Molecular Biology of the Cell*. New York, NY: Garland Publishing.
Berg, P. and Singer, M. (1992). *Dealing With Genes: The Language of Heredity*. Mill Valley, Calif: University Science Books.
Darnell, J.E., Lodish, H.F. and Baltimore, D. (1990). *Molecular Cell Biology*. New York, NY: Scientific American Books.
Kornberg, A. and Baker, T.A. (1992). *DNA Replication*, 2nd ed. New York, NY: W.H. Freeman & Co.
Stryer, L. (1988). *Biochemistry*. New York, NY: W.H. Freeman & Co.
Watson, J.D., Hopkins, N.H., Roberts, J.W., Steitz, J.A. and Weiner, A.M. (1987). *Molecular Biology of the Gene*. Menlo Park, CA: Benjamin/Cummings.
Wells, R.D. and Harvey, S.C. (eds.) (1988). *Unusual DNA Structures*. New York, NY: Springer-Verlag NY Inc.

History of DNA

Crick, F.H.C. (1988). *What Mad Pursuit: A Personal View of Scientific Discovery*. New York, NY: Basic Books Inc. Publishers.
Dubos, R.J. (1976). *The Professor, the Institute, and DNA*. New York, NY: Rockefeller University Press.
Fruton, J.S. (1972). *Molecules and Life*. New York, NY: Wiley-Interscience.
Judson, H.F. (1979). *The Eighth Day of Creation*. New York, NY: Simon & Schuster.
Lederberg, J. (1987). 'Genetic Recombination in Bacteria: A Discovery Account'. *Annual Review of Genetics* **21**: 23–46.
McCarty, M. (1985). *The Transforming Principle*. New York, NY: W.W. Norton & Co.
Olby, R.C. (1974). *The Path to the Double Helix*. Seattle: University of Washington Press.
Portugal, F.H. and Cohen, J.S. (1977). *A Century of DNA*. Cambridge, MA: MIT Press.
Sayre, A. (1975). *Rosalind Franklin and DNA*. New York, NY: W.W. Norton & Co.
Watson, J.D. (1968). *The Double Helix*. New York, NY: Atheneum Publishers.
Watson, J.D. and Tooze, J. (1981). *The DNA Story*. San Francisco, CA: W.H. Freeman & Co.

REFERENCES

Caskey, C.T. (1993). 'Molecular Medicine: A Spin-off from the Helix'. *JAMA* **269**: 1986–92.
Pei, D., Ulrich, H.D. and Schultz, P.G. (1991). 'A Combinatorial Approach toward DNA Recognition'. *Science* **253**: 1408–11.
Watson, J.D. and Crick, F.H.C. (1953). 'Molecular Structure of Nucleic Acids: A Structure for Deoxyribose Nucleic Acid'. *Nature* **171**: 737–8.
Watson, J.D. and Crick, F.H.C. (1953). 'Genetical Implications of the Structure of Deoxyribonucleic Acid'. *Nature* **171**: 964–7.

KENNETH F. SCHAFFNER

THEORY STRUCTURE AND KNOWLEDGE REPRESENTATION IN MOLECULAR BIOLOGY

INTRODUCTION: THE STRUCTURE OF BIOLOGICAL THEORIES

Until quite recently, much of the analysis of theories in the biological and biomedical sciences had subscribed to what I term the "Euclidean Ideal". This notion assumes that the ideal structure of a scientific theory resembles Euclid's approach to geometry: a small number of fundamental definitions and axioms constitutes the essence of a theory, which is then elaborated deductively in the form of theorems which cover a broad (scientific) domain. This view of theory structure obtains fairly strong support in the physical sciences, and in Newton's theory of gravitation and its elaboration in the *Principia* (1942 [1726]), some treatments of Maxwell's electromagnetic theory (see Stratton, 1941), thermodynamics, and quantum mechanics (see von Neumann, 1955). As Beatty (1995) has recently noted, Newton himself in his methodological writings urged that science should follow this Euclidean ideal.

Biologists – especially those biologists seeking a methodological unity with the physical sciences, such as Waddington (1968) – and philosophers of biology, such as the early Michael Ruse (1973), have maintained that the laws and theories of biology have the exact same logical structure as do those of the physical sciences (though recently there have been some changes – see Kitcher (1984), Rosenberg (1985), Culp and Kitcher (1989), van der Steen and Kamminga (1991) and Beatty (1995)). This *unity* view is only supportable if one restricts one's attention to those few – but very important – theories in biology which in point of fact have a very broad scope, and are characterizable in their more simplified forms as a set of "laws" which admit of axiomatization and deductive elaboration. Examples are certain formulations of Mendelian genetics and of population genetics, as well as parts of molecular biology and intermediary metabolism in biochemistry. I maintain that a deeper analysis of even these theories, however, will disclose difficulties with a strong methodological parallelism with the physical sciences (see Schaffner, 1980, 1986, 1993a,b, and Kitcher, 1984). I believe that a close examination of a wide variety of other biological theories in immunology, physiology, embryology, and the neurosciences will suggest that the *typical theory* in the biomedical sciences is a structure of overlapping interlevel causal models.

The models of such a structure usually constitute a series of idealized prototypical mechanisms and variations (mutants) which bear family or similarity resemblances to each other, and characteristically each of these prototypes has a (relatively) narrow scope of straightforward application to (few) pure types. The models are typically *inter*level in the sense of levels of aggregation, containing component parts which are often specified in intermingled organ, cellular, and molecular terms. Stages of temporal development in the models may represent either causally deterministic, causally probabilistic, random (Markovian), or even mixed connections. I argued at length in an earlier article (1980) that this approach to theory, which I termed a "theory of the middle range" (with apologies to R.K. Merton (1968) who first used that term in a somewhat different context), both is pervasive and should be expected to be found in the biomedical sciences. The term "middle range" seemed appropriate for two reasons: first the theories were not broad sweeping general theories but they were not summaries of data either; they were midway between these extremes. Second, in terms of levels of aggregation of the entities in such theories, the theories were not about high level populations evolving in evolutionary time and not about specific DNA sequences or specific enzymes functioning in well defined biochemical pathways, though these molecular entities were frequently included. Rather they were at the level of the organelle, the gene as characterized by functional products, the cell, and the organ. Thus though *inter*level, their levels of aggregation tended to concentrate in the "middle range". I want to emphasize, and will do so in more detail below, that typical *molecular biological theories* – where theories should perhaps be in quotes – such as the operon do conform to this middle range characterization.

Let me motivate this view of theories by citing some of Thomas Kuhn's work on the important role of exemplars, as part of as well as defining of scientific theories.

In his (1970) Postscript to his extraordinarily influential *Structure of Scientific Revolutions* first published in 1962, Kuhn attempted to further refine the concept of 'paradigm' which was so central to the argument in the first edition of his book. (For those of you who do not know, or have forgotten Kuhn's work, a paradigm is what one thinks of as a scientific theory, for example the Copernican theory of the sun-centered universe, or Newton's theory of gravitation.) The paradigm notion had, Kuhn proposed in 1970, two interpenetrating but still rather different aspects, and he suggested that it might be worthwhile to use some new terminology, specifically the idea of a 'disciplinary matrix', to clarify the earlier term 'paradigm'. The first aspect of 'paradigm', which we may describe as a disciplinary matrix *per se*, was similar to what traditional philosophers of science had understood a scientific theory to be, and had two parts, namely (1) symbolic generalizations and (2) models. In his 1962 book Kuhn had made a significant departure from traditional philosophy of science and had introduced "values" – primarily of an epistemic sort – as an important component of a paradigm. This intrinsic value component is again reintroduced as the third element of the disciplinary matrix in his Postscript (1970). A fourth component of the "disciplinary matrix", which Kuhn termed an "exemplar", was, however, viewed by Kuhn as a *distinctly different sense of paradigm*. Kuhn wrote:

For . . . [the fourth sort of element in the disciplinary matrix] the term 'paradigm' would be entirely appropriate, both philologically and autobiographically; this is the component of a group's shared commitments which first led me to the choice of that word. Because the term has assumed a life of its own, however, I shall here substitute 'exemplars'. By it I mean, initially, the *concrete* problem solutions that students encounter as part of their scientific education, whether in laboratories, on examinations, or at the ends of chapters in science texts. To these shared examples should, however, be added at least some of the technical problem-solutions found in the periodical literature that scientists encounter during their post-educational careers and that also show them by example how their job is to be done. More than other sorts of components of the disciplinary matrix, differences between sets of exemplars provide the community finestructure of science. (1970, pp. 186–187) [my emphasis]

Kuhn adds that "the paradigm as shared example is the central element of what [he] now take[s] to be the most novel and least understood aspect of . . . [*The Structure of Scientific Revolutions*]", and in the (1970) Postscript and in some later articles he elaborates further on the concept. For our purposes I want to focus on the problem-solving aspect of exemplars, namely the way in which individuals learning a discipline come to know how to apply the generalizations of the discipline to specific problematic situations. Kuhn writes:

A phenomenon familiar to both students of science and historians of science provides a clue. The former regularly report that they have read through a chapter of their text, understood it perfectly, but nonetheless had difficulty solving a number of the problems at the chapter's end. Ordinarily also, those difficulties dissolve in the same way. The student discovers, with or without the assistance of his instructor, a way to see his problem as like a problem he has already encountered. Having seen the resemblance, grasped the analogy between two or more distinct problems, he can interrelate symbols [constituting the law sketch or scientific generalizations] and attach them to nature in the ways that have proved effective before. The law-sketch, say f = ma, has functioned as a tool, informing the student what similarities to look for, signaling the gestalt in which the situation is to be seen. The resultant ability to see a variety of situations as like each other, as subject for f = ma or some other symbolic generalization, is, I think, the main thing a student acquires by doing exemplary problems. (1970, p. 189)

I believe that this is an important insight of Kuhn's and one that has largely gone unnoticed by most philosophers of science. (Philip Kitcher is an exception, but in this essay I will not have the opportunity to comment on the way he develops his view of practices. Suffice it to say that he and I disagree, at least in a prima facie way, about the role of deductive development of his practices). I view Kuhn's suggestions about exemplars as identifying an aspect of scientific theories that is even more important in the biological and medical sciences than in the physical sciences, since the *law* sketches to which he refers are typically not present in biological theorizing, and quite specific models as well as exemplars bear the burden of the generalizable parts of biomedical theories. Thus it is *not only* in the area of *problem solving* in the biomedical sciences, including molecular biology, that exemplars figure significantly, but also *in the very structure of the theories* in those domains.

The view of theories that I am elaborating in this article and which can be found in some papers I published in 1980 and 1986 (and elaborated on in my 1993a,b) triangulated on a similar position arrived at by a somewhat different route by a committee of the National Academy of Sciences/National Research Council (NAS/NRC) which Dr. Harold Morowitz chaired and which produced the 1985 report on *Models for Biomedical Research: A New Perspective*. That NAS committee's report introduced the notion of the "matrix" of biomedical knowledge:

The workshops demonstrated that the results of biomedical research can be viewed as contributions to a complex body, or matrix, of interrelated biological knowledge built from studies of many kinds of organisms, biological preparations, and biological processes at various levels. (1985, p. 2)

From within this multidimensional matrix, "many-many modelling" occurs, in which analogous features at various levels of aggregation are related to each other across various taxa. The committee report notes:

An investigator considers some problem of interest – a disease process, some normal physiological function, or any other aspect of biology or medicine. The problem is analyzed into its component parts, and for each part and at each level, the matrix of biological knowledge is searched for analogous phenomena. . . . Although it is possible to view the processes involved in interpreting data in the language of one-to-one modelling, the investigator is actually modelling back and forth onto the matrix of biological knowledge. (1985, p. 67)

The committee realized that additional word was required to formulate more clearly this largely intuitive idea of a "matrix" and added:

The development of the matrix and the extraction of biological generalizations from it are going to require a new kind of scientist, a person familiar enough with the subject being studied to read the literature critically, yet expert enough in information science to be innovative in developing methods of classification and search. This implies the development of a new kind of theory geared explicitly to biology with its particular theory structure (Schaffner, 1980). It will be tied to the use of computers, which will be required to deal with the vast amount and complexity of information, but it will be designed to search for general laws and structures that will make general biology much more accessible to the biomedical scientist. (1985, p. 67)

At the final meeting of the committee in the spring of 1985, a consensus appeared to emerge that these conceptions of the matrix and "many-many" modelling could rehabilitate the notion of a "theoretical biology". The old theoretical biology sought for universal theories analogous to Newtonian mechanics and was largely a failure; the "new" theoretical biology was to be reconceived as the discovery and extension of something like middle range theories analogously related via many-many mappings in the matrix.

FURTHER PROBLEMS CONCERNING GENERALIZATION SCOPE AND EXPLANATIONS IN THE MOLECULAR BIOLOGY

There are some additional features regarding generalizations and their scope, as well as explanations in molecular biology, which are closely related to the approach to theory structure discussed thus far. These are perhaps best introduced under the general heading of reduction, about which more can be found in Sarkar's and Beckwith's essays. The comments I offer about this subject will of necessity be somewhat brief, but I think they may help us characterize problems with knowledge representation in molecular biology from another perspective.

Much of the discussion in the philosophy of science about reduction has been about inter-*theoretic* reduction, and has attempted to understand such reduction as a subspecies of Hempelian deductive-nomological *explanation*, sometimes called covering law explanation. Schematically, the deductive-nomological model can be presented as

$$\frac{L_1 \ldots L_n}{C_1 \ldots C_k}$$

$$\therefore E$$

where the L's are universal laws and the C's represent initial conditions. Collectively these constitute the "explanation" or "explanans". E is a sentence describing the event to be explained, and is called the "explanandum". In this model, E follows *deductively* from the conjunction of the law or nomological premises (from the Greek word *normos* which means law) and the statements describing the initial conditions. E here may be a sentence describing a particular event. On the other hand, if the initial conditions are interpreted in a suitable general way, for example as boundary conditions, E may be a law, such as Snell's law in optics or the Hardy-Weinberg law in population genetics.[1]

The *locus classicus* of philosophical analyses of theory reduction can be found in the work of Ernest Nagel (1947, 1961). Nagel envisaged reduction as a relation between *theories* in science. A theory in biology, say, was reducible to a theory in chemistry, if and only if (1) all the non-logical terms appearing in the biological theory were *connectable* with those in the chemical theory, e.g., gene had to be connected with DNA, and (2) with the aid of these *connectability assumptions*, the biological theory could be *derived*, essentially as in Hempelian explanation as characterized above, from the chemical theory (with the additional aid of general logical principles). Later, connectability came to be best seen as representing a kind of "synthetic identity", e.g., gene = DNA sequence (Schaffner, 1967; Sklar, 1967; Causey, 1977). I will refer to this view of reduction or inter-theoretic explanation as *the standard view*.

This standard view has undergone extensive criticism as well as defense over the past 30 years, some of the most influential critiques occurring in the work of Kuhn and Feyerabend in general philosophy of science, and in the work of Hull, Wimsatt, Kitcher, and Rosenberg, among others, within the philosophy of biology. For my purposes in this essay, I want only to dwell on some of the implications of the above view of theory structure for intertheoretic explanations, and in particular I want to consider the consequences of dealing with theories that have (1) interlevel components and (2) generalizations of both broad and narrow scopes of application. Each of these features, it seems to me, will affect the way that we both attempt to explain phenomena in molecular biology, as well as how we seek to represent "theories" that do the explaining (and problem-solving).

What will be the consequences for explanation and reduction of the prima facie fact that in molecular biology we do not have anything *quite* like the "laws" we find so prevalent in physics explanations? (I should add parenthetically that Wimsatt (1976a,b) and Kitcher (1984) have also been concerned with the implications for reduction of this lack of "laws" in molecular biology.) We do find, I have suggested above, some *surrogates* for such laws: There are generalizations of a sort (of varying generality) which we could extract from any molecular biological explanation of some process, such as enzyme induction or muscle contraction, though such generalizations are usually left implicit, typically having been introduced in other (perhaps earlier and more didactic) contexts.[2] Further, these generalizations are typically borrowed from a very wide-ranging set of models (protein synthesis models, biochemical models, etc.). But even when such generalizations, narrow or broad, are made explicit, they need to be supplemented with the details of what might be termed specific "hook-up" connections in the system, in order to constitute an explanation.[3] These are not quite like the "initial conditions" of traditional philosophy of science because they are an integral part of the reducing mechanism, e.g., that the gene sequence in the *lac* system of *E. coli* is i-p-o-z-y-a and that the repressor binds to the DNA operator sequence ACCTTAACACT (Watson et al., 1987, pp. 476–80).

Some of these generalizations we can glean from such molecular mechanisms may have quite a narrow scope, and even broad generalizations typically need to be changed to *analogous* generalizations (or mechanisms) as one changes organism or system. For example, we change to RNA as the material basis of the gene for TMV or the AIDS virus (HIV-1), we appeal to different gene regulation principles as one moves away from prokaryotes to eukaryotes, and even in the same organism we note similarities and differences between skeletal muscle, smooth muscle, and cardiac muscle.

What we appear to have in most of our intertheoretic explanation or "reduction" examples are rather intricate *systems* using both broad and narrow *causal generalizations* which are typically *not* framed in *purely* biochemical terminology, but which are characteristically *interlevel* and *interfield*. This interfield notion is a feature importantly stressed by Darden and Maull in their (1977) writing, and I view it as following naturally from the intertwined interlevel character of biological theories, including those in molecular biology.

This pattern of both broad and narrow interlevel causal generalizations is what we seem to have in many of the examples encountered in current biology and medicine. As an explanation of a phenomenon like genetic dominance or enzyme induction is given in molecular terms, one maps parts of the phenomena into a molecular vocabulary, sometimes by synthetic identities, or sometimes by exhibiting a previously molecularly characterized entity as a causal consequence of a reducing mechanism. Enzyme induction becomes not just the phenomena as Monod encountered it in his early research, but it is *reinterpreted as* a causal consequence of derepressed protein synthesis. Thus, something *like* Nagel's condition of connectability is found but it is laden with new *causal* freight.

Is reduction as characterized above in terms of the examples also in accord with reduction by *derivation* from a theory? Again, I think the answer is generally yes, but with some important caveats. We do not have, I have argued above, a very general set of sentences (the "laws") which can serve as a small set of premises from which we can deduce the conclusion. Rather what we have are *a set of causal sentences of varying degrees of generality*, many of them quite specific to the system in question, and many left implicit. It may be that in some distant future all of these causal sentences of narrow scope, such as in the lac operon system "the *i* gene determines . . . the synthesis of . . . a 'repressor' which blocks synthesis of the β-galactosidase, and which the exogenous inducers remove. . . ." (to quote from the famous Pajama experiment of Pardee et al., 1959, p. 3127), can be explained by general laws of protein chemistry, but this is not the case at present. In part this failure of unilevel explanation is a consequence of our inability to infer the three-dimensional structure of a protein like the repressor from a complete knowledge of the sequence of amino acids which make up this tetrameric protein. Fundamental and very general principles will have to await a more developed science than we will have for some time. Thus the reductans, the explaining generalizations in such an account, will be a *complex web* of interlevel causal generalizations of varying scope, and will typically be expressed in terms of an idealized system of the types discussed, complicated by all of the intertwined interlevel features, along with some textual elaboration on the nature of the causal sequence leading through the system.

This then is a kind of *partial model explanation or partial reduction with largely implicit generalizations*, often of narrow scope, licensing temporal sequences of causal propagations of events through the model. It is not unilevel reduction, i.e., to biochemistry, but it is I think what is characteristically termed a *molecular biological explanation*.[4] The model effecting the explanation or "reduction" is typically interlevel, mixing different levels of aggregation from cell to organ back to molecule. The model or models also may not be robust across this organism or other organisms, it may well have a narrow domain of application, in contrast to what we typically encounter in physical theories.

In spite of all of these qualifications, however, there is no reason why the *logic* of the relation between the reductans and the reductandum cannot be *cast* in deductive form. Typically this is not done, because it requires more formalization than it is worth, but some fairly complex engineering circuits can effectively be represented in the first-order predicate calculus and useful deductions made from the premises, which typically are several dozen in number (see Genesereth and Nillson, 1987, pp. 29–32 and 78–84 for an example).

The picture of explanation or reduction that emerges from any detailed study of molecular biology as it is practiced is thus not an elegant one. There are, in the above characterization, indications that reduction may perhaps better be conceived of, as Wimsatt (1976b) has urged, more along causal/mechanical lines, or Sarkar (1992) has developed in his account which elaborates on Ernst Mayr's notion of explanatory reduction. I think there is some truth in this, but that it only captures one dimension of the relationships that exist between reduced and reducing domains, and that we must appeal to generalizations of the type I have described above along the lines of what John Mackie once termed "laws of working" that we discern operating in such mechanisms. This is a topic which I shall not be able to pursue in the present paper, however, but some discussion of this can be found in my (1993b).

This causal analysis approach to reduction sketched above raises the question as to how it may relate to more classical Nagelian intertheoretic reductions. I will not have space in the present essay to comment on this relation, which is somewhat complex, but an overview of such relations can be found in my (1993a) and more details in my (1993b, chapter 9). Suffice it to say that the causal and classical approaches have different strengths depending in part on the state of codification of the science(s), and in part on the types of questions of interest to the philosopher. For example, questions about the nature of "identity" found between two levels in reduction are, I would argue, better pursued using the classical analysis.

KNOWLEDGE REPRESENTATION AND SIMULATING MOLECULAR BIOLOGICAL KNOWLEDGE

To propose that biomedical theories are a series of prototypical mechanisms and variations (mutants) which bear family or similarity resemblances to each other, that each characteristically has a (relatively) narrow scope of straightforward application to (few) pure types, that the models are typically *inter*level in the sense of levels of aggregation, containing component parts which are often specified in intermingled organ, cellular, and molecular terms, is, I think, a useful first step, but the characterization raises many *new* problems. For example, it is not obvious how such a view of biomedical theories is to be clearly, and perhaps even computationally, represented (see Thagard, 1988, chapter 2), nor how the insight might be further elaborated so as to facilitate new discoveries and problem-solving in the biomedical sciences in general, and in molecular biology in particular. Furthermore, it is not obvious that this view of theory structure really is congruent with what one finds in biology; further development and testing of the concept is called for. It seems to me that one of the best ways to get a handle on these prima facie unwieldy and non-Euclidean surrogates for *theories* is to look to that discipline within Artificial Intelligence called *knowledge representation.*

In an attempt to begin to fill in some additional details regarding knowledge representation of biomedical theories as described above, a preliminary account was developed in (Schaffner, 1987) which adopted the standard techniques of frames and semantic nets from artificial intelligence research. In a chapter in a rather lengthy recent book (Schaffner, 1993b), that work was extended to encompass what is termed in computer science an object-oriented programming approach or OOP for short.

Object-oriented approaches appear to offer some important advantages for knowledge representation in the domains of increasingly complex sciences such as molecular biology. More specifically, large amounts of molecular biology including information about DNA, RNA, and protein sequences, together with representations of the mechanisms of transcription and translation, will be required in any adequate representation system. Moreover, this information contains large variation in the types of mechanisms. The features of modularity, multiple inheritance, abstraction, information hiding, and polymorphism – terms that have fairly clear senses in computer programming but which I will not be able to discuss in this paper (but see Cox, 1986 and Stefik and Bobrow, 1986) – are facilitated by the relatively new style of writing code termed object-oriented programming (OOP) in which they form the essence of the approach. Suffice it to say that these features of OOP mirror the complexity of knowledge found in molecular biology. The GENSIM program developed by Peter Karp (1989) is an example of a program written using this OOP approach, and has achieved reasonable success in representing both the discovery as well as modes of action of the *trp* operon.

THE NEED FOR ADDITIONAL TOOLS TO REPRESENT VARIATION: THE CASE-BASED REASONING APPROACH

Though the use of frames, semantic nets, and OOP techniques constitute powerful knowledge representation tools which can be employed in attempts to represent the type of biological theory structure described above, and which may also capture the complexities of what the NAS report mentioned earlier called the "biomatrix", they are somewhat frustrated by the extensive variation – the analogies rather than identities – found among biological mechanisms. In attempts to represent such variation, I and others have considered explicitly representing it using analogical reasoning techniques (following Winston, 1986 and Carbonell, 1986). I have also explored a "fuzzy set" approach (in Schaffner, 1993b), and in addition have examined the use of *implicit* similarity as represented in *classification hierarchies* composed of frames and objects (Schaffner, unpub.). None of these appeared to squarely face the intertwined issues of representation, abstraction, and appropriate indexing techniques, however, and more recently, in following up some of Schanck's (1982) and Carbonell's (1986) work, I have turned to what is termed "case-based reasoning" (CBR) to examine some very promising alternative procedures. In the remainder of this paper I will outline some of the features of CBR, and then discuss one such system, the PROTOS program, developed by Bareiss (1989) in a bit more detail, indicating why I think it offers additional possibilities for capturing the type of complex variation found in the biomedical sciences in general and in molecular biology in particular. Ultimately, I suspect, any adequate representational system will need to have *many* interacting aspects, perhaps along the lines of what is known in the AI area as a "blackboard" system (see Nii, 1986).

Let me now turn to a discussion of some of the CBR approaches to knowledge representation.

Case Based Reasoning

To characterize the CBR approach in its simplest terms, Schank's student, Janet Kolodner suggests that making a case-based inference includes the following set of steps:
1. Recall a previous case.
2. Focus on appropriate parts of that case.
3. Use those parts of the previous case to derive an appropriate decision for the new case.

Each of these steps requires considerable elaboration. Recall of a previous case is accomplished by probing the memory, and any CBR system will need to have an account, and an algorithm, which will permit this probing. It also goes without saying, perhaps, that some type of a memory structure needs to be implemented before the probing can begin. The "focussing on appropriate parts of the case" step also requires extensive development, and different tacks have been taken by AI theorists such as Carbonell (1986), Kolodner (1985), Ashley and Rissland (1987), and Bareiss (1989). I shall comment on Bareiss' (1989) approach below. Kolodner's step 3, which is said to use the previous solution to solve the new problem, involves making the case-based inference, and again there are various options to pursue. The simplest is to transfer the solution that achieved the goal in the comparison (old) case to the new case. This tack has been employed by a number of authors including Carbonell (1986) and Kolodner (1985). More complex forms of case-based reasoning allow various modifications or allow the use of a multiplicity of cases (e.g., in Ashley's work, 1991). I will discuss Bareiss' approach to this issue in some detail in a moment.

The general value of case-based reasoning for modeling biology in general and molecular biology in particular lies in the fact that biological reasoning, as emphasized earlier, is itself heavily exemplar-based. In addition, CBR has been forced to confront in the context of quite specific problem-solving situations the problems of variation, similarity, and analogical reasoning. Furthermore, by focussing on the issue of memory structures, CBR provides a framework for addressing the enormous amounts of knowledge that contemporary molecular biology is generating.

Though the just-rehearsed account of CBR provides a general outline of some of the tools which the approach can provide, a deeper and more detailed analysis will help us to see the promise – and the problems – more clearly. As already indicated, I want to provide this more detailed account by drawing on some of Bareiss' work on PROTOS.

Concept Representation and PROTOS

A deeper analysis of CBR and also of theory structure can begin with the question of how to best represent concepts with their general and particular aspects. Concept representation has been approached from classical, probabilistic, and exemplar perspectives (see Bareiss, 1989, Appendix A and Schaffner, 1987). In the classical approach, a concept is defined via a series of necessary and sufficient conditions, as in Aristotelian logic. In the probabilistic approach, the features associated with the concept have weights, and some threshold is specified to enable the application of the concept to an individual. This permits a similarity metric to operate in classification. In the exemplar view, often associated with the work of Rosch, a concept is learned by storing examples of the concept, though some examples are viewed as more *prototypical* than others, e.g., a robin is a more prototypical instance of 'bird' than is a penguin. The attractiveness of the exemplar approach is its novelty as well as its congruence with frame-based systems and default reasoning, but it has drawbacks as well (Schaffner, 1987), including the non-monotonicity of default reasoning. An additional difficulty with the exemplar view of concepts has been discussed in detail in the psychological literature (Murphy and Medin, 1985) as well as by Bareiss (1989), namely the problem of conceptual cohesiveness. Not every collection of exemplars constitutes a legitimate concept, and similarity is not a well enough understood or well enough defined notion to be the "glue" which holds together the exemplars. Bareiss' PROTOS program, following the general suggestions of Murphy and Medin (1985), employs prototypical cases but supplements them with "*explanation links*" to provide this "glue".

PROTOS is represented using what is called a semantic net in which the nodes are categories, exemplars, and qualitative features. Figure 1 from Bareiss' book shows a simple example from everyday life – the concept of a chair – and what this looks like. In this semantic net the edges or the arcs are 4-tuple relational links acquired from teacher-provided explanations, thus constituting explanation links that represent domain knowledge. Figure 2 shows that the relational links also include a complex set of *indices* present in PROTOS, which associate (1) exemplar features with categories (these are both (a) the *remindings* and (b) the not shown *censor* [or negative remindings] indices), (2) categories with exemplars (*exemplar* links bearing prototypicality ratings), and (3) exemplars with imperfectly similar examples (*difference* links). An explanation in this system is a set of known relations, shown in Figure 3, (including correlational, definitional, part-to-whole, causal, and functional

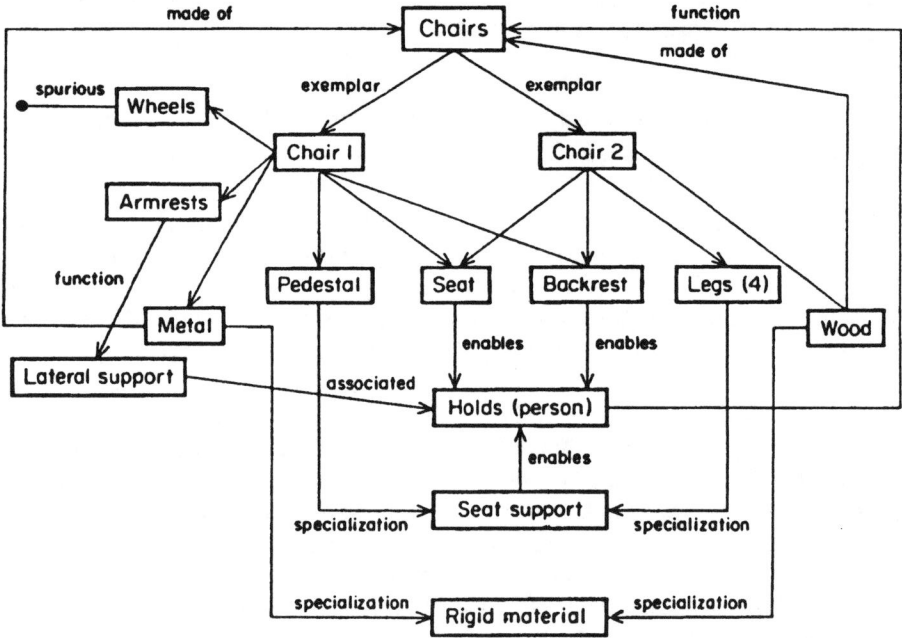

Figure 1. A sample category structure (from Bareiss (1989), reprinted with permission from Bareiss et al. (1988). "Protos: An Exemplar-Based Learning Apprentice", *International Journal of Man-Machine Studies* **29**: 549–561).

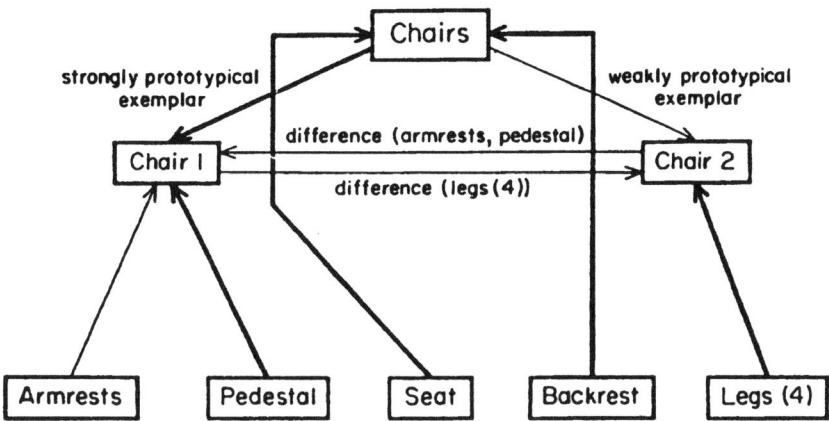

Figure 2. Indices associated with the sample category structure (based on Bareiss (1989), reprinted with permission from Bareiss et al. (1988). "Protos: An Exemplar-Based Learning Apprentice", *International Journal of Man-Machine Studies* **29**: 549–561).

- **Generalization/Specialization**
    - has typical generalization/has typical specialization
- **Part-to-Whole**
    - part of/has part
- **Causal**
    - causes/is caused by
- **Functional**
    - has function/is function of
    - enables/is enabled by
- **Predicate-to-Argument**
    - acts on/is acted on by
    - affects/is affected by
- **Conjunction**
    - and
- **Mutual Exclusion**
    - is mutually exclusive with

Figure 3. Noncorrelational explanation links (from Bareiss (1989) with permission).

connections) connecting a feature (or a feature collection) to another feature or category.

PROTOS proceeds through both classification and learning phases. In its classification phase, remindings and censors operate from the features associated with a new case to generate a set of ordered categories which are possible classifications, i.e., hypotheses. An example from Bareiss' book might illustrate this more clearly. In Figure 4, a series of *features* associated with a patient's audiological exam are listed, and in Figure 5, these features, repeated on the left, are shown as *remindings* of possible hypotheses or classifications for this patient, shown on the right. In this type of application, the classifications or hypotheses are *diagnoses*; they are also *exemplars*. These hypotheses, then, are evaluated until a match to another exemplar is found (a merge), or a new exemplar is created. Figure 6 shows the extent of a match of the case shown in Figure 4 to an exemplar of "possible Menieres" disease. Within categories, exemplars are selected in the order of their prototypicality. The pattern matching process attempts to use domain knowledge (as stored in the explanation links) to explain the equivalence of the new case's features to the exemplar's features. In the learning phase of PROTOS *generalizations* are developed and represented in both implicit and explicit ways. Implicit generalizations reside in the domain knowledge which "increases the featural variation allowable in an acceptable match to an exemplar" (Bareiss, 1989, p. 47). Explicit generalization is represented by ascribing low *importance* values to certain features of an exemplar, and also by replacing (merging) some of an exemplar's features with more general features, again on the basis of explanations which are part of the encoded domain knowledge (but now as represented in fragments of generalization, functional, or

Case: p859OR

Unknown

s_neural(mild,gt4k)   tymp(a)
s_neural(mild,ltlk)   speech(normal)
ac_reflex_u(normal)   air(mild)
ac_reflex_c(normal)   history(vomiting)
o_ac_reflex_u(normal)   history(dizziness)
o_ac_reflex_c(normal)   history(fluctuating)

Figure 4. Raw remindings from the features of Case 1 (from Bareiss (1989) with permission).

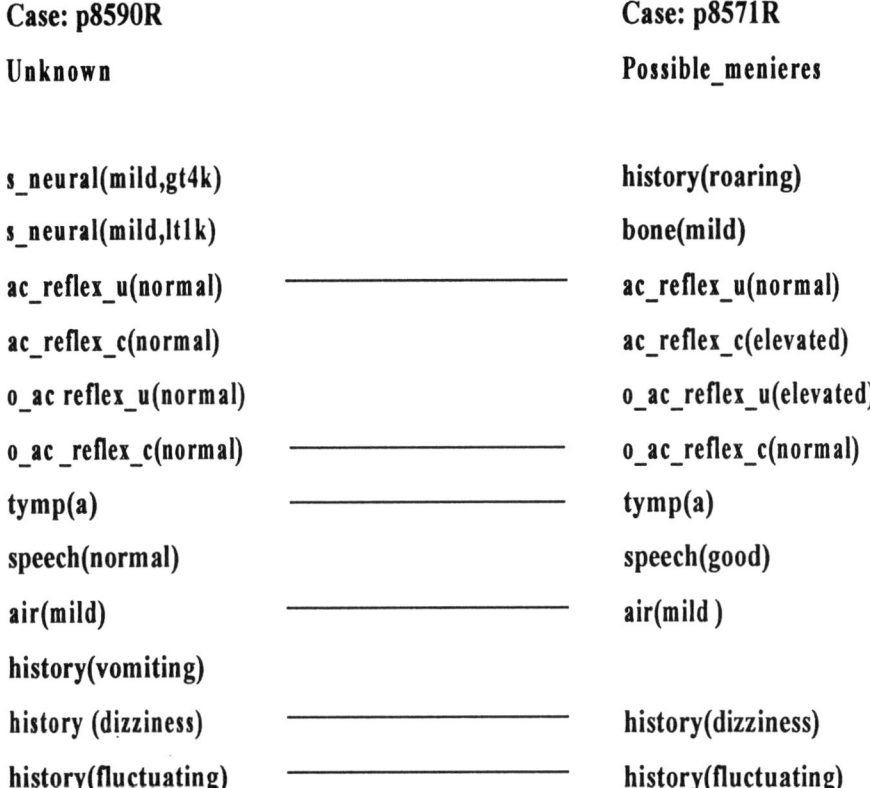

Figure 5. Raw remindings from the features of Case 1 (from Bareiss (1989) with permission).

causal hierarchies – see Bareiss, 1989, p. 49). As suggested from the examples, PROTOS has been tested using audiological diagnosis problems and performs as well as "a human journeyman and experts" (Bareiss, 1989, p. 55). The system is still in its early stages of evolution, and much of the power of the system comes from its interaction with (and correction by) a human "teacher". In its current form, its explanatory capabilities are fairly simple and primitive, though there appear no reasons why, in principle, much more complex explanations could not be developed in PROTOS.

I noted above that PROTOS uses explanation links to supplement similarity relations to hold together exemplars in identifiable groups. This aspect of PROTOS suggests that the special features of explanations appealing to middle range biological theories as discussed earlier in this paper, will need to be carefully considered in any extensions of a CBR approach to biology and molecular biology. Thus one should expect that explanatory links will have

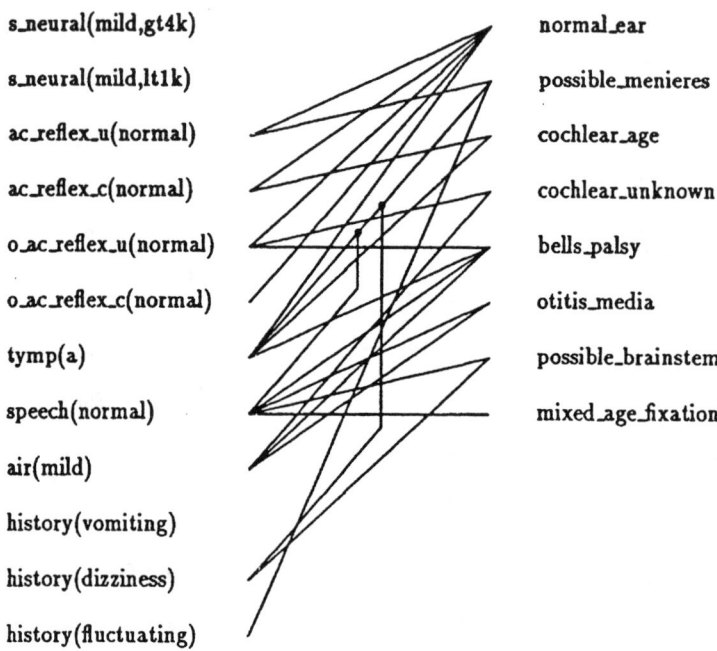

Figure 6. Matching Case 1 to an exemplar of possible_Meniers (from Bareiss (1989) with permission).

to represent the *interlevel* features of standard molecular biological explanations. Furthermore, the explanatory links can be expected to have varying degrees of scope, a feature which will fall out of the CBR approach naturally, but which should be anticipated as such a representation tool is evaluated. What I envision will be encountered, as one begins to represent molecular biological explanations in anything like the detail that one finds in moderately advanced textbooks and in research papers in the area, is something very much akin to the many-many connections emphasized by the NAS committee in their account of the biomatrix. This, in turn, suggests a complexity that will sorely tax both software and hardware capabilities as knowledge representation proceeds.

The PROTOS program discussed above is but one instance of a case-based reasoning program, and others that have been developed also provide promise of useful approaches to representing the types of complexity encountered in the biomedical domain. Several CBR programs have been developed and applied in the clinical domain in addition to PROTOS, including CASEY (Koton, 1988) and MEDIC (Turner, 1988). Even domains as prima facie removed from biomedicine as legal reasoning, in which powerful case-based reasoning programs have been developed (Ashley and Rissland, 1987), may offer insights into how to structure knowledge representation in molecular biology. Several other case-based reasoning programs, such as PARADYME (see Kolodner, 1988) and MBR (Stanfill and Waltz, 1988) have been implemented on the Connection Machine, which can provide the appropriate hardware for the memory demands of this approach to knowledge representation. Though work in the CBR area is of comparatively recent vintage, the approach appears to offer significant promise in assisting knowledge representation research in molecular biology.

CONCLUSION AND SUMMARY

In conclusion, what I have tried to do in this paper is to urge the reader to think about a different approach to theory structure in biology and in molecular biology, an approach which eschews the Euclidean ideal and emphasizes the role of interlevel exemplars functioning as prototypes and related by similarity. I also considered what the implications might be for explanations (and reductions) of this view of theory structure and suggested that explanations will typically be interlevel ones, even in molecular biology, and that they will involve appeals to causal generalizations of both broad and narrow scope. I discussed some approaches to knowledge representation of complex biological domains, and proposed that in addition to the older techniques of using frames and semantic nets, newer approaches such as object-oriented programming or OOP techniques, and especially case based reasoning approaches, should have value in capturing the structure of knowledge in biology and molecular biology.

George Washington University,
Washington, DC, USA

NOTES

[1] Nagel, in developing his account of D-N explanation (1961, pp. 33–7) to apply it to scientific laws, does not detect any serious problems with the extension. Hempel and Oppenheim, in their original (1948) account, were concerned with special problems that such an extension posed, and elaborated on their reservations in what Salmon (1989) terms their "notorious" footnote 33 (Hempel, 1965, p. 273). Salmon indicates Hempel never adequately addressed this problem in any of his subsequent writings.

[2] Thus a molecular geneticist will *assume* extensive knowledge on the part of her readers of DNA, RNA, etc.

[3] I added the term "hook-up" here to distinguish these connections within a system from reduction functions or "bridge laws" that are connections (of a different sort) between reduced and reducing systems.

[4] It may seem paradoxical that molecular biological explanations are not uni-level reductions, but the empirical evidence for this view is overwhelming. Bill Bechtel (personal communication) has argued that thus we are not capturing "reduction" by admitting inter-level features. I would admit that we are not capturing a *uni*-level reduction, but the account given suggests that the higher-level theory is itself not uni-level, and the examples suggest that what occurs is that we move closer to a biochemical explanation, but do not typically achieve it in any full sense (with very few exceptions and those primarily in very simple biochemistry).

REFERENCES

Ashley, K. and Rissland, E. (1987). 'Compare and Contrast, A Test of Expertise'. *Proceedings of AAAI-87*.
Bareiss, R. (1989). *Exemplar-Based Knowledge Acquisition: A Unified Approach to Concept Representation, Classification, and Learning*. San Diego: Academic Press.
Bareiss, R., Porter, B.W. and Wier, C.C. (1988). 'Protos: An Exemplar-Based Learning Apprentice'. *International Journal of Man-machine Studies* **29**: 549–61.
Beatty, J. (1995). 'The Evolutionary Contingency Thesis'. In: Wolters, G., Lennox, J. and McLaughlin, P. (eds.), *Concepts, Theories, and Rationality in the Biological Sciences*. Konstanz: Universitätsverlag Konstanz, and Pittsburgh: University of Pittsburgh Press, pp. 45–81.
Carbonell, J.G. (1986). 'Derivational Analogy: A Theory of Reconstructive Problem Solving and Expertise Acquisition'. In: Michalski, R.S., Carbonnell, J.G. and Mitchell, T.M. (eds.), *Machine Learning*, ch. 14. vol. 2. Los Altos, CA: Morgan Kaufmann.
Cox, B. (1986). *Object Oriented Programming: An Evolutionary Approach*. Reading, MA: Addison-Wesley.
Culp, S. and Kitcher, P. (1989). 'Theory Structure and Theory Change in Contemporary Molecular Biology'. *British Journal for the Philosophy of Science* **40**: 459–83.
Karp, P. (1989). 'Hypothesis Formation and Qualitative Reasoning in Molecular Biology'. Ph.D. Dissertation, Stanford University.
Kitcher, P. (1984). '1953 and All That: A Tale of Two Sciences'. *Philosophical Review* **18**: 335–73.
Kolodner, J. (1988). 'Retrieving Events from a Case Memory: A Parallel Implementation'. In: Kolodner, J. (ed.), *Case Based Reasoning. Proceedings of a Workshop on Case-Based Reasoning*. San Mateo, CA: Morgan Kaufmann.
Koton, P. (1988). 'Reasoning about Evidence in Causal Explanations'. In: Kolodner, J. (ed.), *Case Based Reasoning. Proceedings of a Workshop on Case-Based Reasoning*. San Mateo, CA: Morgan Kaufmann.
Merton, R.K. (1968). 'On Sociological Theories of the Middle Range'. In: Merton, R.K. (ed.), *Social Theory and Social structure*. New York: Free Press.
Morowitz, H. (1985). *Models for Biomedical Research: A New Perspective*. Washington, DC: National Academy of Sciences Press.
Murphy, G.L. and Medin, D.L. (1985). 'The Role of Theories in Conceptual Coherence'. *Psychological Review* **92**: 289–316.
Newton, I. (1972) [1726]. *Principia*, 3rd ed., 2 vols. Cambridge, MA: Harvard University Press.
Nii, H.P. (1986). 'Blackboard Systems'. *AI Magazine* **7**(2): 38–53; (3): 82–106.

Rosenberg, A. (1985). *The Structure of Biological Science*. Cambridge: Cambridge University Press.
Ruse, M. (1973). *Philosophy of Biology*. London: Hutchison.
Sarkar, S. (1992). 'Models of Reduction and Categories of Reductionism'. *Synthese* **91**: 167–94.
Schaffner, K.F. (1980). 'Theory Structure in the Biomedical Sciences'. *The Journal of Medicine and Philosophy* **5**: 57–97.
Schaffner, K.F. (1986). 'Exemplar Reasoning About Biological Models and Diseases: A Relation Between Philosophy of Medicine and Philosophy of Science'. *The Journal of Medicine and Philosophy* **11**: 63–80.
Schaffner, K.F. (1987). 'Computerized Implementation of Biomedical Theory Structures: An Artificial Intelligence Approach'. In: Fine, A. and Machamer, P. (eds.), *PSA-1986*, vol. 2. East Lansing, MI: Philosophy of Science Association.
Schaffner, K.F. (1993a). 'Theory Structure, Reduction, and Disciplinary Integration in Biology'. *Biology and Philosophy* **8**: 319–347.
Schaffner, K.F. (1993b). *Discovery and Explanation in Biology and Medicine*. Chicago: University of Chicago Press.
Schanck, R. (1982). *Dynamic Memory*. Cambridge: Cambridge University Press.
Stanfill, C. and Waltz, D.L. (1988). 'The Memory-Based Reasoning Paradigm'. In: Kolodner, J. (ed.), *Case Based Reasoning. Proceedings of a Workshop on Case-Based Reasoning*. San Mateo, CA: Morgan Kaufmann.
Steen, van der W.J. and Kamminga, H. (1991). 'Laws and Natural History in Biology'. *British Journal for the Philosophy of Science* **42**: 445–67.
Stefik, M. and Bobrow, D. (1986). 'Object-Oriented Programming: Themes and Variations'. *AI Magazine* **6**: 40–62.
Stratton, J.A. (1941). *Electromagnetic Theory*, 1st ed. New York: McGraw-Hill.
Thagard, P. (1988). *Computational Philosophy of Science*. Cambridge, MA: MIT Press.
Turner, R.M. (1988). 'Organizing and Using Schematic Knowledge for Medical Diagnosis'. In: Kolodner, J. (ed.), *Case Based Reasoning. Proceedings of a Workshop on Case-Based Reasoning*. San Mateo, CA: Morgan Kaufmann.
von Neumann, J. (1955). *Mathematical Foundations of Quantum Mechanics*. Princeton: Princeton University Press.
Waddington, C.H. (1968). 'Introduction'. In: Waddington, C.H. (ed.), *Towards a Theoretical Biology*, vol. 1. Chicago: Aldine Press.
Winston, P.H. (1986). 'Learning by Augmenting Rules and Accumulating Censors'. In: Michalski, R.S., Carbonnell, J.G. and Mitchell, T.M. (eds.), *Machine Learning*, ch. 3, vol. 2. Los Altos, CA: Morgan Kaufmann.

DORIS T. ZALLEN

REDRAWING THE BOUNDARIES OF MOLECULAR BIOLOGY: THE CASE OF PHOTOSYNTHESIS*

Establishing the existence of human paternity these days is not a daunting task. Accurate tests are available to detect genetic and biochemical similarities and to use the degree of identity observed to support conclusions about the occurrence of direct connections between the purported parents and their progeny. Establishing the parentage of scientific disciplines – tracking the intellectual, institutional, and instrumentational influences involved – is far more difficult. Nonetheless, I would like to suggest that the current views on the origins of molecular biology are too limited, that molecular biology arose from a broader constellation of research efforts than is currently recognized, and that those areas which today function with the designation "molecular biology" are only a subset of the areas that actually contributed to the molecularization of biology. As a case in point, I would like to examine the work done within plant physiology – specifically, the attempts by plant physiologists to explain the process of photosynthesis. This case will demonstrate that another area of biological study, that of "bioenergetics", or the transformation of energy by living systems, needs to be added to those of "structure" and "information" in the story of the rise of molecular biology.

THE STRUCTURAL AND INFORMATIONAL COMPONENTS OF MOLECULAR BIOLOGY

Much of the work tracing the origins, spread, and conduct of molecular biology, especially as put forth by some of the scientist-participants themselves in their memoirs and in festschrift volumes, has emphasized the role of two major research schools in forming the foci around which what has become the molecular biology enterprise first crystallized, and from which it has been propagated. These are known as the structural and the informational schools.

The structural component refers to work carried out, starting in the 1930s, to determine the detailed atomic architecture of a number of key biological molecules. This work was conducted as the means of understanding the relationship between structure and function at the molecular and subcellular levels, and of developing a greater degree of insight into biological function at higher levels of organization, such as tissues and organs.[1] The major mode of experimental analysis was X-ray crystallography – coupled, later on, with electron microscopy. The chief proponents of this work were the British

* Reprinted from the *Journal of the History of Biology* **26**(1) (Spring 1993), pp. 65–87.

scientists W.T. Astbury and J.D. Bernal, who had received their training from William and Lawrence Bragg, the physicists who developed the technique of X-ray crystallography; in the United States, Linus Pauling and his coworkers at the California Institute of Technology included substances of biological origin in their larger structural chemistry program.[2]

The other source of molecular biology, according to current accounts, was the studies carried out on bacterial viruses (also called bacteriophage or phage) starting in the late 1930s.[3] These were conducted by the "phage group", a loose confederation of scientists, with Salvador Luria, A.D. Hershey, and physicist Max Delbrück playing leading roles in advancing the use of bacterial viruses, and their annual summer course at Cold Spring Harbor serving to bring new recruits to the study of phage. The chief goal of this work, indeed the very reason for choosing bacteriophage as the model organism, was to elucidate one of the secrets of life: the processes whereby the hereditary material serves as a self-perpetuating reservoir of information that produces the characteristic properties of each organism. Major tools in this form of inquiry were genetic crosses involving phage and their bacterial hosts, the use of radioisotope probes, ultracentrifugation, and electron microscopy.

Though the workers associated with each school gave priority to that group's own contributions and were hard-pressed to acknowledge the real significance of the other in originating or giving shape to molecular biology, there seems to have been an accommodation reached that both should be jointly credited in the historical record.[4] There *were* occasions when these two areas of investigation came together. The most spectacular of these was the elucidation of the molecular structure of the genetic material, DNA, by J.D. Watson and Francis Crick,[5] which led within a decade to the decoding of the genetic information contained within the base-pair sequences of the double helix and then to the dissection of some of the pathways by which DNA produces its effect. Experimental programs geared to understanding the molecular mechanisms associated with heredity have become one of the most potent areas of modern biological studies.

Increasingly, molecular biology has come to be viewed as directly connected to and exclusively defined by such work. In the foreword to the collection *Nobel Lectures in Molecular Biology*, David Baltimore states that "molecular biology ... is the study of how DNA, RNA and protein are interrelated".[6] And one continues to see this view of disciplinary origin reaffirmed by other scientists in the biological literature.[7]

It is not surprising that the belief that "molecular biology" is synonymous with the study of the hereditary material and its role in the production of various cellular products has also taken hold in the work of historians, philosophers, and sociologists of science. If this assessment of the origins and focus of molecular biology is correct, then the several excellent studies already produced by scholars should serve well as the framework for further analysis and evaluation. But if the molecularization of biology has occurred in other contexts – so that molecular investigations in genetics turn out to be just one part of a bigger picture – then we would be well advised to expand the current view, or else our insight into the way science is conducted and into mechanisms of conceptual change and disciplinary development in science will be flawed.

ANOTHER STRAND OF MOLECULAR BIOLOGY: BIOENERGETICS

There is no official checklist of the diagnostic markers for what constitutes "molecular biology". However, an examination of factors common to both the structural and informational varieties of molecular biology already described points up the following general characteristics:

(1) It is an area of activity that seeks to understand and explain the key activities of living systems – the so-called secrets of life – through an examination at the molecular level of the structure, functions, and interactions of the relevant molecules.
(2) The biological objects of study are chosen to provide the simplest model system for experimental analysis, and are typically viral, microbial, or isolated molecular species and subcellular units.
(3) Its experimental protocols rely on preexisting technical instruments and procedures of the physical sciences, or modifications of them for use in biological studies.
(4) It makes use of concepts and mathematical approaches developed within the physical sciences – often with the direct involvement of physicists and physical chemists – for the interpretation of biological phenomena.[8]

An examination of work undertaken by plant physiologists starting in the 1920s shows that their efforts to decipher the process of photosynthesis – the conversion of light energy into chemical energy – closely conforms to these criteria. What follows is not meant to be a history of work on photosynthesis, though one deserves to be written, but an attempt to measure work in the bioenergetics of photosynthesis against each of the criteria just given in order to establish that, in the period from about 1920 to 1960, the molecularization of photosynthesis was actually in phase with, and even in advance of, equivalent work in genetics.

When Watson and Crick rushed into the Eagle to announce to their colleagues that they had discovered the secret of life[9] – the secret of how the hereditary material is organized – they were only partially correct. There are other phenomena uniquely associated with the living state that also merit a "secret of life" designation. Not least among these is the process through which the energy of the sun, which reaches earth in the form of sunlight, is converted into chemical energy by living systems: the green plants, algae, and some bacteria. The capacity to carry out photosynthesis is not just a peculiar trademark of a limited variety of organisms, it is fundamental to all life on this planet. The organic compounds, the carbohydrates, that are the major end product of photosynthetic activity are required by all living systems: they provide the energy to drive all metabolic activities, and they provide the carbon skeletons from which the biomolecules required to direct and build the cellular and organismic structures of the living state, including the nucleic acids, can be constructed. Without this conversion of solar energy into chemical energy, life as we know it would not be possible on earth.

The process of photosynthesis has been known for about two hundred years, but for most of this time what was understood about it was rather superficial: chiefly, the fact that in the light and in the presence of the green chlorophyll pigment, plants convert carbon dioxide and water into organic matter, primarily carbohydrate, and oxygen is released. Because of the limited experimental tools available to examine photosynthetic systems, only rough estimates of the uptake and release of gases could be made – showing, for example, the equivalence between the amount of carbon dioxide taken up and the oxygen produced – and no details of the process could be provided. This situation began to change in the 1920s, and with each passing decade investigators were able to penetrate further beneath the surface of reactants and products to identify many of the molecular intermediates involved and to determine their organization into functional assemblies. As the base of knowledge enlarged and instrumentation improved, scientists were able to probe the fundamental events taking place at the reaction centers – those molecular points at which light energy is trapped and the conversion of light energy into chemical energy first occurs.

Most modern chronologies of photosynthesis research[10] start with Otto Warburg's measurements in the 1920s of the quantum yield of photosynthesis – the minimum number of quanta of light, or photons, that are required to yield a molecule of oxygen – which he found to be close to four.[11] These investigations heralded the beginning of attempts to discover the details of the mechanism of the primary light conversion reactions. They also marked the introduction of the use of *Chlorella*, a unicellular green alga, as an organism of choice for pursuing photosynthetic questions.

The prevailing view that the oxygen produced in photosynthesis was derived from carbon dioxide was called into question in the 1920s in France by René Wurmser, and in more widely known work in the 1930s by C.B. van Niel, a Stanford University microbiologist and former student of Albert Jan Kluyver, Delft's famed developer of comparative biochemistry.[12] Van Niel carried out comparative biochemical studies involving a number of different photosynthetic bacteria and green plants. He noted that one type of photosynthetic bacterium, which grows anaerobically using hydrogen sulfide (H_2S) in place of water (H_2O) when fixing carbon dioxide into organic matter, produces sulfur, instead of oxygen, as an end product. The comparison of the general overall equations for green plant photosynthesis and for this type of bacterial photosynthesis allowed van Niel to suggest that it must be the water, not the carbon dioxide, that undergoes photolysis to produce the oxygen. The photolysis of water not only yields oxygen but must also generate a reducing moiety, which could then function, directly or indirectly, in the reduction of the carbon dioxide to carbohydrate.

In 1937, Robin Hill at Cambridge University provided experimental proof for the validity of the photolysis of water as a major event in photosynthesis by showing that isolated chloroplasts evolve oxygen when maintained in an environment deprived of carbon dioxide but with an artificial electron acceptor present.[13] The recognition that it is water that yields the oxygen was as fundamental to allowing an understanding of the photosynthetic process as the later realization that DNA and not protein is the genetic material was to understanding the gene. Hill's experiments accomplished something more: they revealed the ability to separate the "light" and "dark" steps. It was now possible to separate those reactions that require the quanta of light, and where the initial conversion of the energy of the photon into chemical energy useful to the cell occurs, from the dark reactions, those subsequent steps that do not require light – specifically, the biochemical pathway associated with the transformation of carbon dioxide into carbohydrate. Experimenters could then pursue these two different types of events (light and dark) in greater detail in separate research programs.

The major dark reaction sequences were worked out in the period from 1945–1954 by Melvin Calvin, Andrew A. Benson, James Alan Bassham, and their colleagues at Berkeley – thanks, in part, to the availability of radioactive carbon-14, an isotope of carbon with a long half-life, and to improved methods of chromatography.[14] These carbon-dioxide-fixing reactions are localized in the soluble protein fraction of the chloroplast. They require ATP, the major energy source for metabolism, and a strong reductant, NADPH, to drive the synthesis of carbohydrate from carbon dioxide.

Robert Emerson, an American scientist who had been a student of Otto Warburg in Berlin, undertook studies of the events associated with the interaction of light with the photosynthetic apparatus while working at CalTech in the 1930s. Emerson and others used new methods of spectrophotometric analysis they had developed to measure the quantum yield – and found it to be closer to eight, twice as high as the Warburg figure. Emerson also found that these quanta were received by pigment clusters that acted cooperatively in units of about 300 chlorophyll molecules in size.[15] The possible existence of a photosynthetic "unit" narrowed the focus of research from whole cells or whole chloroplasts down to pigment clusters within the chloroplast. By the 1940s and the early 1950s, well in advance of equivalent insight into gene structure and function, various portions of the photosynthetic apparatus had been experimentally analyzed, and attempts had begun to explain the transfer of energy among pigment molecules in a single unit to the reaction center with the use of electromagnetic coupling or resonance transfer models.

By 1960, just before the genetic code was interpreted, several groups of investigators had put forth what is still the most generally accepted mechanism of the photochemical portion of photosynthesis: There are two separate light reactions mediated by two different pigment photosystems.[16] Thus, the photolysis of water in photosystem II produces electrons, which can flow to photosystem I through a series of electron carriers that are alternately oxidized and reduced. Another photochemical conversion in a second set of photosystem I pigments liberates electrons capable of producing a strong reductant, NADPH. The end products of this chain of reactions are the oxygen produced in photosystem II, a reductant (NADPH) produced in photosystem I, and ATP produced as a result of the flow of electrons between the two photosystems. This is the so-called Z-scheme, and it represents a revolution of sorts in photosynthetic thought that has become a "central dogma" for plant physiologists and biophysicists. The NADPH and ATP produced as a result of the light reactions are used in the dark reactions where carbon dioxide is fixed.

In 1961 the chemiosmotic hypothesis advanced by Peter Mitchell related the formation of ATP to the specific organization of the molecules comprising the electron carrier chain in the membranes of the chloroplast.[17] The generation of a proton gradient as a result of proton translocation across membranes during photosynthetic chain electron transport in chloroplasts (and, similarly, respiratory chain electron transport in mitochondria) sets up the high energy state that can drive the formation of ATP.

Work since the sixties has provided considerably more molecular detail and has permitted the incorporation of structural considerations into models of photosynthetic function. The photosynthetic reaction centers and their associated light-harvesting antennae of pigment molecules have been probed using such techniques as picosecond absorption spectroscopy, electron spin resonance spectroscopy, and X-ray crystallography.[18] A detailed picture is now emerging, not only of the transfer of energy among the light-harvesting pigments, but of the molecular arrangement at the reaction centers – and of the associated events that characterize the primary photochemical reaction and bring about the conversion of the energy of the photon to usable cellular chemical energy.

This is a brief and sketchy account, and it would be incorrect to leave the impression that the molecular mechanisms of photosynthesis have been completely elucidated. They have not, and, despite recent work that is permitting the initial photochemical conversion to be understood in more precise detail, much remains to be done. But it would be correct, I believe, to view what has been accomplished experimentally, and the degree of insight that has been obtained into structural and functional relationships at the molecular level of one of the most fundamental of biological processes, as meeting the first criterion, listed above, for "molecular biology".

Though plant physiologists have mainly been interested in the photosynthetic processes of higher green plants, the biological materials chosen for many of these studies have been single-celled microorganisms. The unicellular green algae, such as Warburg's *Chlorella*, have been the mainstay of this work. The ease of manipulation of microbial cells compared to plant tissue, the more uniform illumination that can be achieved using such cells, and the rapid and accurate gas exchange measurements using manometric techniques with these cells made the algae ideal substitutes for the structurally more complex and experimentally more cumbersome leaf tissue. In the mid-1950s, just about the time that the first cell-free systems for protein synthesis were being worked out,[19] Daniel Arnon and his group at Berkeley demonstrated that isolated chloroplasts were indeed capable of carrying out all the photochemical, phosphorylation, and carboxylation reactions associated with photosynthesis.[20] This finding made the chloroplast, not the cell, the locus of experimental attention. As smaller and smaller photosynthetic regions or pigment systems were studied, both types of materials – the chloroplast preparations and the algal cell cultures – were fractionated into their membrane subunits. The photosynthetic bacteria were also widely used.[21] There are significant differences between green plant and bacterial photosynthesis: the bacteria do not evolve any oxygen, and they appear to be devoid of the equivalent of photosystem II. Yet, there are advantages to using bacteria as model systems for all types of photosynthesis, especially for studies on the organization of molecular components at the reaction center sites. This emphasis on the use of simple model systems, even if such systems may differ from what obtains in the intact organismic state, closely conforms to the second criterion for molecular biology mentioned above.

Criterion 3 stresses the role of the experimental tools of the physical sciences in programs of molecular biology research. Given the types of experimental questions raised in attempting to decipher the events attendant on photosynthetic processes – questions such as the nature of the light-absorbing pigments, the efficiency of light quanta in stimulating the photochemical conversion, the mechanism of the photoconversion, the sequence and timing of the subsequent electron transport events – it is not surprising that the experimental tools used in physics research became central to studies of photosynthesis. In 1945, Eugene Rabinowitch, in chronicling the advancing state of knowledge on photosynthesis, pinpointed both plant physiologists and physicists as the appropriate audience for his compendium *Photosynthesis and Related Processes*: plant physiologists so that they could appreciate "the experimental results obtained by various physical methods", and physicists so that they could better understand the physiology of photosynthesis and how it ties in with the overall life processes of plants.[22]

Chief among the physical tools was that of optical spectroscopy, in which the absorption of light passing through a sample is used as an indicator of the identity of the molecular species present and a measure of their abundance. The changes in absorbance can also be used to monitor physical and chemical changes in the sample. At about the same time that X-ray crystallography was being pressed into service for the study of biomolecular structure, the development of photoelectric spectrophotometry for studying photosynthetic pigments and their reactions was already taking place. Frederick P. Zscheile, Jr., a plant physiologist working in the laboratory of Thorfin R. Hogness, a physical chemist at the University of Chicago, helped apply photoelectric spectroscopy to photosynthesis work in the mid-1930s.[23] In the photoelectric spectrophotometer, a photoelectric cell replaces a photographic plate as the light detector. Light striking the photocell is converted to electrical current, which can be measured accurately and rapidly. The importation of spectroscopy into biological work, though not directed toward plant studies, was made a major programmatic objective by the Rockefeller Foundation in its efforts in the 1930s and 1940s to transform biological laboratory practice. One of its projects involved a collaborative effort between the Physical Laboratory at the University of Utrecht in the Netherlands directed by Leonard Ornstein (a theoretical physicist turned experimentalist) and the Microbiological Laboratory at Delft headed by A.J. Kluyver. These two groups created a center for new spectroscopic approaches to the study of photosynthesis that has produced many of the techniques and trained the leading investigators in the field.[24]

For a long time there was no commercially available equipment of sufficient sensitivity for use by biologists. Those wishing to make use of spectrophotometers had to design and construct the equipment within each individual laboratory. Robert Emerson and William Arnold in 1931 developed an experimental system that allowed them to separate the dark and light photosynthetic reactions using flashing light. In order to get the light to flash at the appropriate rate and to be of sufficient intensity to induce a photochemical effect, a novel design was employed using neon tubes and complex circuitry. The paper that reports these results has an elaborate materials-and-methods section so that readers could understand and, if desired, copy the experimental apparatus. It could not be purchased anywhere.[25] Until commercially produced units became available, such as the Beckman DU spectrophotometer in 1941, researchers had to build the equipment themselves, connect all the component parts, calibrate, trouble-shoot, and do the repairs. This feature produced the right conditions for interactions between physical scientists and biological scientists. At the very least, it demanded that the plant physiologists themselves be well trained in physics and physical chemistry so that they could usefully incorporate these necessary approaches into their work.

Not only hardware but, as indicated in criterion 4 above, conceptual and quantitative tools from the physical sciences were imported into molecular biology. With the recognition of the existence of photosynthetic units – those collections of pigment molecules with a reaction center – the problem of transfer of energy amongst these molecules became important, and here the theoretical work in physics on resonance transfer was applied to photosynthesis. Even before Max Delbrück had turned his attention to bacteriophage genetics,[26] James Franck, who had won the Nobel prize in physics for his work on the quantized transfer of energy during collisions of electrons with atoms, sought to extend "the understanding of excitation and photodissociation from diatomic molecules to liquids and solids and finally to the process of photosynthesis in plants".[27] Many other physicists also entered into work on the various aspects of energy transfer, looking at theoretical models to explain the transfer of energy among chlorophyll molecules or from accessory pigment molecules to chlorophyll, as well as the primary photochemical conversion itself.

There are many parallels between the scientific work in molecular genetics and that in the bioenergetics of photosynthesis. Both are concerned with revealing a key secret of life; both have a "central dogma" that guides the development of research programs; both have successfully identified fundamental units at the molecular level and have advanced highly productive models to relate structure to function; both have emphasized – and continue to require – the use of physical and chemical tools to probe these fundamental units; and both have involved physicists who have been intrigued by the way this biological subject has provided a means of extending their own ideas – based on inanimate material – to living systems, and who, in many cases, have collaborated directly with biologists as research partners. By all of these markers, then, studies of photosynthetic bioenergetics should qualify as molecular biology. Other areas of bioenergetics, including the production of mechanical energy for contractile activity or of electrical energy for sensory function, are likely to qualify as well. However, it must be left to others to test these fields for their molecular biology markers.

THE HISTORICAL PLACEMENT OF BIOENERGETICS WITHIN MOLECULAR BIOLOGY

It has been argued above that bioenergetics, particularly as related to photosynthesis, meets all the criteria for being regarded as a form of molecular biology. This section will explore how molecular biology was originally perceived by the scientific community, to show that in its earliest uses in institutional and professional settings, molecular biology as a category of scientific work never excluded – and often expressly included – areas of research centered on bioenergetics. Three examples will be described below.

The term "molecular biology" is generally regarded as having first been used by Warren Weaver in the 1938 *Report of the Rockefeller Foundation*. In the Foundation's annual reports, the president typically provided a comprehensive initial summary, after which each division described the overarching themes of its work and provided a brief sketch of the projects supported that year in line with those themes. In President Raymond D. Fosdick's report for 1938 there is a section entitled "Molecular Biology", which includes the following text describing what that term means, undoubtedly supplied for the report by Warren Weaver:

> Using many of the delicate and exact instruments which the physicist devised in his attack on organic material, [the biologists] are beginning the description and analyses of biological phenomena, not in terms of cells as units, but in terms of genes and the other critically important subdivisions of cells; and recently even in terms of molecular structure and forces. ... A new biology – molecular biology – has begun as a small but significant salient in the whole field of biological research. ... [I]t is by means of the new tools and techniques developed in many instances by the physical sciences that the door to a biology of molecules has only recently been opened.[28]

After enumerating the many different tools and techniques provided from the physical scientists, the summary went on to caution:

> It should not be expected, of course, that continual fragmentation will of itself necessarily reveal the true inner meaning of life processes. A living organism may well be something more than the sum of its parts.[29]

Rather than indicating a whole new effort by the Foundation, what Weaver and Fosdick were referring to with this term was a program of research support that had been under way for several years – a program that favored experimental biology projects employing the tools of the physical sciences to explore vital biological processes, mainly human biological processes. In the Natural Sciences section of that same *Annual Report*, Weaver described more fully the recent grants awarded under the Molecular Biology heading. Only one of these, a grant to Astbury at Leeds, could be considered in the "structural" tradition; one, to Curt Stern at the University of Rochester, to explore "problems related to hereditary changes and genic structure" by attempting to induce mutations in fruit flies with radioactive phosphorus, could be considered to come close to the "informational" category.[30] Other genetically based projects (for pigeon immunogenetics, for dog cystinuria, and for producing standardized mouse colonies) were placed in their own separate category and not collected under the Molecular Biology grouping. The bulk of the support for Weaver's molecular biology, however, was for several different types of physiological and metabolic studies of cells, especially of the energy-transforming nerve and muscle cells. Most of the support given in 1938 for molecular biology, then, could be regarded as falling more closely within a "bioenergetic" tradition than any other. Among other Natural Sciences

Division-sponsored projects started before 1938 that would also have qualified for molecular biology status from the Foundation was the spectroscopy project set up at Utrecht. So the meaning attached to Weaver's use of "molecular biology" could, and did, encompass molecular bioenergetics, including photosynthesis.

The Foundation also dedicated a substantial degree of funding to selected academic institutions as a means of establishing cooperative efforts by biologists and physical scientists in the pursuit of molecular studies of biological systems. Caltech was a major locale for the receipt of this funding, and biophysics was included in the program of support. Photosynthesis researcher Robert Emerson was one such beneficiary of the Foundation's largesse while he was working on photosynthetic mechanisms in Thomas Hunt Morgan's biology division in the Kerckhoff Laboratories from 1930 to 1937.[31] Weaver was well aware of this work, pointing out in 1933 that "the department has, so far, heavily emphasized work in biophysics, biochemistry and genetics".[32] However, despite Emerson's geographic proximity to the genetics workers, his degree of contact with them during that time (and any reciprocal influences that may have occurred as a result) appears to have been minimal.[33]

In 1952, the term "molecular biology" also came to be used by the newly inaugurated National Science Foundation (NSF) in the United States. The principal body within NSF charged with disbursing funds for research in the biological and medical sciences chose to eschew a traditional division by type of organism, deciding instead to award grants in "functional" areas.[34] One of the functional areas so identified was molecular biology, and among the kinds of work that were considered to properly fit in here were studies on structure-function relationships in chloroplasts.[35] By the NSF guidelines, then, studies of photosynthetic processes were a valid form of "molecular biology". An examination of the projects receiving funds in the early years shows that work on photosynthesis was indeed supported by molecular biology funds. Before the demise of the microbiology division (it was abandoned because it was felt to represent a morphological and not a functional category), some photosynthetic work on algae, which would otherwise have been placed within the molecular biology grouping, was accommodated there as well.[36] From its start, NSF had sought to promote photosynthesis research: it was cited as one of the three fields of interest, along with immunochemistry and protein structure and synthesis, "that are of great value to the progress of man and to his social and biological welfare and . . . hold out significant and fundamental value scientifically".[37]

Admittedly, the decisions made by public and private funding agencies may depend more on practical and political factors than on philosophic ones, and they cannot be regarded as definitive evidence of the correctness of any scientific label. But they do show the range of acceptable meanings – of areas of research that were seen to fit – when the term was first used.

What meaning were the laboratory scientists themselves attaching to molecular biology? The *Journal of Molecular Biology (JMB)*, the first journal identifying molecular biology as a central research focus and using it in its title, appeared in 1959 with John Kendrew as editor-in-chief. In the "Instructions to Authors" section of that first issue, the subject matter appropriate for publication is described as follows:

The Journal of Molecular Biology will publish papers on the nature, production and replication of biological structures at the molecular level and its relation to function. Suitable subjects are sub-cellular organization: molecular genetics: structure and replication of viruses: molecular structure of muscle, nerve and other tissues: structure of proteins, nucleic acids, carbohydrates, lipids, etc. and their synthetic analogues, as investigated by X-rays, light absorption and other methods: *problems of inter- and intra-molecular energy transfer.*[38]

The area staked out here as lying within the realm of molecular biology goes beyond DNA central dogma and related questions to include structure and function of specialized cells (not just bacteria and viruses), all macromolecules (not just DNA and protein), and problems of energy transfer (not just problems of molecular structure and information). Even though the *JMB* changed its emphasis in later years – preferring not to publish papers on lipids, for example – it still continues to consider "energy transfer" as falling within its purview.[39]

A survey of papers and letters published in the first years of the *JMB* shows that the great preponderance of papers were related to DNA and protein structure, function, and interaction, as well as to the regulation of gene function. The editorial board and the advisory board were made up of researchers experienced in dealing with these questions. Melvin Calvin, who won the Nobel prize for working out the cyclic, dark reactions for fixing carbon dioxide, was the only person with credentials in photosynthesis on either the advisory or the editorial board. But there were some papers on photosynthesis published in the *JMB*: ten appeared over the first seven years and fourteen volumes of publication. Two papers came from Calvin's own group, one on the "physical separation of metabolic pools of chemically identical substances participating in both respiration and photosynthesis in the *Chlorella*"[40] and the other on "the molecular organization responsible for the energy conversion process".[41] After 1965, only an occasional paper on photosynthesis can be found in an entire volume, even with eight journal volumes coming out each year, and most of these papers deal with the characterization of the DNA and the protein-synthesizing system found in the chloroplast. While this is not an impressive showing – some reasons for which will be offered in the next section – it does demonstrate that bioenergetics in general, and photochemical conversions in particular, were seen by the journal most closely associated with the field as a valid area within the molecular biology agenda.

CONCLUSION

If the work carried out to gain a detailed understanding of the process of photosynthesis, and probably other types of bioenergetic conversions as well, fulfills the criteria of a molecular biology, and if the groups funding this research and those who worked in the laboratory regarded it as such, why has it been necessary for me to argue here that bioenergetics should always have been counted as part of – indeed, may have been in the forefront in establishing – the molecular biology tradition? Why have the investigators themselves not insisted on correcting the historical record that has emerged thus far? I would like to offer two reasons: one institutional, the other scientific.

First, the scientists working on the detailed examination of photosynthetic mechanisms were perfectly well integrated into their research environments – whether these were university departments of botany, biology, plant physiology, or agricultural science, or whether they were privately funded efforts to figure out ways to expand the food supply or to use solar energy as a new source of power. The goal of gaining a detailed picture of photosynthesis was appreciated and encouraged by their colleagues in allied areas, even if their forms of analysis and the data they collected were not understood. The most classically trained botanists, for example, focusing on plant systematics, still knew that the ability to carry out photosynthesis was a fundamental feature of the organisms on which they were concentrating, and that their own research efforts could be enriched by knowing more about the ways in which photosynthesis occurred and what it demanded of or contributed to whole organism function. Work on the molecular basis of photosynthesis was regarded as useful, indeed essential, in these research environments. And there were many well-established outlets for professional interaction and publication: specialist journals such as *Plant Physiology, Enzymologia*, and *Biochimica et Biophysica Acta* were available for reporting work from the biological side, and journals such as the *Journal of Biological Chemistry* and the *Journal of Physical Chemistry* published the work with a more physical or physical-chemical cast.[42] Thus, those studying the molecular basis of photosynthesis could continue to work productively in their usual institutional settings under their preexisting professional designations of biochemist, biophysicist, plant cell biologist, or even botanist.

In contrast, many scientists working on molecular genetics had a less sympathetic reception among their colleagues, who felt that the way the whole organism functioned – and the developmental changes that occurred – required explanations of interactions above the cellular level and could not be accurately extrapolated from processes studied exclusively at the molecular level. The molecular geneticists tended to meet far more resistance to their genre of work among their professional colleagues than did those working on energy transformations. One way to resolve the tensions created by conflicts between molecular and organismic views is to create separate and more congenial environments for molecular studies of genetic mechanisms: separate units, or even whole new departments, with new labels like "molecular biology", were an obvious solution. Journals such as *JMB* were not needed by the molecularly oriented plant physiologists, but were vital to molecular geneticists to support their efforts to carve out and validate their niches in genetics.

The second reason that those working on photosynthesis and (though this needs to be examined further) other areas of bioenergetics do not form a visible entity within current molecular biology has to do with the success that has been achieved in the last four decades in studies in molecular genetics. As Watson and Crick noted in their landmark paper in 1953, their proposed structure of DNA with its striking feature of complementary base-pairing immediately pointed to possible mechanisms of replication. It also opened up new pathways for thinking about information transfer to the protein level, as well as about the regulation of gene function. A number of enormously fruitful areas of investigation could be staked out and exploited, while nonproductive routes such as plasmagene hypotheses could be quickly abandoned. And with these experimental successes has come a shifting of additional resources into the field of molecular genetics. What happened in 1953, then, was not the beginning of molecular biology, but the catapulting of a particular branch of molecular biology – molecular genetics – into prominence.

In contrast to this, the knowledge of the structure of chlorophyll, worked out by Hans Fischer in 1940,[43] and even the ability to synthesize it in the laboratory, accomplished by R.B. Woodward in 1960,[44] did not make obvious the next stages of experimental or theoretical work, nor did they help separate more-useful from less-productive models. The same could be said to be true for the alpha-helix, which did not create a clear road-map for further experimental studies. The complex nature of photosynthetic processes makes their analysis particularly difficult, and minority views – such as that of Warburg, that even one quantum of light could bring about the transformation of carbon dioxide into carbohydrate – were able to coexist with other models in the photosynthesis literature for long periods of time. In recent years there has been a shifting of financial resources away from both basic and applied work in photosynthesis.[45] Applied research, in particular, has been diminished, as efforts to improve the food supply have tended to proceed through genetic means rather than by trying to manipulate the photosynthetic process itself. The success attained by those working under the "molecular biology" label has even led the plant physiologists who originally had no need to rely on this term to begin now to use it themselves to describe their own work.[46]

The spectacular successes achieved in molecular genetics have relegated other areas of molecular biology to the shadows. If those of us concerned with the history and philosophy of science fail to take account of these other areas – especially, as I have argued here, bioenergetics – we will be left with a skewed sense of the molecularization of biology and of the ways in which new scientific disciplines are actually generated. A more inclusive view of molecular biology will allow us to appreciate the many other patrons, such as the Kettering Foundation and the Fels Foundation, who created opportunities for cross-disciplinary sharing between physics and biology.[47] It will permit us to assess the contributions of the many other laboratories, such as the Carnegie Institution's Department of Plant Biology or the Biophysical Laboratory at Utrecht, where biological studies at the molecular level were initiated and pursued. It will enable us to consider the part played by different patterns of scientific training in encouraging molecular modes of thinking: physiologists, for example, were given more training in physics, mathematics, and chemistry than were geneticists educated at the same time. It will facilitate the evaluation of the role of other forms of instrumentation – ultraviolet and visible spectroscopy, spin resonance, manometry, or the oxygen electrode – in providing new forms of access to the molecular level, or even assessing their eventual impact on genetic research. Otherwise, we run the risk of proceeding along an unprofitable scholarly pathway: gaining solely a view of molecular genetics, only to lose the whole world of molecular biology.

ACKNOWLEDGMENTS

This work was aided by useful discussions with Toby Appel, Robert Knox, David Robinson, René Wurmser, and participants at the Boston University Workshop on the Philosophy and History of Molecular Biology (April 1991), as well as by helpful comments on the manuscript by Richard Burian, Marjorie Grene, Ann La Berge, Muriel Lederman, and Jennifer Zallen. Adam H. Serchuk ably assisted in the content analysis of the *Journal of Molecular Biology*. Hospitality provided by the History of Science and Technology Group at Imperial College of Science, Technology and Medicine and by the Wellcome Institute for the History of Medicine in London is also greatly appreciated.

Virginia Polytechnic Institute and State University,
Blacksburg, VA, USA

NOTES

[1] As described in W.T. Astbury, 'Adventures in Molecular Biology', *Harvey Lect.* **46** (1950–51), 3: "It implies not so much a technique as an approach, an approach from the viewpoint of the so-called basic sciences with the leading idea of searching below the large-scale manifestations of classical biology for the corresponding molecular plan. It is concerned particularly with the *forms* of biological molecules, and with the evolution, exploitation and ramification of those forms in the ascent to higher and higher levels of organization. Molecular biology is predominantly three-dimensional and structural – which does not mean, however, that it is merely

a refinement of morphology. It must of necessity enquire at the same time into genesis and function."

[2] Linus Pauling, 'Fifty Years of Progress in Structural Chemistry and Molecular Biology', *Daedalus* **99** (1970), 988–1014.

[3] This view is most forcibly expressed in the collection put together to mark Max Delbrück's sixtieth birthday: *Phage and the Origins of Molecular Biology*, ed. J. Cairns, G.S. Stent, and J.D. Watson (Cold Spring Harbor, NY: Cold Spring Laboratory of Quantitative Biology, 1966).

[4] An exchange in the literature between John Kendrew and Gunther Stent illustrates the separate descriptions of molecular biology – structural and informational – held by those biologists associated with each type of work. See J.C. Kendrew, 'How Molecular Biology Started', *Sci. Am.* **216** (1967), 141–4; and Gunther S. Stent, 'That Was the Molecular Biology That Was', *Science* **160** (1968), 390–5.

[5] James D. Watson and F.H.C. Crick, 'A Structure for Deoxyribose Nucleic Acid', *Nature* **171** (1953), 737–8.

[6] David Baltimore, 'Foreword', in *Nobel Lectures in Molecular Biology, 1933–1975* (New York: Elsevier, 1977), p. viii.

[7] One example of this is the following statement: "The birth of molecular biology occurred in 1953 when Watson and Crick, using data obtained by Rosalind Franklin and Maurice Wilkins, proposed the now well-known structure of DNA" (Helen M. Berman, 'How Eco RI Recognizes and Cuts DNA', *Science* **234** (1987), 1482–83).

[8] It should be emphasized here that each one of these features is shared by *both* the structural and informational forms of molecular biology. Item (1), for example, is central both to Astbury's own description of molecular biology from a structural school perspective (see above, n. 1) and to that of the informational school, with the latter's prime attention being given to one particular molecule: DNA; see G.S. Stent, *Molecular Biology of Bacterial Viruses* (San Francisco: Freeman, 1963). Item (2) derives from the fact that, for each of these schools, experimental questions, coupled with the requirements of the instruments employed, caused attention to be concentrated either on isolated biomolecules, usually in a crystalline state, or on the least complex biological entity displaying the hereditary phenomena of interest, such as bacteriophage or protein-synthesizing systems composed of the requisite subcellular components. Items (3) and (4), respectively reflect the two schools' dependence on experimental techniques from the domain of the physical sciences – X-ray diffraction, electron microscopy, ultracentrifugation, and the like – to examine biological material; and the movement of some physicists and physical chemists to leading roles in biological research, bringing the quantitative and analytical tools of the physical sciences with them to the biological setting – as, for example, J.D. Bernal and M. Wilkins to the study of biomolecular structure, or M. Delbrück to the analysis of phage genetics. The criteria cited here are also analogous to those advanced for molecular biology in L. Kay, *The Molecular Vision of Life: Caltech, the Rockefeller Foundation, and the Rise of the New Biology* (New York and Oxford: Oxford University Press, 1992), p. 4.

[9] James D. Watson, *The Double Helix* (New York: New American Library, 1969), p. 126.

[10] Particularly helpful reviews of the recent history of photosynthesis can be found in Jack Myers, 'Conceptual Developments in Photosynthesis, 1924–1974', *Plant Physiol.* **54** (1974), 420–6; C. Stacy French, 'Fifty Years of Photosynthesis', *Ann. Rev. Plant Physiol.* **30** (1979), 1–26; D.I. Arnon, 'Photosynthesis 1950–1975: Changing Concepts and Perspectives', in *Photosynthesis I: Photosynthetic Electron Transport and Photophosphorylation*, ed. A. Trebst and M. Avron (Berlin and Heidelberg: Springer-Verlag, 1977), pp. 7–56; and Govindjee and R. Govindjee, 'Introduction to Photosynthesis', in *Bioenergetics of Photosynthesis*, ed. Govindjee (New York: Academic Press, 1975), pp. 1–50.

[11] O. Warburg and E. Negelein, 'Über den Energieumsatz bei der Kohlensäureassimilation', *Z. physikal. Chem.* **102** (1922), 235–66; O. Warburg and E. Negelein, 'Über den Einfluss der Wellenlänge auf den Energieumsatz bei der Kohlensäureassimilation', *Z. physikal. Chem.* **106** (1923), 191–218.

¹² See René Wurmser, 'Le rendement énergétique de la photosynthèse chlorophylliene', *Ann. Physiol. Physiochim. Biol.* **1** (1925), 47–63; idem, *Oxydations et réductions* (Paris: Presses Universitaires de France, 1930). More recently Wurmser has called attention to this work in idem, Letter to the Editor, *Photosynth. Res.* **13** (1987), 91–3. Van Niel's contributions are summarized in C.B. van Niel, 'Photosynthesis of Bacteria', *Cold Spring Harbor Symp. Quant. Biol.* **3** (1935), 138–150; and idem, 'The Bacterial Photosyntheses and Their Importance for the General Problem of Photosynthesis', *Adv. Enzymol.* **1** (1941), 263–328.

¹³ R. Hill, 'Oxygen Evolution by Isolated Chloroplasts', *Nature* **139** (1937), 881–2; idem, 'Oxygen Produced by Isolated Chloroplasts', *Proc. R. Soc. London*, Ser. B, **127** (1939), 192–210. This photolysis, which Hill called the "chloroplast reaction", has become generally known as the "Hill Reaction".

¹⁴ Among the papers that describe this work are M. Calvin and A.A. Benson, 'The Path of Carbon in Photosynthesis', *Science* **107** (1948), 476–80; W. Stepka, A.A. Benson, and M. Calvin, 'The Path of Carbon in Photosynthesis II: Amino Acids', *Science* **108** (1948), 304; M. Calvin and A. Benson, 'The Path of Carbon in Photosynthesis IV: The Identity and Sequence of the Intermediates in Sucrose Synthesis', *Science* **109** (1949), 140–2; M. Calvin, 'The Path of Carbon in Photosynthesis', *Harvey Lect.* **46** (1952), 218–25.

¹⁵ R. Emerson and W. Arnold, 'The Photochemical Reaction in Photosynthesis', *J. Gen. Physiol.* **16** (1932), 191–205.

¹⁶ The papers that first put this idea forth are R. Hill and Fay Bendall, 'Function of the Two Cytochrome Components in Chloroplasts: A Working Hypothesis', *Nature* **190** (1960), 136–7; Bessel Kok and George Hoch, 'Spectral Changes in Photosynthesis', in *A Symposium on Light and Life*, ed. William D. McElroy and Bentley Glass (Baltimore: Johns Hopkins Press, 1961), pp. 397–423; and L.N.M. Duysens, J. Amesz, and B.M. Kamp, 'Two Photochemical Systems in Photosynthesis', *Nature* **190** (1961), 510–11.

¹⁷ P. Mitchell, 'Coupling of Phosphorylation to Electron and Hydrogen Transfer by a Chemiosmotic Type of Mechanism', *Nature* **191** (1961), 144–8.

¹⁸ See, for example, J.R. Norris, R.A. Uphaus, H.L. Crespi, and J.J. Katz, 'Electron Spin Resonance of Chlorophyll and the Origin of Signal I in Photosynthesis', *Proc. Nat. Acad. Sci. U.S.A.* **68** (1971), 625–8; D.C. Youvan and B.L. Marrs, 'Molecular Mechanisms of Photosynthesis', *Sci. Am.* **256** (1987), 42–8; and W.W. Parson, 'Electron Transfer in Reaction Centers', in *Chlorohylls*, ed. H. Scheer (Boca Raton, FL: CRC Press, 1991), pp. 1153–80.

¹⁹ For the first report from animal systems (rat liver), see P. Siekevitz, 'Uptake of Radioactive Alanine in Vitro into the Proteins of Rat Liver Fractions', *J. Biol. Chem.* **195** (1952), 549–65. For bacterial systems (*Escherichia coli*), see J.H. Matthaei and M.W. Nirenberg, 'Characteristics and Stabilization of DNAase-sensitive Protein Synthesis in *E. coli* extracts', *Proc. Nat. Acad. Sci. U.S.A.* **47** (1961), 1580–88; and M.W. Nirenberg and J.H. Matthaei, 'The Dependence of Cell-Free Protein Synthesis in *E. coli* upon Naturally Occurring or Synthetic Polyribonucleotides', ibid., pp. 1588–602.

²⁰ D.I. Arnon, 'The Chloroplast as a Complete Photosynthetic Unit', *Science* **122** (1955), 9–16.

²¹ The photosynthetic microbes used in this research are the green bacteria (Chlorobacteriaceae) and the purple sulfur (Thiorhodaceae) and nonsulfur purple (Athiorhodaceae) bacteria.

²² Eugene Rabinowitch, *Photosynthesis and Related Processes* (New York: Interscience Publishers, 1945), p. v.

²³ F. P. Zscheile, Jr., 'An Improved Method for the Purification of Chlorophylls a and b', *Bot. Gaz* **95** (1934), 529–62; idem, 'Toward a More Quantitative Photochemical Study of the Plant Cell's Photosynthetic System', *Cold Spring Harbor Symp. Quant. Biol.* **3** (1935), 108–15.

²⁴ D.T. Zallen, 'The Rockefeller Foundation and Spectroscopy Research: The Programs at Chicago and Utrecht', *J. Hist. Biol.* **25** (1992), 67–89.

²⁵ Robert Emerson and William Arnold, 'A Separation of the Reactions in Photosynthesis by Means of Intermittent Light', *J. Gen. Physiol* **15** (1932), 391–420.

[26] Max Delbrück, well known as a founder of the phage-based informational school of molecular biology, had also given serious consideration to directing his research efforts to studies of photosynthesis. In the mid-1930s, he held meetings with other scientists (including Hans Gaffron, Eugene Rabinowitch, and C. Stacy French) devoted to photosynthesis topics at his house in Berlin. See French, 'Fifty Years of Photosynthesis' (above, n. 10), pp. 6–7. That he did not convert this interest into a viable research program has been attributed to his dislike for Otto Warburg's overbearing and inflexible behavior in scientific matters, especially those related to quantum yield measurements in photosynthesis and their implication for basic photosynthetic mechanisms. See E.P. Fischer and C. Lipson, *Thinking about Science: Max Delbrück and the Origins of Molecular Biology* (New York and London: W.W. Norton, 1988), p. 69. However, he did maintain his interest in photosynthesis throughout his lifetime – for example, taking van Niel's famed microbiology course at Stanford, which included photosynthesis as a central topic, and attending (apparently without any official role) a major conference on photosynthetic mechanisms held in 1960; see 'Additional Participant List' in McElroy and Glass, *Symposium on Light and Life* (above, n. 16), p. xi. Delbrück's first graduate student at Caltech, Roderick Clayton, would go on to become a leading investigator of photosynthetic bioenergetics.

[27] H.G. Kuhn, 'Franck, James', *Dict. Sci. Biog.* **5**, 118.

[28] *Annual Report* (New York: Rockefeller Foundation, 1938), pp. 40ff.

[29] Ibid., p. 43.

[30] Ibid., p. 222.

[31] See Kay, *Molecular Vision* (above, n. 8); E. Rabinowitch, 'Emerson, Robert', *Dict. Sci. Biog.* **4**, 362–3.

[32] Rockefeller Foundation Archives (hereafter cited as RFA), Warren Weaver officer's diary, September 8–10, 1933, RG 12.1, Rockefeller Archive Center, North Tarrytown, N.Y.

[33] In Emerson's papers, published during that time, he expresses close links with the electrical engineering and physics faculties, rather than with other groups within the biology division. Weaver's comments in several diary entries indicate that a measure of isolation existed, related in part, perhaps, to Emerson's personality and scientific training. During a visit in 1933, Weaver recorded that "R. Emerson . . . concentrates his interest in photosynthesis. He is more of a lone worker, probably not as effective as B[orsook]" (Warren Weaver officer's diary, October 23, 24, 25, 1933, RFA, RG 12.1); and in 1936, "S[poehr] says R. Emerson at CIT [California Institute of Technology, or Caltech] is shy, modest, and has no advertising capacity whatsoever. However, S[poehr] considers E[merson] a sound and important worker. S[poehr] thinks E[merson] is still a little too much under the dominating influence of his experience with Warburg, but that he will emerge" (Warren Weaver officer's diary, March 12, 1936, RFA, RG 12.1). Also, Emerson left Caltech to work at the Carnegie Laboratory of Plant Biology at Stanford just as the Rockefeller investment in the Beadle/Ephrussi collaboration was first beginning to bear fruit in understanding gene function, and as Max Delbrück arrived to carry out his first forays into genetics on a Rockefeller Foundation fellowship.

[34] This plan was based on ideas first put forth by Paul Weiss for the National Academy of Sciences: Paul Weiss, *Within the Gates of Science and Beyond: Science in Its Cultural Commitments* (New York: Hafner, 1971), ch. 5, pp. 79–85. In Weiss's view, molecular biology was "concerned with the elementary compounds, their interactions, transformations and the attendant energy balance"; to the cellular biology category went "the coordination of molecular events underlying orderly structure and function"; the "laws and mechanisms of heredity" were placed in the "*genetic biology*" category (p. 88).

[35] A.T. Waterman, 'The National Science Foundation and the Life Sciences', *Publ. Health Rep.* **69** (1954), 380. According to the NSF definition, "*Molecular Biology* can be categorized or defined as a study of the structure, syntheses and molecular properties of biologically important molecules and kinetic studies of their reactions. Involved are such studies as: Physicochemical

studies of purified natural substances such as proteins, nucleic acids, lipids, carbohydrates, etc., identification and structure of particulate matter such as mitochondria, *chloroplasts* [my emphasis], chromosomes, viruses, enzyme structure and kinetics – chemistry of coenzymes, electrochemical phenomena, membranes and fibers, solid and liquid state phenomena, reaction of proteins – long range forces, mathematical approaches to biological problems." In Toby Appel, *A History of the National Science Foundation* (in preparation) ch. 2: 'Beginnings: Biology at NSF, 1950–1952'.

[36] It is here that Emerson's work on "Carbon Dioxide Exchange during the Induction Period of Photosynthesis" was placed, along with Wolf Vishniac's project on 'Enzymatic Reactions in Photosynthesis and Chemosynthesis' in 1952, as was Jack Myers's project on the blue-green algae: *Second Annual Report of the National Science Foundation, Fiscal Year 1952* (Washington, DC: GPO, 1952), p. 47; *Third Annual Report of the National Science Foundation, Fiscal Year 1953* (Washington, DC: GPO, 1953), p. 77.

[37] William V. Consolazio to Fernandes Payne, 'Programs for Molecular and Genetic Biology: The State of the Biological Sciences Division', January 15, 1953, Annual Reports of the Biological and Medical Sciences Division, Historical Files, NSF, Washington, DC. NSF targeted work in this field not only by supporting individual research programs but by sponsoring a major international conference on photosynthesis at Gatlinburg, Tennessee, in 1951 and by financing the National Research Council's Committee on Photobiology.

[38] 'Instructions to Authors', *J. Mol. Biol.* **1** (1959), after p. 409 (emphasis added).

[39] See, for example, 'Instructions to Authors', *J. Mol. Biol.* **220** (1991), last unnumbered page; under item 1, suitable subjects for publication include "Organelle structure and function, motility, transport and sorting of macromolecules, *energy transfer* [my emphasis], growth control, genetics of development. . . ."

[40] V. Moses, O. Holm-Hansen, J.A. Bassham and M. Calvin, 'The Relationship between the Metabolic Pools of Photosynthetic and Respiratory Intermediates', *J. Mol. Biol.* **1** (1959), 22.

[41] Kenneth Sauer and Melvin Calvin, 'Molecular Orientation in Quantasomes I: Electric Dichroism and Electric Birefringence of Quantasomes from Spinach Chloroplasts', *J. Mol. Biol.* **4** (1962), 451.

[42] *Plant Physiology* was first published in 1924, *Enzymologia* in 1936, and *Biochimica et Biophysica Acta* in 1947. In 1973, *Enzymologia* changed its name to *Molecular and Cell Biology*. The *Journal of Biological Chemistry* started publication in 1905, joining the *Journal of Physical Chemistry*, which had commenced publication in 1896.

[43] H. Fischer and H. Orth, *Die Chemie des Pyrrols* (Leipzig: Akad. Verlagsges., 1937), vol. II, part 1; and H. Fischer and A. Stern, *Die Chemie des Pyrrols* (Leipzig: Akad. Verlagsges., 1940), vol. II, part 2.

[44] R.B. Woodward, 'The Total Synthesis of Chlorophyll', *J. Am. Chem. Soc.* **82** (1960), 3800–2.

[45] H. Kende, P.K. Stumpf and J.L. Key, 'The State of Plant Biology: Views from the Other Side of the Fence', *Cell* **56** (1989), 914–5.

[46] In 1988, for example, the *Annual Review of Plant Physiology* changed its name to the *Annual Review of Plant Physiology and Plant Molecular Biology*.

[47] See, for example, Ondess L. Inman, *The C.F. Kettering Foundation for the Study of Chlorophyll and Photosynthesis* (Yellow Springs, OH: Antioch College, 1937).

RICHARD M. BURIAN

UNDERAPPRECIATED PATHWAYS TOWARD MOLECULAR GENETICS AS ILLUSTRATED BY JEAN BRACHET'S CYTOCHEMICAL EMBRYOLOGY

INTRODUCTION

The present paper puts forward some controversial views about the history, historiography, and nature of molecular biology. These views arise out of series of case studies, now in progress. Because of space limitations, however, I shall develop here only part of one of those case studies in significant detail. Accordingly, I will utilize the first third of this chapter to develop some of the major claims about the formation and nature of molecular biology to which the larger series of case studies leads. I will then present parts of a case study dealing with aspects of the cytological and embryological work of a particularly interesting Belgian biologist, Jean Brachet, recently deceased. This work, originally morphological and biochemical in character, but eventually characteristic of molecular biology, illustrates very nicely a number of the points I wish to make. At the end of the paper I will return to the framework claims that this case study illustrates and supports.

FRAMEWORK

Molecular biology is unquestionably a highly theoretical field or group of fields. It is safe to say that molecular biologists deal in large part with distinctly theoretical entities. Yet I hold, controversially, that *molecular biology contains no central unifying theory*.[1] Molecular biology, I contend, is an agglomeration of fields and disciplines the accomplishments of which depend, in large part, on the ability to describe and manipulate theoretical entities (many of which, but by no means all, can now be indirectly visualized) and to alter or control their behavior. Molecular biology includes highly theoretical disciplines such as molecular genetics and the new immunology. But there is not, I claim, a central theory that binds together the many distinct (sub)disciplines of molecular biology. To be sure, some theorists argue that the so-called central dogma, according to which information flows from nucleic acid to protein, but not *vice versa*, counts such a theory. But they are wrong about this, as I will argue on another occasion. What serves to unify the distinct sorts of work in molecular biology together is *not* a central theory, but the use of an immense battery of techniques and a general approach to explaining – and altering – organismic function by reference to, and use of, an *omnium gatherum* of detailed molecular mechanisms. This point is nicely reinforced by Doris Zallen's chapter on photosynthesis (this volume).

In brief, molecular biology is less like Newtonian mechanics than it is like auto mechanics: What it studies are *mechanisms* and it *uses* those mechanisms to intervene in nature. Thus, the subject matter of molecular biology is detailed mechanisms. Characteristically, it studies those mechanisms "all the way down" to the molecular level. But it also works, as I will emphasize in the conclusion, "all the way up" to organismic structure, function, development, and evolution. Thus, although it makes significant use of laws, it does so more-or-less incidentally; *its central research objectives do not include the articulation of general laws*. Furthermore, there are no general laws that belong, characteristically, to molecular biology (see Beatty, 1995). This is not to deny the relevance of fundamental theory or of fundamental laws to molecular biology – who could deny it? Nonetheless, and this is my central point, *there is no large-scale theory peculiar to molecular biology*.

Despite the lack of an overarching theory, a Newtonian or quantum mechanics of its very own, molecular biology has become a unifying discipline in virtue of the power of its techniques, its ability to extrapolate from the molecular to higher levels, and its synthesis of problems of form and function at the molecular level. This synthesis of form and function is a central, ill-understood, and historically important feature of molecular biology. Among other things it explains the degree of truth – and of error – in Erwin Chargaff's and Seymour Cohen's mordant witticism that molecular biology is the practice of biochemistry without a license. We will return to this matter below.

One way of gaining a greater appreciation of the position just staked out is to explore some of the less appreciated pathways that contributed to the formation of molecular genetics. Even the limited case study presented below shows convincingly that the conventional histories do not take proper account of the amazing breadth of the disciplines, problems, techniques, lines of investigation, conceptual approaches, and so on that, in fact, made crucial contributions to the formation of molecular genetics as a discipline (let alone all of molecular biology).

One reason for the common but mistaken perception that work like that which I will discuss had little to contribute to molecular biology is that molecular biologists have marginalized a number of hypotheses and research gambits of the forties and fifties (and some of their key proponents), dropping them from their textbooks and informal histories. Kendrew's and Stent's notorious classification of 'the' two approaches (or schools?) making the primary contributions to the formation of molecular biology – the 'structural' and 'informational' schools – illustrates the point nicely; many important contributions that were crucial to the founding of molecular biology simply are not included in the Procrustean bed of this categorization.[2] Since historians of science are only beginning to pay extensive attention to the breadth of work in this period in the history of biology, many important figures, areas of research, and lines of work have not yet been adequately treated. There will surely be major changes in the historiography of molecular genetics when the balance is redressed. One example of a line of work worth extensive study, alluded to below and discussed briefly by Scott Gilbert (this volume) and Jean-Paul Gaudillière (1992), is the so-called plasmagene hypothesis. A fuller set of examples, some of which will be mentioned in passing below, would demonstrate some of the many ways in which the study of the molecular mechanisms of cellular defense, of morphogenesis in micro-organisms, of bioenergetics and respiration, of the kinetics of nutrition, and of the form of molecules and organelles, *inter alia*, contributed to the foundations of molecular biology.[3] Such work dissolved old discipline boundaries and led to a major reconfiguration of the biological sciences.

Within this framework the present limited discussion of Brachet's work uncovers only a few of the lacunae in the traditional accounts of the genesis of molecular biology. My discussion will emphasize the contributions of lines of work and traditions not usually discussed in treatments of the phage school, Watson and Crick, X-ray crystallography, Benzer's fine structure analysis of the gene, Jacob and Monod's work on the operon, and similar reasonably well-studied aspects of the founding of molecular biology. The work to be discussed here has simply not entered the historiographical lore of molecular biology. I shall emphasize especially the importance of certain sorts of physiological work to the founding of molecular biology, work which, I believe, provided crucial interconnections between studies of molecular form and function on the one hand and studies of cellular and organismic form and function on the other. A history of molecular genetics or, more generally, of molecular biology that restricts itself to the so-called structural and informational traditions cannot adequately account for the ability of molecular biology to make the connection to cellular and organismic function.

NUTRITIONAL STUDIES

Before turning to Brachet, it will help to set the tenor of my historical claims by recalling an often-overlooked setting for Beadle and Tatum's well-known studies of the early 1940s on nutritionally deficient mutants of *Neurospora*. Most readers of this chapter will know that many major developments of the forties and fifties took those studies as a point of departure. Beadle, Tatum, and their colleagues exploited techniques involving the use of mutagenized *Neurospora*, grown on minimal (i.e., nutritionally unsupplemented) media and, alternatively, on nutritionally supplemented media. This enabled them to detect mutants unable to digest specific sources of carbon because they lacked certain wild-type enzymes required to metabolize essential intermediate substances. The patterns of enzyme loss led Beadle and Tatum to the famous proposal that eventually came to be known as the one gene – one enzyme hypothesis.[4] This hypothesis rested mainly on the recognition that typical genetic lesions in *Neurospora* blocked a single well-defined enzymatically-controlled step required for the biosynthesis or degradation of a particular substance.

A seldom-recognized part of the background for Beadle's work, barely mentioned in the historical literature on molecular biology or by molecular biologists, traces back to the period of his joint program of research on *Drosophila* with Ephrussi in Paris and Berkeley in the mid 1930s. At that time he became closely familiar with André Lwoff's work on growth factors and nutritional competence in bacteria, ciliates, and other micro-organisms.[5]

In this research, Lwoff and his colleagues dissected some of the biochemical pathways required for the digestion of various carbon sources in micro-organisms and characterized specific steps that specific organisms could not perform when they depended on a host or food source to provide a substance or manifested some sort of metabolic block or enzymatic deficiency. They showed that closely related organisms often differ in their ability to carry out a single synthetic or degradative step (or, more commonly, a small series of related steps) required to utilize a particular carbon source; they held that the ability to carry out each of these steps was controlled by the presence or absence of an enzyme or enzymes. Enzymes, they claimed, exhibited "genetic continuity" – i.e., they arose from other enzymes or enzyme precursors by some sort of duplication and template system rather than being manufactured *de novo*.[6] Lwoff and his co-workers also provided a technique for identifying the biochemical lesions of micro-organisms incompetent to grow on minimal media – the study of the growth kinetics of cultures derived from well-characterized initial stocks raised on biochemically defined media. These techniques, highly adapted, are prominent in the work of Beadle and Tatum in the early forties.

This account of one source contributing to the Beadle-Tatum research program in no way minimizes the originality or importance of their work on the genetic control of nutritional competence in *Neurospora*; rather, it shows that that work draws on a physiological tradition seldom recognized as contributing to the formation of molecular genetics. Yet the tools and techniques developed in this tradition provided important criteria for determining the number, the specificity, and the chemical roles of enzymes involved in particular biosynthetic or degradative pathways – information that became critical for determining whether the presence or absence of a particular enzyme was under genetic control.[7] That very same tradition, as Jean Gayon, Doris Zallen, and I have argued, played a central role in the career of one of Lwoff's protégés, Jacques Monod (see Burian, 1990; Burian and Gayon, 1991; Burian and Gayon, in preparation; and Burian et al., 1988).

The general point cannot be pursued in detail here; it is that physiological, specifically nutritional, work of this sort depends heavily on kinetic studies of growth and metabolism. Physiological work along such lines provided a crucial background for many early studies contributing to what is now called molecular biology. For one example, the work of Lindegren, Monod, Spiegelman,[8] and others on so-called enzymatic adaptation (later, enzyme induction), i.e., the competence of cells to switch from one carbon source to another without genetic change, is essentially physiological and kinetic in character. Many of the founding figures of molecular biology, including Ephrussi, Hershey, Lederberg, Lindegren, Luria, Monod, Pontecorvo, Spiegelman, and many others drew on a background in nutritional studies and allied work – a fact that I believe is not widely appreciated and the importance of which is not widely understood. This background helps explain the centrality of the many studies of chemical kinetics in the early days of molecular biology – an importance that is very hard to understand on accounts that trace the foundations of molecular biology back primarily to the interaction of structural and informational traditions.

A COMMENT ON BIOCHEMISTRY

This section prepares the way for a discussion of some of Jean Brachet's work and identifies the error in treating molecular biology as the practice of biochemistry without a license. For both purposes it is important to discuss some of the cytochemical work that contributed to the foundations of molecular genetics.

It will help to begin by stating one of the morals of this part of the story. A central distinction between molecular biology and biochemistry, at least as the latter was practiced in the forties and fifties, turns on the chemical, as opposed to biological, orientation of biochemists. This meant that biochemists often treated the chemical formula and isomeric structure of a molecule as very nearly the whole story, so that physical binding of enzymes or proteins to membranes, or location of an enzyme on or in an organelle, was thought to be, at best, of secondary importance. The image of the biochemistry of the day as treating the cell as a bag of enzymes is not, after all, *entirely* a caricature. It has *something* to do, for example, with the resistance to Peter Mitchell's chemiosmotic hypothesis.[9] In this respect, aspects of biological form or structure were of secondary or tertiary interest to many biochemists in the period when molecular biology was getting off the ground. In particular, many biochemists resisted the ideas that proteins or enzymes would function differently in different cell compartments and that one had to take account of their interactions with membranes or organelles. Accordingly, many biochemists were not particularly concerned with the localization of particular biochemical interactions on particular structures.[10] In contrast, the need to work out such localization was a commonplace to most of the physiologically oriented biologists who participated in the formation of molecular biology.

For example, in characterizing the cytochromes (enzymes crucial to respiration) in the 40s and 50s, the distinction between membrane bound and soluble cytochromes, eventually crucial to molecular biologists and biochemists studying mitochondrial function, was irrelevant to most biochemists' attempts to analyze respiration.[11] In molecular biology, as it eventually came to be practiced, it became critical to understand the molecular interlocking of forms – literally shapes – that determined the breaking and creating of bonds, the anchoring of a molecule on a membrane and so on. This is part of the distinctive wedding of form and function in molecular biology in contrast to earlier biochemistry. Consider, for example, the changes of shape that oxygenation causes in the hemoglobin molecule, the conformal effects of the breaking of hydrogen bonds in analyzing enzymatic interactions or allosteric transitions, the complex configuration of molecules required, for example, to manufacture ATP in photosynthesis, or recent findings about the conformal peculiarities of membrane-binding domains, and so on. Indeed, *form* and *change of form*, treated along these lines, came to be the keys to the analysis of molecular mechanisms. The pathways by which such issues came to play a central role in molecular biology are of great interest – particularly since they connect also to the analysis of the form and structure of various cellular organelles.

BRACHET'S CHEMICAL EMBRYOLOGY AND THE DETERMINATION OF NUCLEIC ACID FUNCTION[12]

I turn now to the work of Jean Brachet (1909–1988), whose use of cytochemical techniques mainly in the service of understanding embryological development and its regulation led him, as we shall see, to make fundamental contributions to molecular genetics. Partly because of his orientation toward embryological, morphological, and biochemical questions, Brachet has not figured centrally in standard histories of molecular biology. Judson (1979) describes part of his work on nucleic acids and protein synthesis from the late thirties to 1947 on five pages of *The Eighth Day of Creation*. At p. 236 he puts his finger on one of the key reasons that Brachet is neglected; he cites Crick's complaint about the messiness of [embryological, physiological, and biochemical] work on topics such as RNA turnover. Part of the mythology of molecular biology is that its founders did everything possible to get around the messiness, complication and muddle of approaches such as Brachet's (and others who were more muddled). The contrast between embryological lovers of complexity and molecular biologists' drive for simplicity is, indeed, a central part of the folklore of molecular biology.[13] The absence of much of Brachet's work (only a small part of which is covered below) from standard histories may result partly from the fact that, in this respect, his contributions to molecular biology came from the wrong side of the tracks.

Brachet's work on chemical embryology began as an extension of his father, Albert's, research on causal embryology. The part of the son's work that is of immediate interest here stems from his effort to understand the synthesis, localization, and physiological role of nucleic acids in embryonic development. As early as 1933,[14] he obtained results with virgin sea urchin eggs, suggesting that "yeast" (or pentose) nucleoproteins were present exclusively in the cytoplasm and that small quantities of "thymus" nucleic acid were present in the nucleus. (In spite of the anachronism, for convenience I shall use the modern labels, *RNA* and *DNA* respectively, for the two nucleic acids from here on.) Noting that the amount of RNA nucleoprotein in the sea urchin embryo's cytoplasm decreased in rough synchrony with the increase in nuclear DNA after fertilization, he hypothesized that the pentoses in the RNA served as a reserve of precursors that were transformed into DNA during embryonic development.

The techniques Brachet used to localize RNA and DNA were largely cytochemical.[15] He employed Feulgen, Unna and toluidine blue stains – the second of which usually stains RNA red and DNA green. These were combined with the use of deoxyribonuclease and (after 1938) ribonuclease to confirm that the color produced was, indeed, due to the presence of the suspected nucleic acid. Over the years, these techniques were refined and cross-checked, and the findings obtained by their use integrated with more sophisticated techniques and hypotheses.[16] The spirit of his enterprise, however, is already clear in the paper of 1933.

In the 1933 paper, Brachet noted that cells producing large quantities of protein or enzymes had high concentrations of RNA, suggesting that RNA might somehow be connected with protein synthesis. This suggestion, at first based on rather diffuse evidence, came to play a larger role in Brachet's subsequent discussions of the physiology of nucleic acids.

By 1942,[17] he was able to demonstrate high concentrations of RNA in the ergastoplasm (i.e., basophilic cytoplasm, especially what later came to be recognized as endoplasmic reticulum) and in nucleoli, but was also able to show that up to 10% of the nucleic acid in nuclei is RNA. By 1944, the topic of the synthesis, localization and physiological role of nucleic acids was important enough for Brachet to devote 56 of the 500 pages of his *Embryologie chimique*[18] to it. Although he did not yet identify genes with nucleic acid, he argued, against Koltzoff (1939), that the proportion of DNA in chromomeres (which, like many cytologists of the day, he thought might correspond to genes) was constant throughout the cell cycle, so that DNA might be a constant component of genes (pp. 70–1). He demonstrated a relationship between the quantity of RNA and the amount of synthetic activity in various cells, noting, *inter alia*, that an increase of RNA is required for specialized cells to begin facultative production of their distinctive products and that secretory functions are expressed in proportion to the amount of RNA in the endoplasmic reticulum. He showed that RNA in the endoplasmic reticulum is associated with microsomes, as isolated by Albert Claude (1941, 1943) using ultracentrifugation. The smallest microsomes maintain a constant ratio of nucleic acid to protein, though they also form larger units containing more protein.

There are better articulated hints of new findings in a paper presented in 1948 on "Plasmagenes in Development" at a symposium on "Units Endowed with Genetic Continuity".[19] The plasmagene paper makes some important additions to Brachet's account of nucleic acid physiology. The smallest microsomes, or ribonucleoprotein granules, have a remarkably constant chemical constitution. Their importance in protein synthesis is shown by the fact that in cells making a lot of a particular product, e.g., hemoglobin, granules with varying amounts of the product attached to them can be isolated. The granules exhibit a sort of growth, starting from their basal size.[20] They behave like a "sort of nucleoprotein germ for diverse molecules – proteins, which may or may not have enzymatic activity, and lipids" (Brachet, 1949, p. 156). Brachet reports parallel findings for respiratory enzymes and mitochondria. Furthermore, he offers evidence that "the synthesis or the multiplication of the particles occurs under nuclear control" (p. 157).

The resultant picture is consistent with, but does not decisively favor, an account of biosynthesis that treats the ribonucleic granules as plasmagenes. Such an account would involve nuclear synthesis of granules with specific competences in the cytoplasm – or nuclear regulation of their activation. If the granules were self-reproducing, which is what was meant by "endowed with genetic continuity" in this context, and if they, in turn, synthesized (or controlled the synthesis or specificity of) proteins and other cellular products, one would obtain the outline of a general solution to the problem of protein synthesis. A huge number of details were missing – and, in their absence, Brachet was duly skeptical of the plasmagene story – but at least the picture was consistent with the available information and provided a clear direction for further research.

By 1950, Brachet recognized a complication for the plasmagene hypothesis. He recognized that at least one more fraction of RNA, critical for protein synthesis, remaining in the supernatant after ultracentrifugation, is present in cells that are actively producing some protein or ready for major growth. There are some complications yet to be resolved in understanding the historical sequence of his treatment of this RNA. One of the problems is its relation to so-called "soluble RNA". In American usage by the later fifties, "soluble RNA" was produced at pH 5; at least in Zamecnik's laboratory, but I think quite generally, it appears that only transfer RNA counted as soluble RNA on the western side of the Atlantic.[21] I have yet to puzzle out Brachet's usage in 1950; judging by the quantities he measured, however, there probably were transfer, messenger, and non-membrane-bound polysomal RNAs (though without any recognition of the distinctions among them) among the RNAs remaining in the supernatant, for he determined that "a large portion of [RNA], which may exceed 50 per cent of the nucleic acid present in the extracts of

UNDERAPPRECIATED PATHWAYS TOWARD MOLECULAR GENETICS   75

frog eggs and embryos, as well as of chick embryos, is not sedimented by ultracentrifugation" (Brachet, 1950, p. 864).[22]

By 1952, he had a picture of protein synthesis that looked like a first approximation of what Crick later termed the "central dogma" of molecular biology (Crick, 1970), arguing, in effect that nuclear DNA makes RNA and that nucleic acid provides the specificity for proteins, but not *vice versa*. To be sure, he had, at this point, no notion of the structure of DNA or of a code, but as the concluding Diagram of Brachet (1952), reproduced here as Figure 1, shows, a moderately accurate and modern picture of protein synthesis was beginning to emerge from his cytochemical and kinetic studies.

Although, in act, the RNAs in the supernatant whose importance Brachet stressed in 1950 included an unholy mess of different RNAs, during the next few years he was able to show that they are critical to the determination of protein specificity and to separate them into a variety of classes. Furthermore, Brachet already knew that ribonuclease stops protein synthesis but left microsomes intact. Sometime before 1960 he also knew that addition of ribosomes to a cell stimulates protein synthesis, but non-specifically. (The precise date, which I do not yet know, is critical, as should be clear from note 21.) That

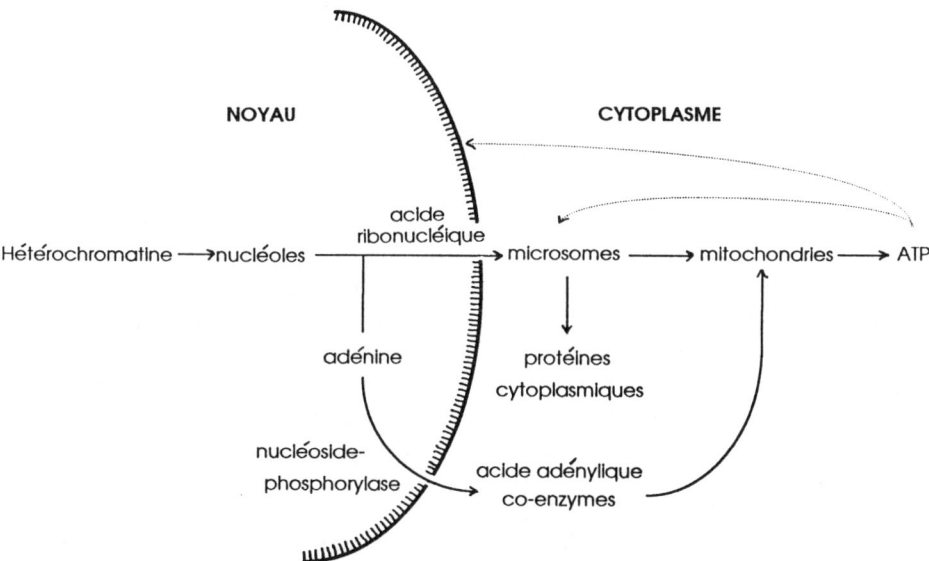

Figure 1. Figure redrawn from p. 115 of Brachet (1952), showing the scheme of protein synthesis at which he had arrived by that date. Note the feedback control loops and the approximation to subsequently-accepted details of the mechanisms of protein synthesis as well as the obvious similarity to the doctrines underlying the claim that nuclear DNA makes RNA and the "central dogma" that information is transferred only in the direction from nucleic acid to protein (Crick, 1970).

is, he knew that *which* proteins a cell produces depends on the non-centrifugable RNAs of the host cell and not on the source from which the injected ribosomes were derived.[23] Once this was clear, it was obvious that if RNA is the intermediary between the genes and the production of protein, it was not ribosomal RNA, but the soluble RNA or some other RNA fraction that was responsible for the specificity of protein synthesis. To fill in the case study properly, it is essential to determine the precise timing of Brachet's determination of these results.

One special investigation deserves particular mention, for it provides an especially elegant illustration of the bearing of cytochemical investigations on fundamental claims of molecular genetics. The studies in question concerned the extraordinary giant unicellular alga, *Acetabularia mediterranea*. This organism, though composed of a single cell, grows to over two centimeters in length, with a long thin stalk and a foot (rhizoid) containing the nucleus. When it is ready to reproduce, it forms an umbrella-like cap in which its spores are located. As Hämmerling had reported long ago (Hämmerling, 1934), *Acetabularia* can live virtually indefinitely – for a period of many months – after its nucleus is excised. Even more striking, and the subject of a large number of studies by Brachet and his colleagues (reviewed in Chantrenne, 1961), is the fact that for at least two months after its nucleus has been excised, a capless individual can still regenerate an umbrella. Thus, *Acetabularia* can carry out a major morphogenetic step in the absence of a nucleus. Nonetheless, if an enucleate fragment is supplied with the nucleus of a related species, *the cap assumes the morphology of the donor of the nucleus* rather than that of the recipient. It follows that the genetic determinants of the umbrella are provided by the nucleus, but stored for a long period in the giant cell.

Since Brachet could test whether RNA and DNA are manufactured by enucleated *Acetabularia*, this organism played an important role in working out nucleo-cytoplasmic relations, the extent of nuclear control of morphogenesis, and the role of the nucleus and other organelles in the manufacture of nucleic acids. Suffice to say that by 1951 he was able to demonstrate that protein synthesis, dependent on cellular RNA, continued for months after the enucleation of the cell (Brachet and Chantrenne, 1951), but that over time the proteins produced were less frequently nuclear in origin and more frequently chloroplastic.[24]

As early as 1951, Brachet and his colleagues showed that cap formation in enucleated *Acetabularia* involved genuine protein synthesis. The enucleate fragments of the alga were thus able to utilize genetic information, derived from the nucleus but stored in the cytoplasm, to make protein, although only for a certain period of time. By 1960, Brachet was able to distinguish fairly cleanly between RNAs produced by the nucleus and those produced by chloroplasts and mitochondria. This enabled him to argue that "cytoplasmic ribonucleic acid carries the genetic information originating from the genes, and controls, for a certain time in any event, synthesis of specific protein" (Brachet, 1960b, p. 197) and that "a true synthesis of chloroplastic ribonucleic acid occurs in . . . anucleate *Acetabularia*; this synthesis takes place at the expense of the other cytoplasmic ribonucleic acid fractions" (*ibidem*).

The string of largely physiological investigations underlying these results clearly led to doctrines belonging centrally to molecular genetics, doctrines that are commonly believed to have been derived almost exclusively by work along other lines. Because there was a great deal of interaction among the scientists pursuing different investigative pathways (to use F.L. Holmes's term), it is very difficult, perhaps impossible, to isolate the contributions of one style of investigation from those of others. But this account of Brachet's work shows, I think, that a great number of investigative pathways were triangulating on rather similar conclusions. This supports a moral toward which I have been building: no one research group, indeed, no small set of research groups or of investigative pathways, is privileged as the key to obtaining the crucial results from which molecular genetics got its start. Indeed, if I am right about the extensive list of disciplines, approaches, techniques, groups, and hypotheses that made central contributions to the foundation of molecular biology and about the character of work in this domain, no research group or small set of research groups played an indispensable role in the founding of molecular biology – not even Watson and Crick, Jacob and Monod, or the phage group. The myth of the founder(s) is, as always, a powerful and important one and it has played an extremely important role in the transformation of molecular biology into the compound of disciplines and practices (and into the center of power) that it has become – but its historical accuracy (not its powerful role in forging disciplinary identity) is clearly challenged, indeed, undermined, by careful historical analysis.

HISTORICAL CONCLUSION

With this, let us turn from our narrative to the larger themes with which this chapter began. As should now be clear, Brachet played an important and underappreciated role in the founding of molecular biology. The same can be said for Lwoff's early work on the utilization of carbon sources by micro-organisms and of a large number of other programs of research into structure, function, and their interplay at various biological levels. Brachet, partly in parallel with Caspersson and Schultz, opened up an important pathway toward the recognition that the nucleic acids play a crucial role in protein synthesis; he was one of the first to show that RNA (and not just the RNA in ribosomes) had to be an intermediate between the nucleus and its protein products. He demonstrated the non-specificity of ribosomal function and provided important clues within a non-informational tradition regarding the existence and specificity of what we now know as mRNA. He took major steps toward a molecular analysis of the composition and organization of such nucleoprotein structures as ribosomes and chromosomes in spite of the fact that he did not examine molecular structures directly in the manner of, say, Watson and Crick or simple systems in the style of Delbrück. Furthermore, he provided important evidence that the specificity of protein synthesis, ultimately controlled by chromosomal DNA, is intimately correlated with the supernatant RNA left after centrifugation.

True, in his early work Brachet had the polarity of some DNA-RNA interactions wrong because he though that cytoplasmic RNA provided the raw material for the biosynthesis of DNA. Nor is it hard to find other well supported hypotheses in Brachet's work that we now know to be wrong. The same is true for virtually all historically important scientists. In spite of such well-founded mistakes, however, (and partly because of them!) his research and that of his collaborators constitutes a major contribution to the elucidation of transcription and translation. His research is clearly an important part of the story of the founding of molecular biology. The portion of Brachet's work examined here centered on the physiology of the nucleic acids, including their role in protein synthesis and the differences in their behaviors in different cellular compartments. His larger purpose was to understand morphogenesis and the control of development and differentiation; in this respect his findings in molecular genetics, crucial as they were for that discipline, were ancillary to his main research program. In this respect, he does not fit the 'standard' profile of a founder of molecular biology.

However, that may be, his use of biochemical and physiological techniques to localize and determine the functions of nucleic acids typifies the sorts of work that have turned up time and time again in my historical research – work in ciliatology, physiological genetics, molecular morphology, developmental biology, bacterial genetics, biochemistry of respiration, and on and on. It strains credulity to force such research into the procrustean beds of 'the' two founding traditions of molecular biology – informational and structural – as these are usually characterized in the standard histories. In this respect, Brachet is typical of the figures I have been studying: his entry into molecular biology derives from the marriage of his interest in physiological problems – in his case, those of causal and chemical embryology – with sets of techniques that produced reliable results at the molecular level. Such physiological interests provided major tools for forming and transforming molecular biology, contributing to some of the most fundamental results of that field. The importance of such work has not been recognized in most histories of molecular biology.

Putting the point in a more historiographical mode, the myth of the founder(s) plays a crucial role in cementing a new discipline or area of work. Agreement on whom to count as the founders of a discipline helps set the style, modes of thinking, and the technical tools of that discipline. It provides the discipline with a definite profile, an identity. But acceptance of certain individuals or groups as founders also obscures the pathways by means of which its central early accomplishments were achieved, for the myth requires that the contributions of non-founders be ranked as secondary in comparison with those of the founders. In a certain sense, this becomes a self-fulfilling prophecy, for contributions parallel to those made by the recognized founders, but made by different individuals or groups, are treated as setting the background for what was done in the new discipline or as ancillary, secondary, or as providing after-the-fact support. Once such views have entered thoroughly into the culture of a newly-formed discipline they acquire considerable force and are extremely difficult to dislodge. This, I believe, is exactly what has happened to a considerable body of the work of Jean Brachet – work that provided crucial techniques to molecular biology, helping to marry the analysis of function to the localization of molecules and the analysis of molecular structures. An expanded appreciation of the contributions of many figures (of whom Brachet is only one) to the body of techniques, doctrines, and practices that became molecular biology will, I am confident, give us a deeper appreciation of the nature of this super-discipline.

PHILOSOPHICAL CONCLUSION

I shall close with a brief discussion of the transforming role that the entry of molecular biology into physiology, immunology, genetics, developmental biology, and evolutionary biology is playing in these and virtually all the traditional botanical, zoological, and microbiological disciplines. This will allow me to state one of the larger philosophical points deriving from case studies like that presented here.

In my generation, training in the philosophy of science focused mainly on theories. It is theories, or so we were taught, that unify fundamental disciplines, that provide the conceptual core and the unifying glue of basic science. The analysis of theories was supposed to provide solutions to such urgent philosophical problems as that of reductionism in science and metaphysics; a discipline's claims would count as reductionist or emergentist *not* in virtue of the techniques that it employed, but in virtue of the ontology of theoretical entities to which its theories were committed or in virtue of the parameters that entered into the underived laws of those theories.

In recent years there has been a significant movement away from this theory-centrism, exemplified by works such as Hacking (1983) and many others. One laudable consequence of this shift is renewed emphasis on the details of scientific experimentation and a wide range of scientific practices. The analysis of molecular biology presented here (and parallel analyses being developed by many other historians and philosophers) fits well with these trends. Indeed, if my account of molecular biology is correct even as a zeroth approximation, a theory-centered philosophical analysis must be fundamentally wrong at least for this science.

The unifying power of molecular biology stems in good part from the reorganization of biological disciplines that it is bringing about. As I have argued, this reorganization is not based on a grand theory, but on an immensely powerful battery of techniques for getting at the interrelated families of complex mechanisms found in all sorts of organisms. These mechanisms marry molecular form – but also the form of organelles, cells, components or organs, and so on – to biological function. The modes of attack on those mechanisms employed by biologists breach all the old boundaries – boundaries that separated zoologists from botanists, drosophilists from maize workers, microbiologists from geneticists, biochemists from organismic biologists. There will, of course, never be a complete unification of any sort among the diverse disciplines involved, but molecular biology is dramatically changing the ways in which scientists can communicate with one another, the character of the barriers between them, the techniques they can employ for particular tasks, and the overlap in the problems they recognize and consider focal.

Finally, a brief promissory note regarding reductionism will show that there is philosophical bite in these changes. I implied – though it will take another paper to spell it out properly – that contemporary molecular biology differs from the older style of biochemistry in ways that make molecular biology far less reductionistic than biochemistry was. As Schaffner has explicitly argued (1993, this volume, and elsewhere), a crucial feature that distinguishes molecular biology from such fields as biochemistry is that *molecular biology requires the interaction of mechanisms, of structures and functions, at a great many levels.* One cannot understand the functioning of the immune system without treating whole cells – T-lymphocytes for example – as integrated structural-functional units. Structural integration at many levels turns out to be a prerequisite for reliable molecular function in organisms of all sorts. The analysis of this structural integration must proceed at multiple levels simultaneously, not just at the molecular level. In this respect, molecular biology has become far less reductionistic than many, perhaps most, biologists and philosophers thought it would be a brief decade or two ago.

In fact, I *do* believe that organisms are mechanisms all the way down – provided the many-leveled aspect of structural integration is properly taken into account. The workings of most (but not all) of those mechanisms are best understood by use of the tools of molecular biology. For this reason, the old division of biological disciplines on organismic vs. functional lines will probably not endure much longer. But even if we see a return to biology as a sort-of unified discipline, as I think we will, *the unification in question, rather than being based on some grand synthetic theory, will revolve around the techniques for investigating multi-leveled mechanisms.* One particular matter is of key importance in understanding many of the molecular mechanisms of particular interest: *in many cases, it is only in the context of higher level structures, functioning properly, that molecular mechanisms operate 'correctly'*. One small symptom of this fact is the variation in rates of molecular evolution according to the higher-level structures in which the molecules of interest play a functional role. Considerations solely at the molecular level are not likely to yield a full understanding of this variation in evolutionary rates. Indeed, the integration of work across levels is a key to the nature of molecular biology. The structure and progress of the resultant science(s) cannot fail to be of interest to scientists, historians, sociologists, and philosophers.

ACKNOWLEDGMENTS

Work on this paper was conducted during the tenure of a grant from the National Endowment for the Humanities, a residential fellowship at the National Humanities Center, and with support from Supplemental Grants from Virginia Polytechnic Institute and State University. The paper was substantially improved by discussions with Ruth Alscher, Jean Gayon, Muriel Lederman, Doris Zallen and with colloquium audiences at Virginia Tech, the University of Chicago, and the Boston Colloquium for the Philosophy of Science. I also benefited from criticisms by an anonymous referee and by Sahotra Sarkar. I am grateful for all of this help.

Virginia Polytechnic Institute and State University,
Blacksburg, VA, USA

NOTES

[1] This position may be controversial, but it is by no means original. Various allied views, some indirectly related, may be found in many recent analyses in the philosophy of biology. An incomplete sample includes Bechtel (1986, 1993), Darden and Maull (1977), Hull (1972, 1973, and subsequent work), Sarkar (1989, 1992), and Wimsatt (1974, 1985, 1986, and other work).

[2] Thus, Olby, who organized *The Path to the Double Helix* in part around the contributions of the informational and structural schools (see chs. 15 and 16 of Olby 1974), does not refer to Brachet in this book.

[3] The need to focus on a single case study here rather than a battery of half-a-dozen or more is a serious drawback, for a single case study can only illustrate, but not provide, the rich support available for this stance. Nonetheless, I hope I say enough to make it clear that there are many examples like the one central to this paper.

[4] It is hard to find early clear statements of this hypothesis. Useful formulations, though ambiguous, are found in Beadle (1945) and Tatum and Beadle (1945). A preliminary treatment of this case, with references to some further primary literature may be found in Burian (1993a). Lily Kay and Norman Horowitz have each stressed this point in personal communication. Cf. Kay (1992), ch. 7, esp. pp. 198–9 and Horowitz (1990, esp. pp. 5–6, and 1991) for some further details.

[5] The early portions of this work are summarized in Lwoff (1932); later parts of it, probably

more familiar to Beadle, are presented in Lwoff (1943); cf. the discussion in Burian and Gayon (1991).

[6] The terminology of genetic continuity may mislead current readers. In this context it meant that the enzyme had to be produced from a pre-existing enzyme or a generalized precursor of the same nature as enzymes. More generally, genetically continuous structures or entities meant structures like cells, nuclei, and chromosomes that could only arise out of pre-existing structures of the same sort. There were many candidates for this status: genes, plasmagenes, viruses, kinetosomes, chromosomes, nucleoli, nuclei, mitochondria, chloroplasts, enzymes, and many more. Some workers held that biology should be built on the study of genetically continuous entities since organisms, themselves genetically continuous, are constructed, perhaps indirectly but primarily, by combination of parts that are genetically continuous. It is worth noting that Lwoff employed this terminology in reference to organelles (specifically kinetosomes) as early as 1929 and soon after to enzymes; cf. Chatton and Lwoff (1929). For Lwoff, a major break from this treatment of genetic continuity arose with his discovery that viruses are not, in this sense, genetically continuous (see Lwoff 1953, 1957).

[7] It is highly unlikely that Tatum's biochemical contributions to his collaboration with Beadle had any direct connection with the work of Lwoff. What is likely, however, is that Beadle's knowledge of Lwoff's work influenced, directly or indirectly, the problems that he posed and to which Tatum applied extensions of the techniques he had learned and developed in Wisconsin.

[8] For more on Spiegelman's work see Gilbert, this volume, and Gaudillière (1992).

[9] Cf. Allchin (1991, 1993) and Weber (1991). As F. L. Holmes has pointed out repeatedly in discussion, much of the biochemistry of the 20s and 30s was more concerned with intracellular structures and their roles in mediating biochemical processes than was typical in the 50s or of Mitchell's opponents in the 60s.

[10] As is so often true, there are ironic twists to the pathway by which biochemistry arrived at this state. Indeed, such research programs as that of F.G. Hopkins, a founding father of British biochemistry, placed considerable emphasis on morphology and supra-molecular structure and was based on a far more dynamic vision of cellular processes and the tasks of biochemistry than came to be common in biochemistry by mid century. Further historical research into this aspect of the development of biochemistry is desirable.

[11] This point was stressed in series of interviews between P.P. Slonimski, a founder of mitochondrial genetics, and R. Burian, J. Gayon, and D. Zallen. (Transcripts available from RB on request.) There were, of course, biochemists who emphasized the importance of intact mitochondria for certain aspects of respiratory chain reactions (e.g., Britton Chance – see, e.g., Chance and Williams 1956). Lehninger (1965) – a classic text – still found it necessary to proclaim the need to bring together "not only the basic enzymology of biological oxidations and phosphorylations which appear to constitute the major activities of the mitochondrion, but also many other aspects, such as ion transport, membrane contractile phenomena, control and integration mechanisms, the molecular structure of the mitochondrial membrane, and the origin and biogenesis of these structures" (p. viii). The main body of the book bears the stigmata of an early unsuccessful attempt to achieve the integration of these diverse phenomena with the enzymology of oxidation and phosphorylation.

[12] A portion of the text of this section parallels that of a section of Burian (1994).

[13] E.F. Keller (in press) provides a strong, somewhat idealized version of the contrast between the drive for simplicity associated, on standard accounts, with the founding of molecular biology and the insistence on complexity among embryologists during the transition from traditional embryology to developmental biology. Her account makes it clear why many scientists and historians would not count Brachet as a molecular biologist, at least during the period under examination here.

[14] Brachet (1933). All quotations from Brachet's French language articles have been translated freely by RB.

[15] It should be remarked that embryologists who employed biochemical techniques and most cytochemical workers *were* concerned with form, structure, and localization to a greater degree than most biochemists.
[16] The stages of the refinement are indicated in Brachet (1947a).
[17] Brachet (1942). In this article Brachet cites his own earlier findings and similar findings published by Caspersson in 1940 and 1941.
[18] Brachet (1944); cf. ch. VI, pp. 194–250.
[19] Brachet (1949), also discussed in Burian (1990).
[20] These results are amplified by Brachet's collaborator, H. Chantrenne (Chantrenne 1947) and discussed in Brachet (1947a), n. 9.
[21] For the work of Zamecnik's group, see the recent publications of Hans-Jörg Rheinberger (1993 and much other work; see also Burian 1993b). S. Sarkar has suggested (pers. commun.) that mRNA cannot be excluded as a component of soluble RNA under the relevant technical conditions. I am sceptical whether significant quantities of mRNA were included in the supernatant in Zamecnik's treatments of soluble RNA (sRNA), but I do not know enough of the relevant biochemistry or of the protocols employed in other laboratories to rule out this suggestion. In any event, the distinction between mRNA and tRNA was not, of course, clearly drawn until 1959–60, and then initially only to an inner circle (see Judson 1979, ch. 7, esp. pp. 421ff. for the relevant chronology), so all parties to the puzzles about the role of ribosomes in protein synthesis and the functions of sRNA were left with unresolved problems about the different classes of RNAs that might be involved until at least then.
[22] Brachet (1950). The centrifugation was at 100,000 G for less than an hour. An initial report of this finding in frog eggs is given at p. 24 of Brachet, (1947b).
[23] Brachet (1960a); these results are taken from ch. 1, sect. 4, 'The Role of the Microsomes and Ribonucleoprotein Particles in Protein Synthesis'.
[24] The evidence for this last claim was only suggestive until sometime after 1963, when CsCl density centrifugation of the DNAs produced in enucleate *Acetabularia* confirmed beyond question the source of the RNAs produced at different times after enucleation. Cf. Green et al. (1967).

REFERENCES

Allchin, D. (1991). *Resolving Disagreement in Science: The Ox-Phos Controversy, 1961–1977*. Ph.D. Dissertation, University of Chicago.
Allchin, D. (1993). 'New Realities From Experimental Chimeras (Notes for a chemosmotic history)'. Paper presented at the International Society for the History, Philosophy, and Social Studies of Biology.
Beadle, G. (1945). 'Biochemical Genetics', *Chemical Reviews* **37**: 15–96.
Beatty, J. (1995). 'The Evolutionary Contingency Hypothesis'. In: Wolters, G., Lennox, J. and McLaughlin, P. (eds.), *Concepts, Theories, and Rationality in the Biological Sciences*. Konstanz: Universitätsverlag Konstanz, and Pittsburgh: University of Pittsburgh Press, pp. 45–81.
Bechtel, W. (1986). 'The Nature of Scientific Integration'. In: Bechtel, W. (ed.), *Integrating Scientific Disciplines*. Dordrecht: Nijhoff, pp. 3–58.
Bechtel, W. (1993). 'Integrating Sciences by Creating New Disciplines: The Case of Cell Biology'. *Biology and Philosophy* **8**: 277–300.
Brachet, J. (1933). 'Recherches sur la synthèse de l'acide thymonucléique pendant le développement de l'oeuf d'Oursin', *Archives de Biologie* **44**: 519–76.
Brachet, J. (1942). 'La localisation des acides pentosenucléiques dans les tissus animaux et les oeufs d'Amphibiens en voie de développement'. *Archives de Biologie* **53**: 207–57.

Brachet, J. (1944). *Embryologie chimique*. Paris: Masson; Liège: Desoer.
Brachet, J. (1947a). 'Nucleic Acids in the cell and Embryo'. *Symposia of the Society for Experimental Biology* **1**: 207–24.
Brachet, J. (1947b). 'The Metabolism of Nucleic Acids'. *Cold Spring Harbor Symposia in Quantitative Biology* **12**: 18–27.
Brachet, J. (1949). 'L'hypothèse des plasmagènes dans le développement et la différenciation'. In: *Unités biologiques douées de continuité génétique*. Paris: CNRS, pp. 145–63.
Brachet, J. (1950). 'The Localization and the Role of Ribonucleic Acid in the Cell'. *Annals of the New York Academy of Sciences* **50**: 861–9.
Brachet, J. (1952). *Le rôle des acides nucléiques dans la vie de la cellule et de l'embryon*. Liège: Desoer; Paris: Masson.
Brachet, J. (1960a). *The Biological Role of Ribonucleic Acids*. Amsterdam: Elsevier.
Brachet, J. (1960b). 'Ribonucleic Acids and the Synthesis of Cellular Proteins'. *Nature* **186**: 194–9.
Brachet, J. and Chantrenne, H. (1951). 'Protein Synthesis in Nucleated and Non-Nucleated Halves of *Acetabularia mediterranea* Studied with Carbon-14 Dioxide'. *Nature* **168**: 950.
Burian, R. (1990). 'La contribution française aux instruments de recherche dans le domaine de la génétique moléculaire'. In: Fischer, J.-L. and Schneider, W. (eds.), *Histoire de la génétique*. Paris: ARPEM, pp. 247–69.
Burian, R. (1993a). 'Unification and Coherence as Methodological Objectives in the Biological Sciences'. *Biology and Philosophy* **8**: 301–18.
Burian, R. (1993b). 'Technique, Task Definition, and the Transition from Genetics to Molecular Genetics: Aspects of the Work on Protein Synthesis in the Laboratories of J. Monod and P. Zamecnik'. *Journal of the History of Biology* **26**: 387–408.
Burian, R. (1994). 'Jean Brachet's Cytochemical Embryology: Connections with the Renovation of Biology in France?' In: Debru, C., Gayon, J. and Picard, J.-F. (eds.), *Les sciences biologiques et médicales en France 1920–1950*, vol. 2 of *Cahiers pour l'histoire de la recherche*. Paris: CNRS Editions, pp. 207–20.
Burian, R. and Gayon, J. (1991). 'Un évolutionniste bernardien à l'Institut Pasteur? Morphologie des Ciliés et évolution physiologique dans l'oeuvre d'André Lwoff'. In: Morange, M. (ed.), *L'Institut Pasteur: Contribution à son histoire*. Paris: Editions de la Découverte, pp. 165–86.
Burian, R. and Gayon, J. (in preparation). 'D'un possible style française de recherche en matière de hérédité'.
Burian, R., Gayon, J. and Zallen, D. (1988). 'The Singular Fate of Genetics in the History of French Biology, 1900–1940'. *Journal of the History of Biology* **21**: 357–402.
Chance, B. and Williams, G.R. (1956). 'The Respiratory Chain and Oxidative Phosphorylation'. *Advances in Enzymology* **17**: 65–134.
Chantrenne, H. (1947). 'Hétérogénéité des granules cytoplasmiques du foie de Souris'. *Biochimica et Biophysica Acta* **1**: 437–48.
Changrenne, H. (1961). *The Biosynthesis of Proteins*. London: Pergamon.
Chatton, E. and Lwoff, A. (1929). 'Les infraciliatures et la continuité génétique des systèmes ciliares récessifs'. *Comptes rendus de l'Academie des Sciences* **188**: 1190–2.
Claude, A. (1941). 'Particualte Components of Cytoplasm'. *Cold Spring Harbor Symposia in Quantitative Biology* **9**: 262–71.
Claude, A. (1943). 'The Constitution of Protoplasm'. *Science* **97**: 451–6.
Crick, F.H.C. (1970). 'Central Dogma of Molecular Biology'. *Nature* **227**: 561–3.
Darden, L. and Maull, N. (1977). 'Interfield Theories'. *Philosophy of Science* **44**: 43–64.
Gaudillière, J.-P. (1992). 'J. Monod, S. Spiegelman et l'adaptation enzymatique. Programmes de recherche, cultures locales et traditions disciplinaires'. *History and Philosophy of the Life Sciences* **14** (1992): 23–71.
Green, B., Heilporn, V., Limbosch, S., Boloukhere, M. and Brachet, J. (1967). 'The Cytoplasmic

DNA's of *Acetabularia mediterranea*'. *Proceedings of the National Academy of Sciences, USA* **58**: 1351–8.
Hacking, I. (1983). *Representing and Intervening*. Cambridge: Cambridge University Press.
Hämmerling, J. (1934). 'Über formbildende Substanzen bei *Acetabularia mediterranea*, ihre räumliche und zeitliche Verteilung und ihre Herkunft'. *Archiv für Entwicklungsmechanik* **131**: 1–81.
Horowitz, N. (1990). 'George Wells Beadle (1903–1989)'. *Genetics* **124**: 1–6.
Horowitz, N. (1991). 'Fifty Years Ago: The Neurospora Revolution'. *Genetics* **127**: 631–5.
Hull, D. (1972). 'Reduction in Genetics – Biology or Philosophy?' *Philosophy of Science* **39**: 491–9.
Hull, D. (1973). 'Reduction in Genetics – Doing the Impossible'. In: Suppes, P. (ed.), *Logic, Methodology, and Philosophy of Science, IV*. Amsterdam: North Holland, pp. 619–35.
Judson, H. (1979). *The Eighth Day of Creation: Makers of the Revolution in Biology*. New York: Simon and Schuster.
Kay, L. (1992). *The Molecular Vision of Life: CalTech, the Rockefeller Foundation, and the Rise of the New Biology*. New York: Oxford University Press.
Koltzoff, N. (1939). *Les molécules héréditaires*. Paris: Hermann.
Lehninger, A.L. (1965). *The Mitochondrion*. New York: W.A. Benjamin.
Lwoff, A. (1932). *Recherches biochimiques sur la nutrition des Protozoaires*. Paris: Masson.
Lwoff, A. (1943). *L'évolution physiologique*. Paris: Hermann.
Lwoff, A. (1953). 'Lysogeny'. *Bacteriological Reviews* **17**: 269–337.
Lwoff, A. (1957). 'The Concept of a Virus: The Third Marjory Stephenson Memorial Lecture'. *Journal of General Microbiology* **17**: 239–53.
Olby, R. (1974). *The Path to the Double Helix*. London: Macmillan.
Rheinberger, H.-J. (1993). 'Experiment and Orientation: Early Systems of in Vitro Protein Synthesis'. *Journal of the History of Biology* **26**: 443–72.
Sarkar, S. (1989). 'Reductionism and Molecular Biology: A Reappraisal'. Ph.D. Dissertation, University of Chicago.
Sarkar, S. (1992). 'Models of Reduction and Categories of Reductionism'. *Synthese* **91**: 167–94.
Schaffner, K.F. (1993). 'Theory Structure, Reduction, and Disciplinary Integration in Biology'. *Biology and Philosophy* **8**: 319–48.
Tatum, E. and Beadle, G. (1945). 'Biochemical Genetics of Neurospora'. *Annals of the Missouri Botanical Garden* **32**: 125–9.
Weber, B. (1991). 'Glynn and the Conceptual Development of the Chemosmotic Theory: A Retrospective and Prospective View', *Bioscience Reports* **11**: 577–617.
Wimsatt, W. (1974). 'Complexity and Organization'. In: Schaffner, K. and Cohen, R. (eds.), *PSA 1972*. Dordrecht: Reidel, pp. 67–86.
Wimsatt, W. (1985). 'Forms of Aggregativity'. In: Donagan, A., Perovich, M. and Wedin, M. (eds.), *Human Nature and Natural Knowledge*. Dordrecht: Reidel, pp. 259–93.
Wimsatt, W. (1986). 'False Models as Means to Truer Theories'. In: Nitecki, M. and Hoffman, A. (eds.), *Neutral Models in Biology*. New York: Oxford University Press, pp. 23–55.

LILY E. KAY

LIFE AS TECHNOLOGY: REPRESENTING, INTERVENING, AND MOLECULARIZING*

In December 1945, soon after the end of World War II and in anticipation of postwar expansion of science, Linus Pauling submitted to the Rockefeller Foundation (the principal patron of molecular biology) a research plan for life science at CalTech that was remarkable in its scope and structure. It was a twenty-five page grant proposal – unusually long for pre-NSF and NIH genres – asking for $6 million to be distributed at the rate of about $400,000 per year over a period of about 15 years (this would be equivalent to about $30 million in current dollars) to support research on the fundamental problems of biology.[1] The other striking feature of the proposal was the emphasis on group projects organized around specific technologies; on instrumentation as the driving force and dominant conceptual framework of life science research. These structural and cognitive features constituted a turning point in the history of biology – and not only at CalTech. For as I have argued elsewhere, Pauling's vision and CalTech's program had an impact well beyond its temporal and spatial confines – it shaped the cognitive and social development of molecular biology for decades (see Kay, 1987).

To molecular biologists today, Pauling's technology-centered grant would hardly seem remarkable. Of course, they would argue, one can only plan experiments around available apparatus, and of course state-of-the-art techniques define the vanguard of research, and the more the better. But applying such a common-sense view to CalTech's project is deceptive in its simplicity, for it obscures some deep issues that not only are crucial to our understanding of the history and philosophy of biology, but that are also central to our assessment of some recent developments in molecular biology.

In particular I have in mind the discourse about current genetic engineering technologies, developments which are frequently characterized as a natural consequence of pure and theoretical research in the 1950s, 1960s, and 1970s. I wish to propose the reverse: that from its inception, the molecular biology program was defined and conceptualized in terms of technological capabilities and possibilities. Representations of and within the new biology were predicated on interventions that, in turn, aimed from the start at reshaping vital processes, definitions and concepts that formed a continuum of representing and intervening. By examining the 1945 proposal in its broader cognitive and ideological context, I will show in this paper that most of the CalTech plan

---

* Reprinted with kind permission from *Rivista di Storia della Scienza* (sec. II) **I**(7)(June): 85–103 (1993).

was organized around either existing or projected instruments and techniques that were a priori restricted to investigating only properties of submicroscopic dimensions. Thus I argue that the project was caught in a circularity. First, practitioners of molecular biology defined properties of life, health, and disease in terms of macromolecules, then grounded their program in representational strategies that can be understood, verified, legitimated, and perpetuated only through technologies restricted to this realm. Such circular knowledge-claims regarding the nature of life, including the more recent promises of genetic engineering, must be evaluated in light of this circular reasoning.

THE ANALYTICAL FRAMEWORK

Before embarking on narrative and analysis I wish to provide the structure, tools, and some working definitions. The use of the terms "representing and intervening" is not novel. These are the foci of analysis of Ian Hacking's 1983 book in which he proposes that we redirect our priorities from studying theory construction to focusing on experimental practice, that we view reality as intervention and see science as doing, rather than merely knowing. Briefly to summarize his argument, he pins the problematics of realism on representation. Realism is a philosophical problem because reality is an attribute of representation, and yet we have evolved alternative styles of representation, or different pictures of the world. In the physical sciences such representations consist of elaborate systems of modelling, structuring, theorizing, calculating, and approximating. Hacking, believing as he does in the plasticity of scientific truths, supports the argument that models often carry a greater weight in science than do theories.

But rather than get trapped in the old realism-antirealism duel – with combatants each latching on to something in the nature of representation that will vanquish the other – Hacking sees they way out by turning to intervening, to reality as causation and change. He counts as real what we can use to intervene in the world to affect something else, or what the world can use to affect us. He traces the interlocking of representing and intervening to the Baconian program of the 17th century, to the birth of an autonomous experimental tradition, whose primary aim was to manipulate and control nature for the utility of man and to collapse the dichotomy of the natural and artificial. Accordingly, since the primary aim of science is to change the world rather than understand it or make sense of it through change, Hacking urges that when analyzing science we should focus on intervention, and on the interlocking of intervention and representation (see Hacking, 1983, pp. 130–146; Merchant, 1980, chapters 7–9).

Hacking, Paul Forman, and Evelyn Fox Keller have examined the implications of this thesis for the project of pure science, and for the long-cherished ontological cleavage between pure and applied. Looking at cases of 20th-century physics, they have concluded that a built-in interventionist drive in basic research has predetermined the types of applications (see Hacking, 1986; Forman, 1987; Fox Keller, 1992; Pauly, 1987). Thus, for example, they show that military presence – the pervasive culture and the institutional structure of military contracts – shapes basic research in subtle cognitive ways, so that the way questions are selected and experiments are designed is likely to be militarily productive, or rather destructive. This general line of thinking – that interventionist goals shape representational strategies – has informed my own analysis of the growth of molecular biology as an interventionist program, a project whose applications were already inscribed in its design.

We now turn to the use of the term molecular biology, a term that has been problematized by both scientists and historians of science. In order to avoid potential confusions, a few working definitions of molecular biology will be useful. If by molecular biology we mean strictly DNA molecular genetics, then most of the 1940s activities of life science would of course fall outside this domain. On the other hand, if one means the Rockefeller Foundation's sponsored program by that name in the 1930s–1950s, then this definition would include many areas of life science that are not typically associated with molecular genetics (e.g., biophysics and immunochemistry). As Pnina Abir-Am has pointed out, retroactively and in light of the high status of molecular biology, life scientists who would have never identified themselves as molecular biologists were all too eager to reconstruct their careers as elements of the molecular biology success story (see Abir-Am, 1980, 1985).

I do not wish to position myself within this debate, but rather to offer a set of criteria that together explain how I use the term molecular biology.

1. In terms of programmatic statements, the new biology, as T.H. Morgan has said, stressed the unity of life phenomena common to all organisms, rather than focusing on their diversity. Thus the new biology would focus, for example, on respiration or reproduction as central biological (in contradistinction to biochemical) problems, regardless of whether the object of study were a mammal or a bacterium (see Morgan, 1928).

2. Based on this logic, it became much more convenient to study fundamental vital phenomena on their minimalist levels. Thus the new biology increasingly employed simple biological systems – primarily bacteria and viruses – as phenomenological probes, or as conceptual models. (This line of thinking led to Monod's notorious dictum that what is true for the bacterium is true for the elephant.)

3. In cleaving life processes from their host organisms, the molecular biology program aimed at discovering general physico-chemical laws governing vital phenomena, and in so doing distanced its concerns from interactive processes occurring within higher organisms, between organisms (e.g., symbiosis), and between organisms and their environments. Concomitantly, this

physiochemical approach negated historical explanations in biology, developmental and evolutionary accounts of life processes – the arrow of time. The new biology acknowledged only mechanisms of upward causation, ignoring the explanatory role of downward causation.

4. In the process of defining life in terms of fundamental physiological mechanisms, molecular biology ultimately narrowed its focus principally to macromolecules. And before the mid-1950s, this meant primarily the so-called "giant protein molecules", thus defining the locus of life phenomena at the submicroscopic region between $10^{-6} \div 10^{-7}$ cm. This region was the functional domain of the new biology.

5. This domain could be investigated only with complex and sophisticated apparatus, specifically designed to investigate life strictly at this range of dimensions – electron microscopes, ultracentrifuges, electrophoresis, spectroscopy, X-ray diffraction. These instruments, in turn, became organizing principles for research.

6. Because of the complexity of the apparatus and the intricacies of the techniques, research problems were often defined around instruments and were increasingly organized as team projects, with teams consisting of principal investigators, post-doctoral fellows, and graduate students. Doing molecular biology thus entailed structural changes in the organization of departments and laboratories.

This composite offers a reasonable contour of molecular biology, at least until the mid-1950s, when DNA replaced proteins as the master molecule. But my analysis today will focus primarily on the fourth feature: the dimensional characterization of life. It is crucial to grasp that the submicroscopic region of $10^{-6} \div 10^{-7}$ cm had been privileged already in the 1940s as the essence of life, the locus of representation of life and coterminous with it. The representations of this locus – theories and models – were predicated on several modes of intervention. Significantly, this region was seen already int he 1940s as the locus of genetically-engineered life.

These cognitive strategies and structural reconfigurations were facilitated through the powerful resource base and effective institutional mechanisms of the Rockefeller Foundation. As Robert Kohler and Pnina Abir-Am have shown, under the leadership of mathematical physicist Warren Weaver, the Rockefeller Foundation's Molecular Biology Program was conceptualized primarily in terms of technologies applied from the physical sciences to biology; so much so that the director of the Rockefeller Institute, Simon Flexner, in opposing the program in 1934, pointed out that instruments and techniques do not make for a new biology (see Kohler, 1979; Abir-Am, 1982). But in ways that Flexner had not envisioned, it did just that.

The instruments were designed to intervene in the biological universe; but the intervention had a broader level of significance: the social universe. Several scholars have argued that the Rockefeller Foundation's investment in the new biology was not motivated primarily by a commitment to basic research for its own sake. As is well known by now, the Foundation's main interests always lay in the human sciences, and its support of research was never an end in itself. It always aimed at finding rational bases for solving the recalcitrant social and economic problems of industrialized society, seeking to address the root causes of social dysfunction. The Foundation's programs were designed to neutralize the pernicious effects of what Chicago sociologist and Foundation adviser William F. Ogburn had termed "cultural lag", and the lag of social engineering behind science and technology (see, Yoxen, 1981; Haraway, 1979; Pressman, unpub.; Kay, 1992).

The molecular biology program, originally named "psychobiology", was part of the Foundation's new agenda called the new "Science of Man". Its goal was to coordinate the biological sciences, social sciences, and medical sciences toward a comprehensive rationalization of human behavior in quest of social control. The study of heredity as the deterministic component of human behavior, which later became molecular genetics, was central to this program. Thus, to extract the full significance of the relation between representing and intervening, we must view the promotion of the technocratic approach to biology in conjunction and resonance with the technocratic approach to social reform. Just like Bacon's program, the new biology should be understood on *two* nested levels of intervention: the ultimate level of intervention, affecting social change on the macro level, and the proximate level of intervention, the technological control of life on the micro level (or the macromolecular level).

This cognitive and institutional emphasis on physico-chemical methods and the near obsession with the primacy of technology in biology are what made the California Institute of Technology an ideal site for implanting the new biology. It was no mere historical accident that some of the main American research school coalescing as molecular biology – T.H. Morgan's group, Max Delbrück, George Beadle, and Linus Pauling – developed at CalTech. Having pioneered interdisciplinary programs in other fields of the physical sciences, CalTech's institutional mechanisms for interdisciplinary cooperation were already in place. Its strength in science and engineering provided the compatible context; and as important, the absence of competing biological traditions – agriculture, medicine, and evolutionary biology – insured an uncontested implementation of the vision. In the mid-1940s, Pauling's towering presence at CalTech, his X-ray crystallography studies of molecular architecture – both inorganic and organic molecules – his work on the biological role of the hydrogen bond, and his bold forays into immunochemistry became central to the Rockefeller Foundation's investment in the new biology. As Max Mason (president of the Rockefeller Foundation) had already said in the 1930s, CalTech's project under Pauling was "at the center of the program of study of vital processes furnishing aid in the sciences underlying human behavior".[2]

Pauling's priorities eventually came to shape the molecular biology program at CalTech and beyond.

THE CONSTRUCTION OF LIFE AS TECHNOLOGY

We now return to Linus Pauling and his 1946 research proposal, in which he expounded both his visions and prescriptions for the future course of biology. "We believe that the science of biology is just entering into a period of great and fundamental progress, similar to that through which physics and chemistry had passed during the last thirty-five years", he proclaimed, looking down the Comtean ladder. This unfortunate cognitive lag, according to Pauling, could finally be addressed since the methods of physics and chemistry have become sufficiently powerful to mount a successful attack on the central problems of biology.[3]

Pauling went to great lengths to outline these central problems. The fundamental problems of biology and their solutions, he believed, were best understood in terms of the dimensional characterization of life. Forty years ago, Pauling explained, the dark forest of the unknown stretched from somewhere below 10^{-4} cm – the limit of the visible microscope – back indefinitely into the region of smaller dimensions. But technologies of physics and chemistry had recently begun illuminating the dark molecular landscape. It was Pauling's conviction that the answers to the most basic problems of biology – the nature of growth, gene and cell replication, enzyme action, physiological activity of hemoglobin, drugs, hormones, vitamins, and neurological functions – were all hidden within the folds of the giant protein molecules. "The answers to all of these problems are hiding in the remaining unknown region of the dimensional forest, mostly in the strip between 10^{-6} and 10^{-7} cm; and it is only by penetrating into this region that we can hope to track them down."[4] The dimensional characterization of life was not new to Pauling. Researchers since the nineteenth century had viewed life as a continuum from molecules to higher organisms, wondering where the demarcation line between the inanimate and animate should be drawn. Some, like Wendell Stanley, regarded "giant virus molecules" as the "twilight zone of life" (see Kay, 1986). But no biology program before has privileged the lowest order of magnitude of structure as the exclusive domain of explaining life, health, and disease.

What is significant in this historical context is that Pauling, a physical chemist and an outsider to the dominant intellectual traditions of biology – traditions that he neither understood nor respected – could boldly enter a new field and define what was interesting, important, and worth doing in biology, reducing it to the narrow strip between $10^{-6} \div 10^{-7}$ cm; and that he actually succeeded in setting a lasting trend. He compared the advances down the ladder of dimensions in molecular biology to the progress up the scale of magnitudes in astrophysics. Except that unlike astrophysics, which depended principally on the powerful reflecting telescope, no single technology could solve the great problems of biology.

Biology for Pauling was the combinatorial and convergent effect of many technologies. The explorations in biology, as Pauling elaborated, could only be accomplished by the combination of the versatile tools of the physical sciences. Each approach, X-rays, ultracentrifuge, light-scattering techniques, biochemical assays, isotope tracers, or the electron microscope, each contributed only one piece of the molecular puzzle. It was the sum of these techniques, through successive approximations, that could yield insights into nature's animate secrets. The crucial feature here is that this molecular representation of nature was not merely *aided* by various interventionist techniques. What is new is that in Pauling's technological vision, biological representations were predicated exclusively on intervening. The strip between $10^{-6} \div 10^{-7}$ cm consisted of invisible entities that could be only "visualized" through several modes of manipulations. In fact, he stressed that the search for new and powerful methods would be one of the aims of the program.

The research program in molecular biology consisted of 15 different group projects: X-ray studies of proteins, chromatography, molecular weight and shape of protein molecules, electron microscopy, protein chemistry, enzyme chemistry, nucleic acid chemistry, immunochemistry, serological genetics and embryology, chemical genetics, virus studies, microbiology, general physiology, metabolism, and biophysics. What is striking, especially to a historian of biology, is the cognitive construction of biology as a relay system of technologies having a high degree of specificity with respect to particular problems. With the exception of serological genetics and embryology, microbiology, and to some extent also general physiology, each project either emphasized the need for developing more sensitive instruments or techniques, or stressed the benefits of supplying tools which would provide higher resolution of the macromolecular phenomena.

Thus to get at some of the basic explanations of biology, one had to amplify the power of X-ray analyses of proteins, as well as the methods of carrying the calculations generated by the Paterson Fourier patterns. This would be followed by the determination of the amino-acid composition and structure of proteins. Here "great progress would result form the development of a powerful and reliable method of analyzing mixtures of amino acids and peptides", linear and centrifugal stills, radioactive and non-radioactive isotopes, mass spectrography, and further power of resolution of chromatographic analysis (techniques that were greatly improved while working on war projects).[5]

X-ray and amino acids studies would be complemented by studies of molecular properties: ultracentrifuge, the Tiselius electrophoresis apparatus, light-scattering techniques, and birefringence flow. And these in turn would be coordinated with the electron microscope project. Here Pauling extrapolated from observations about the size and shape of virus molecules, structure of bacteria, the nature of muscle and tendon, to the biological fine-structure to be obtained with increased resolving power (100 ÷ 20 Å). One continues down the list to find that even biochemical genetics is promoted as promising exceedingly powerful methods of accomplishing metabolic syntheses. This area, under George Beadle's leadership, had been flourishing with the support of military contracts and the food and drug industries during the Second World War; Beadle's Neurospora turned out to be an effective system for the biosynthesis of vitamins and amino acids.[6]

The core of interdisciplinary program of the new biology would consist of research on the giant protein molecules, the main substance of life. As is well known, in the mid-1940s most life scientists adhered to the view that proteins were the primary biological determinants: that biological specificity – including gene action – was determined by proteins, and that proteins acted as templates for the synthesis of other proteins. DNA was thought to have an important but secondary role. CalTech's program epitomized the protein paradigm in the life sciences; and Pauling's own project on X-ray crystallography of proteins, was at the core of this protein program. Protein structure, he explained, was the master key to all fundamental biological problems: growth, reproduction, and neural functions.

In the context of his studies of protein structure, Pauling was the first to introduce into biological research the building of scaled-up molecular models, a practice which exemplifies Pauling's general view of the relationship among theory, models, and reality. Perhaps best understood as pragmatism, Pauling's approach to natural phenomena centered on the premise that theories were essentially convergent approximations, explanatory schemes designed to encompass diverse scientific observations. To be sure, Pauling was not a philosopher-scientist and was not given to introspection. Inferences regarding his epistemological stand have been extracted from several discussions in his scientific papers, and from a rare moment of self-reflection when Pauling described his approach to science through a contrast with the immunologist Karl Landsteiner. "I found", Pauling reflected, "that Landsteiner and I had a much different approach to science: Landsteiner would ask, 'What do these experimental observations force us to believe about the nature of the world?'and I would ask 'What is the most simple, general, and intellectually satisfying picture of the world that encompasses these observations and is not incompatible with them?'" These were nearly Pauling's exact words when introducing his immunochemistry program in 1940 (see Pauling, 1970, p. 1005; Pauling, 1940).

This same approach guided Pauling's model-building studies of protein structure, an approach he labeled molecular architecture. Molecular models generated several possible pictures of the world, encompassing observations and calculations; those possessing the highest degree of convergence became a representation of biological reality. Molecular architecture was an epistemology, a technology, and a metaphor, which Pauling propagated among scientists and the lay public.

In a rather typical article on Pauling in a 1949 issue of *Science Illustrated* entitled 'The Atomic Architect', a smiling Pauling is photographed holding up a molecular model, predicting that within twenty years medicine would be an exact science if only medical researchers were to become atomic architects. He explained that the architecture of proteins was the key to understanding and controlling life. In fact, he proclaimed, he hoped to create life in the laboratory by designing self-reproducing protein molecules. The article concluded that "if anyone within the next twenty-five years becomes master of this second creation, it very probably will be Pasadena's wizard of atomic architecture" (see Science Illustrated, 1949). Here was a clearly articulated vision of science as doing, science as technology effecting change in the world.

Furthermore, one sees here intervention not merely as an instrumentality of representing, but also an instrumentality of design and control. There is little doubt that Pauling envisioned the region between $10^{-6} \div 10^{-7}$ as the basic level of explanation as well as the fundamental level of intervention: the same kind of technologies that would be used to represent these macromolecular phenomena would also be mobilized to manipulate matter and control process, to effect biological change, to create artificial life.

The reference to medicine was not mere rhetoric. Just at that time Pauling's group was completing the work on sickle-cell anemia. No single project could have been more effective in swiftly confirming the grand claims of molecular architecture as a blueprint of life, and in privileging the molecular level as the fundamental domain of individual and social health. Just published in 1949 in *Science* magazine, the work bore the provocative title, 'Sickle-Cell Anemia, a Molecular Disease', and the article demonstrated how physical chemistry and sophisticated technology accounted for the molecular basis of a genetic disease (see Pauling et al., 1949; Kay, 1988). The project captured the essence of the interlocking of molecularizing, representing, and intervening, as methodology and scientific ideology.

The team of physical chemists and medical researchers examined blood samples from 30 patients: 15 with sickle-cell anemia, 8 with sicklemia – a milder form of the disease – and 7 normal adults. They extracted the hemoglobin from the red cells of each specimen and conducted a series of physico-chemical tests to see how the molecules from the three sources differed. Since the three species of hemoglobin differed in shape, they expected to detect differences in size and weight, but studies with the analytical ultracentrifuge which measures molecular weights revealed no difference.

The team then turned to the study of electrochemical properties of the hemoglobin molecules, using the Tiselius electrophoresis apparatus – a giant machine built at CalTech in the early 1940s – that measured differential rates of migration of charged molecules in an electric field. They subjected the three species of hemoglobin to varying magnitudes of electric forces in solutions of various degrees of acidity and alkalinity, until a point was reached where the effect was striking. At that point (pH 6.9) the normal hemoglobin migrated to the positive electrode, while the sickle-cell hemoglobin migrated to the negative electrode; the hemoglobin sample from patients with the milder form of the disease behaved like a mixture of normal and sickle-cell hemoglobinmolecules in roughly equal proportion. The boundary between health and disease was marked by a small difference in electric charge, a difference which a few years later would be traced to a replacement of just a single amino acid.

Just as the project neared completion, genetist James V. Neel from Ann Arbor found that sickle-cell anemia was a manifestation of a homozygous condition; that is, both parental genes contributed to the offspring's sickle-cell trait. Sicklemia, then, was a result of heterozygous condition, with only one parental gene contributing to the sickle-cell trait. In fact, Neel published his results in the same issue of *Science* in which Pauling's article appeared (see Neel, 1949).

Pauling arrived at the same genetic interpretation based on physico-chemical data. He concluded, in agreement with Neel, that the manufacture of abnormal sickle-cell hemoglobin is controlled by a gene, which, when present in double dose, causes the red cells of the individual to produce only the abnormal hemoglobin. These hemoglobin molecules, when not combined with oxygen, clamp on to one another to form long rods, which then aggregate to deform the cell. This was the first demonstration of how molecular architecture depended on the genetic protein blueprint, the first example of what Pauling called a "molecular disease". With Pauling's own enthusiastic promotion, both in scientific circles and in the popular media, the work was regarded as a spectacular achievement, its significance reaching far beyond explaining this particular medical syndrome. It served to validate the broad claim that the etiology of disease and health was be sought and found in the molecular-technological domain, in the strip between $10^{-6} \div 10^{-7}$ cm.

Equally significant is the relatively tepid reception of competing explanations and alternative frameworks of defining life, health, and disease. In the 1950s, researchers recognized that in the its heterozygous form the gene responsible for the mutant sickle-cell hemoglobin also conferred resistance to malaria. While being a genetic deficiency, it was also an adaptive advantage for black Africans living in malarial regions. It is quite instructive that this finding did not benefit from enthusiastic promotions. Evolutionary explanations complicated the definition of genetic deficiency and fitness, and diluted the impact of Pauling's neat conception of molecular disease.

Furthermore, the molecular definitions of health, disease, and life bracketed out a host of biological processes known at the time, mechanisms that when included in the explanatory framework would certainly modify the accounts of behavior of macromolecules in vivo, and would surely alter the purely technological representation of molecular life. The work of Walter Cannon on the mechanism of homeostasis, coordination of intracellular events, processes of differentiation and development, a host of other biological phenomena subsumed under the subject of biological organization, or the works of Paul Weiss at the University of Chicago, would have challenged the exclusive explanatory power of mechanisms of upward causation. Within the molecular biology program, however, interactive biological processes linking macromolecules to cellular, organismic, and environmental events were largely deemed as epiphenomena.

The promotion of Pauling's spectacular results with the sickle-cell anemia project helped legitimate the growing trend of regarding life as a genetically directed activity of molecules, a trend that placed the study of vital processes in the hands of biochemists and biophysicists equipped with sophisticated molecular probes. At the same time this approach validated the Molecular Biology Program of the Rockefeller Foundation, which provided the framework, and the financial and institutional resources for the research at CalTech and in other institutions. More importantly, Pauling's work encouraged a mode of thinking which was articulated in the 'Science of Man' agenda, the goal he shared with the Foundation's officers, of mapping the molecular pathways of man's soma and psyche, and of rationalizing human behavior through molecular knowledge.

About a decade later, in a 1958 television broadcast entitled 'The Next Hundred Years', Pauling described his vision of a scientific utopia attained through a detailed knowledge of the molecular structure of man.[7] The study of sickle-cell anemia, he stated, set a precedent for that kind of approach. Recounting the biochemical and genetic aspects of the discovery of that first molecular disease "discovered in our laboratory", Pauling postulated that there were "thousands, tends of thousands molecular diseases. My colleagues and I now working on the molecular basis of mental diseases." Pauling believed that mental deficiencies, like other physiological abnormalities, were genetically determined molecular abnormalities.[8]

His vision of the nearing golden age was a move from mere palliative action: biology turning molecular, medicine maturing into an exact science, and social planning becoming rational. "It will not be enough just to develop ways of treating the hereditary defects . . .", he said. "We shall have to find some way to purify the pool of human germ plasm so that there will not be so many seriously defective children born . . . We are going to have a institute birth control, population control. . . ."[9] Pauling's interventionist concepts of social control resonated with those of the Rockefeller Foundation.

Pauling, by then representing the mainstream rather than the vanguard, was receiving nearly a million dollars in grants from the Ford Foundation for biochemical studies of mental deficiency. In his early papers in psychobiology he explained that the new program was just a natural progression of the studies of the previous two decades. The progress in molecular biology, he noted, had related mainly to somatic and genetic aspects of physiology, rather than to psychic aspects.

"We may now have reached the time", he proclaimed, "when a successful molecular attack on psychobiology, including the nature of encephalonic mechanisms, consciousness, memory, narcosis, sedation, and similar phenomena, can be initiated." (See Pauling, 1961, p. 15.) Indeed by the mid-1960s, Pauling had coined a new term and annunciated his concept of "orthomolecular psychiatry", proposing a treatment of mental diseases that would involve providing the optimal molecular environment of mind, including the introduction of nucleic acid into cells to correct genetic abnormalities – DNA by then had replaced proteins as the "master molecule".[10] With the gene problem deemed basically "solved" conceptually, the brain was defined as the last frontier to be explored with molecular probes; the debates on inherited mental attributes recast in the language of molecular genetics gained new currency. The more recent reincarnation of that project, the aspiration to control the sequences coding for human behavior, is now inscribed in the Human Genome Initiative.

CONCLUSION

This historical analysis has traced some of the lineages of recent technologies of life, disease, and behavior control. Rather than subscribe to the received view that current genetic engineering represents the application phase following a period of pure science, and that its pledges are the fruits of basic research in molecular biology from the 1940s to the 1960s, the paper has presented an alternative interpretation. This case study of a dominant research program has shown that the goal of engineering life was inscribed into the molecular biology program from its inception.

Consequently, the knowledge-claims of genetic engineering technologies with respect to the nature of life, health, and disease must be evaluated in light of this inscription, for this inscription constitutes a circularity which I have analyzed through the criteria of representing and intervening. First the great problems of biology were defined strictly in terms of macromolecules: the essence of life resided in the strip between 10^{-6} and 10^{-7} cm, and was coterminous with it. Then, this region could only be represented through technologies designed exclusively to probe only these dimensions. In turn, measurements and observations generated by these new technologies and formulated in molecular language were privileged as the explanations of life, with the ultimate aim of intervening in natural processes. Yet based, at least in part, on a tacit and unwarranted extrapolation to higher-level organization, this intervention ultimately aimed at affecting changes at a higher biological level, on the level of the organism and medical practice.

But this development has a more general and a more profound significance. Laboratories did not produce knowledge in isolation from a more pervasive structure of representing and intervening. The Rockefeller Foundation's program was explicitly an interventionist agenda aimed at producing a science for social control. I am not positing a simplistic argument of causality or even directionality. The Foundation did not impose on scientists a particular research agenda; the Foundation's officers depended at every turn on their scientific advisers – researchers like Pauling or Beadle – to inform them about cognitive priorities and technological capabilities. These scientists, in turn, sought to promote their own research goals and technologies. I have advanced an argument of cognitive and ideological resonance. The Foundation's technocratic vision of social control and its representational strategies were articulated on the discursive level of programs and polices; the scientists' technocratic vision of life was represented at the bench. No sharp demarcation can be drawn between the sociological context of discovery and epistemological context of justification.

What we witness here is a dialectical process of representing and intervening, in which the molecularization of life was empowered by the synergy between laboratory and boardroom, through the commensurability of social and scientific ideologies.

Massachusetts Institute of Technology,
Cambridge, MA, USA

NOTES

[1] Rockefeller Archive Center (hereafter RAC), RG 1.1, 205D, Box 4.23; Grant Proposal, Pauling to Weaver, December 4, 1945.
[2] RAC RG 1.1, 205D, Box 5.71; Mason to Gunn, December 18, 1993.
[3] RAC RG 1.1, 205D, Box 4.23; Grant Proposal, p. 1.
[4] RAC RG 1.1, 205D, Box 4.23; Grant Proposal, p. 2.
[5] RAC RG 1.1, 205D, Box 4.23; Grant Proposal, 'Detailed Statements', pp. 2–9.
[6] RAC RG 1.1, 205D, Box 4.23; Grant Proposal, 'Detailed Statements', pp. 16–7. See Kay (1989).
[7] California Institute of Technology Archives, Historical File, Box 88, Pauling File; 'The Next Hundred Years', KRCA-Channel 4, December 13, 1958.
[8] California Institute of Technology Archives, Historical File, Box 88, Pauling File; 'The Next Hundred Years', KRCA-Channel 4, December 13, 1958, pp. 10–1.
[9] California Institute of Technology Archives, Historical File, Box 88, Pauling File; 'The Next Hundred Years', KRCA-Channel 4, December 13, 1958, p. 12.
[10] Ford Foundation Archives, Report Section, Grant File PA56-223, Reel 2741; 'Orthomolecular Psychiatry', 1967, pp. 1–8; Pauling 1962, Section 5, Attachments.

REFERENCES

Abir-Am, P. (1980). 'From Biochemistry to Molecular Biology: DNA and the Acculturated Journey of the Critic of Science Erwin Chargaff'. *History and Philosophy of the Life Sciences* 2: 360.

Abir-Am, P. (1982). 'The Discourse of Physical Power and Biological Knowledge in the 1930s: Reappraisal of the Rockefeller Foundation's "Policy" in Molecular Biology'. *Social Studies of Science* **12**: 341–82.

Abir-Am, P. (1985). 'Themes, Genres, and Orders of Legitimation in the Consolidation of New Disciplines: Deconstructing the Historiography of Molecular Biology'. *History of Science* **23**: 73–117.

Forman, P. (1987). 'Behind Quantum Electronics: National Security as Basis for Physical Research in the United States, 1940–1960'. *Historical Studies in the Physical and Biological Sciences* **18**: 149–229.

Hacking, I. (1983). *Representing and Intervening*. Cambridge: Cambridge University Press.

Hacking, I. (1986). 'Weapons Research and the Form of Scientific Knowledge'. *Canadian Journal of Philosophy*, Suppl. 12: 235–50.

Haraway, D. (1979). 'The Biological Enterprise: Sex, Mind, and Profit from Human Engineering to Sociobiology'. *Radical History Review* **20**: 206–37.

Kay, L.E. (1986). 'W.M. Stanley's Crystallization of the Tobacco Mosaic Virus, 1930–1940'. *Isis* **77**: 450–72.

Kay, L.E. (1987). *Cooperative Individualism and the Growth of Molecular Biology at the California Institute of Technology, 1928–1953*, Ph.D. Dissertation, The Johns Hopkins University.

Kay, L.E. (1988). 'The Tiselius Electrophoresis Apparatus and the Life Sciences, 1930–1945'. *History and Philosophy of the Life Sciences* **10**: 51–72.

Kay, L.E. (1989). 'Selling Pure Science in Wartime: The Biochemical Genetics of G.W. Beadle'. *Journal of the History of Biology* **22**: 73–101.

Kay, L.E. (1992). *The Molecular Vision of Life: CalTech, the Rockefeller Foundation and the Rise of New Biology*. New York: Oxford University Press.

Keller, E.F. (1992). 'Critical Silences in Scientific Discourse: Problems of Form and Re-Form'. *Secrets of Life, Secrets of Death: Essays on Language, Gender, and Science*. New York: Routledge, pp. 73–92.

Kohler, R.E. (1979). 'Warren Weawer and the Rockefeller Foundation Program in Molecular Biology: A Case Study in the Management of Science'. In: Reingold, N. (ed.), *The Sciences in the American Context: New Perspectives*. Washington DC: Smiths. Instit. Press, pp. 236–70.

Merchant, C. (1980). *The Death of Nature*. San Francisco: Harper and Row.

Morgan, T.H. (1928). 'Study and Research in Biology'. *Bulletin of the California Institute of Technology* **36**: 87.

Neel, J.V. (1949). 'Inheritance of Sickle-Cell Anemia'. *Science* **110**: 64–6.

Pauling, L. (1940). 'A Theory of the Structure and Process of Formation of Antibodies'. *Journal of the American Chemical Society* **62**: 2643–57.

Pauling, L. (1961). 'A Molecular Theory of General Anesthesia'. *Science* **134**: 15–21.

Pauling, L. (1962). 'Academic Address'. In: Rinkel, M. (ed.), *Biological Treatment of Mental Illness*. New York: L.C. Page, pp. 31–7.

Pauling, L. (1970). 'Fifty Years of Progress in Structural Chemistry and Molecular Biology', *Dedalus* **99**: 990–1108.

Pauling, L., Itano, H.A., Singer, J.S. and Wells, M.C. (1949). 'Sickle-Cell Anemia, a Molecular Disease'. *Science* **110**: 543–8.

Pauly, P.P. (1987). *Controlling Life: Jacques Loeb and the Engineering Ideal in Biology*. New York: Oxford University Press.

Science Illustrated. (1949). 'Linus Pauling Atomic Architect', January 1949, pp. 39–40.

Yoxen, E. (1981). 'Life as Productive Force: Capitalising the Science and Technology of Molecular Biology'. In: Young, R.M. and Lavidow, L. (eds.), *Studies in the Labour Process*. London: CSE Books, vol. 1, pp. 66–122.

SCOTT F. GILBERT

ENZYMATIC ADAPTATION AND THE ENTRANCE OF MOLECULAR BIOLOGY INTO EMBRYOLOGY

INTRODUCTION: THE PROBLEMS OF SYNTHESIS

There are many problems whenever a synthesis is attempted between two divergent disciplines. This is especially true when the two disciplines have an historical enmity and when the assumptions and axioms of the disciplines are at opposite ends of the continuum that characterizes the sciences. Genetics and embryology are two such divergent disciplines that are presently being united through molecular biology, and there are great differences between the genetic and embryological sciences. Moreover, to effect this integration, genetics and embryology are being placed into the common language of molecular biology. The molecularization of embryology has come about gradually and grudgingly. It is certainly far from complete, and it may never be completed. Phenotype-level embryologists (those concerned primarily with developmental anatomy and morphogenesis) fear for the integrity of embryology as a discipline and fear the lack of funding if they don't construct cDNA libraries of their favorite organs (see Malacinski, 1991; Gilbert, 1992). They see the handwriting on the wall, and it is full of As, Cs, Gs, and Ts. The molecularization of embryology has come about from several sources: cancer research, studies of globin synthesis, the isolation of ribosomal RNA genes, and analysis of embryonic lethal mutations being some obvious paths.[1] This essay seeks to look at some of the reasons for this delay in the molecularization of embryology and to see how research into bacterial metabolism enabled molecular biology to establish a major foothold in embryology.

There are at least four major dimensions to any disciplinary synthesis: conceptual, professional, philosophical, and aesthetic. All these dimensions are being negotiated in the attempts to synthesize genetics with embryology, and this has caused many embryologists to worry that the result won't be so much a *synthesis* as a *take-over* of embryology by the powerful and already molecularized geneticists. The first and most obvious dimension is the *conceptual* dimension. In this case, the problem concerns: What is the relationship between the genotype and the organismal phenotype? how does the DNA of the cell nuclei generate the living, eating, mating organism? Our models of this relationship have changed dramatically over the past fifty years. In the 1930s, many embryologists thought that the genes merely put the finishing touches on the organism, but that cytoplasmic proteins controlled the major events of ontogenesis. Others (such as E.E. Just) were convinced that there was no relationship between the genes and the organismal phenotype. The nucleus merely sequestered the unused morphogenetic determinants (Sander, 1986; Gilbert, 1988). Several geneticists, however, claimed that differentiation and morphogenesis were just epiphenomena of gene activity and that development was equivalent to differential gene expression. In any synthesis of molecular genetics and embryology, the new trinity of DNA, RNA, and Protein has to be reconciled with the older trinity of Ectoderm, Mesoderm, and Endoderm. This is the conceptual problem that will be the focus of this paper.

The second dimension to any synthesis is the *professional* dimension. Genetics and embryology diverged in the 1920s and 1930s. The mechanism of this separation has been addressed elsewhere (Gilbert, 1978, 1988; Sander, 1986; Allen, 1986) and will be mentioned but briefly here, for this was not an amicable separation. At the turn of the last century, genetics and embryology had been joined in a common science of heredity. The speculations of August Weismann, W.K. Brooks, Theodor Boveri and others each assumed that the hereditary and developmental determinants were the same. What was inherited was a mode of development. Even Mendel's term for his factors, *Formbildungelementen*, displays this union. T.H. Morgan, an embryologist, had initiated experiments to show that the cytoplasm, not the nucleus, determined the sexual phenotype of the organism. His results, reported first in 1911, however, showed that the chromosomes were the important determinants not only of sex, but of other traits as well. Although originally presented in embryological terms, Morgan's laboratory gradually refined the gene concept, and by 1926, Morgan formally separated genetics – the transmission of nuclear genes – from embryology, the expression of those genes (Morgan, 1926a,b).

By the 1930s, genetics and embryology had their own rules of evidence, their own paradigmatic experiments, their own favored organisms, their own professors, their own journals, and most importantly, their own vocabulary. Since then, genetics has become explained in terms of molecular biology,, and embryology has become developmental biology. Any resynthesis that attempts to explain development in terms of molecular biology has to overcome entrenched disciplinary boundaries.[2]

Thirdly, there is the *philosophical* dimension to the synthesis of embryology and molecular biology. They are as far apart on the philosophical spectrum as they can be. The geneticists of the 1930s and the molecular biologists of the 1980s tended to be reductionists. Historians (Roll-Hansen, 1978; Allen, 1985) and philosophers (Wimsatt, 1984; Darden, 1991) have documented the reductionist philosophy of the Morgan school of genetics, and the same philosophy is seen in their molecular descendents. If you know the rules for the "genetic program", the development and evolution of the organism can be understood. In this way, "genetic program" become synonymous with mechanism of development. Embryologists, on the other hand, have a reputation for being holistic thinkers (Harwood 1993; Gilbert and Faber, in press). They have historically stressed that "the organism in its totality is as essential to an explanation of its elements as an explanation of its elements are to an explanation of the organism" (Haraway, 1976). Embryologist Paul Weiss (1968) thought it philosophically untenable that the organization of the embryo could arise without pre-existing order. "The true test of a reductionist system", he said "is whether or not an ordered unitary system . . . can, after decomposition into a disorderly pile of constituent parts, resurrect itself from the shambles by virtue solely of the properties inherent in the isolated pieces". He graphically illustrated this point (Weiss 1962) by showing a photograph of an intact chick embryo, a chick embryo that had been blended through a homogenizer, and a chick embryo whose homogenized components had been centrifuged. The problem for reductionists, he maintained, was how to get that chicken back.

Pnina Abir-Am (1991) has shown the difficulties that Joseph Needham experienced in his attempt fo try ot reconcile the holistic embryology of the 1930s with reductionist chemistry. The chemists saw the constantly changing embryo as poor material from which they could isolate and characterize their enzymes. Embryologists saw Needham's biochemistry as a reductionist threat. Any snythesis of molecular biology and embryology must be able to reconcile reductionism with holism or else subjugate one by the other.[3]

Another philosophical difference concerns epigenesis and preformationism. E.B. Wilson said in 1925 that "Heredity is effected by the transmission of a nuclear preformation which in the course of development finds expression in a process of cytoplasmic epigenesis." Note that he was using the old term – heredity – which incorporates both genetics (which stresses the transmission of preformed genes) and embryology (which stresses the epigenetic changes that create new cell types and organs from the mitotic descendents of the fertilized egg). How can these identical preformed genes create such divergent cell types? Several embryologists doubted they could. In 1941, N.J. Berrill, presiding over the first Growth Society meeting (which was to become the Society for Developmental Biology), defined genes as "statistically significant little devils collectively equivalent to one entelechy". Similarly, the contemporary developmental biologist Lauri Saxén (1973) has claimed that "Our present idea of progressive differentiation actually is not far removed from this classical homunculus concept. Thus, all the information required to build a complete organism is already present within the zygote and development is seen as a progressive expression of this genomic information." He satirized this view by comparing a "'homunculus' in the sperm as illustrated by the 16th century animalculists" with "the present view of the 'homunculoid' information in a germ cell". Susan Oyama (1985) and embryologist H.F. Nijhout (1990) have also commented extensively on the similarity in the modern use of "genetic information" and "genetic program" with older concepts of entelechy and preformation.[4]

Fourth, there is an *aesthetic* dimension. Embryology has a tradition of celebrating the complexity and diversity of life. Molecular biology has a tradition of celebrating life's underlying unity and simplicity. Embryologist Berrill, for instance, (1961) writes repeatedly of "the amazing diversity of developmental performances" and "the complex reality" of embryonic development. Molecular biologists such as Monod claim (quoted in Jacob, 1988) that the elephant is constructed on the same principles and using the same materials as *E. coli*. The aesthetic of molecular biology is abstract formalism. Like abstract art, molecular biology seeks to get past the apparent diversity of nature to reveal an underlying unity "more real than the real". Just as an abstract painter might represent a table by a line without concern as to whether the table is oak, plastic, metal, red, or white, so molecular biology has traditionally ignored species differences to discover the underlying unities of living organisms. The traditional aesthetic of embryology has been naturalism. Every species developes in a different manner, and generalizations from one species to another are very risky. So aesthetically, embryology is to molecular biology as a Michaelangelo statue is to a Brancussi. Needham (1932) was probably correct when he depicted biology as contested ground between Aristotle and Plato.

EMBRYOLOGY AND THE GENETIC REDEFINING OF DEVELOPMENT

One of the reasons that molecular biology has taken so long to enter embryology is the longstanding fear among embryologists that genetics – whether it be classical genetics or molecular biology – is trying to take over their discipline and bring with it all its reductionism and lack of appreciation for the complexity and species differences.[5] The fear that developmental biology might be "taken over" by genetics is nearly as old as the separation of genetics from embryology in the 1920s.

The remarkable success of genetics in the 1920s and 1930s caused it to become the pre-eminent way to study inheritance, and it redefined the other disciplines in genetic terms. The study of inheritance became genetics, which Morgan defined as the discipline concerned with the transmission of nuclear genes (Morgan, 1926a,b), and Morgan's exclusion of cytoplasm from the realm of inheritance was soon viewed as dogma (Sapp, 1987). Embryology was redefined as the study of changes in *gene expression* over time (Morgan, 1934), and evolution was redefined as changes in *gene frequency* over time (Dobzhansky, 1937). Thus, evolution and embryology, which had traditionally been sciences of the phenotype, were given new, genotypic, definitions.

These new definitions went against the prevailing paradigms of these fields. Evolution had been the province of paleontologists who reconstructed ancient skeletons and phylogenies. Similarly, few embryologists had concerned themselves with questions of gene expression. The predominant problem of embryology from the 1700s through the 1950s was the creation of ordered form, morphogenesis, not differentiation (Haraway, 1976; Lenoir, 1982; Fischer and Smith, 1984). Morphogenesis was a whole-embryo question, differentiation a cellular question. The genetic redefinition of embryology collapsed the morphogenesis question into a subset of the cell differentiation question. To geneticist Richard Goldschmidt, this was axiomatic. He wrote in 1939, "Development is, of course, the orderly production of pattern, and therefore after all, genes control pattern." Similarly, Sturtevant (1932) told the International Congress of Genetics:

One of the central problems of biology is that of differentiation – how does an egg develop into a complex organism? That is, of course, the traditional major problem of embryology; but is also appears in genetics in the form of the question, "How do genes produce their effects?"

Note how embryology's main question is said to be differentiation, not morphogenesis. Note also how genetics can now take over this question. Embryology is being reduced to genetics. And why not, since the geneticists will write that they have the superior methodology? After this separation of genetics from embryology, the geneticists defined themselves against embryology. They often depicted embryology as being an intellectual backwater and saw themselves as the true heirs of Darwin and the question of heredity.[6] William Bateson, who was fond of biblical rhetoric, cast the split in terms of faith and truth. Embryology had lost the faith, while genetics carried it forward. In his essay of 1922, 'Evolutionary Faith and Modern Doubts', he claimed, "Morphology having been explored in its minutest corners, we turned elsewhere . . . The geneticist is the successor of the morphologist." As the victorious geneticists wrote the history, this historiography became the received doctrine. We have been told that genetics grew as embryology withered on the vine. Even embryologists started believing it. One contemporary developmental biology book presents the history of developmental biology in terms of 'Embryology: Losing the Faith' and 'Genetics: Keeping the Faith'.

But while the geneticists were making their great discoveries into the mechanism of hereditary transmission, the embryologists were also having their own golden era. Ignoring genetics altogether, embryologists embarked on the program which Joseph Needham (1936) christened *Gestaltungsgesetze*, "the rules of morphological order". Here, the transplantation experiments of Spemann, the Mangolds, Holtfreter, Hamburger, Hörstadius, Harrison, Witschi, Lewis, Child, Willier, and Rawles set new experimental standards for embryologists and provided astounding new insights into how organs were constructed. The evidence coalesced into a concept called the morphogenetic field, and this morphological unit, rather than the gene, was seen as being the fundamental unit of development (and in the case of planaria, inheritance) (Huxley and De Beer, 1934; Weiss, 1939). Indeed, the concept of gene expression is absent in the major embryology books of the 1920s through the 1940s (see Gilbert, 1988). Although experimental embryology had successfully separated itself from the earlier traditions of developmental anatomy, it remained a phenotypic science, and it identified itself as a science concerned with cytoplasmic changes. As Frank R. Lillie wrote in his critical review of 1927, "The germ exhibits the duality of nucleus and cytoplasm; the geneticist has taken the former for his field, the embryologist the latter."

However, the nuclear envelope proved to be a permeable barrier. More and more, geneticists began to see some form of differential gene activity as the cause for embryogenesis. Jumping over the nuclear boundary, they claimed embryology as part of their domain as well. Morgan's group (Dobzhansky, Sturtevant, Schultz, and the visiting Waddington) and the German groups (Kuhn's laboratory, Goldschmidt's laboratory) began studying the mutations which altered the basic patterns of insect development (Harwood, 1993; Kohler, 1994). Goldschmidt (1939, p. 1) saw the development as being identical with "physiological genetics". He claimed that *geneticists* must explain embryology because the embryologists were not capable of doing so. In a later statement that reflects this boundary dispute, Goldschmidt wrote (1955) that "geneticists will continue to worry about the problem of genetic action and take the risk of climbing over the fence erected by some jealous embryologists, who, while claiming the kingdom for themselves, do not set out to till its soil". C.H. Waddington (1939) began reintroducing embryology into English-language genetics textbooks by stating. "Now that the mechanism of inheritance is known, in its main outlines at least, it is possible to tackle the next question, of how the genes affect the developmental process which connect the fertilized egg into the adult organism." If the embryologist were not going to discuss embryogenesis in terms of gene activity, the geneticists would.

But the embryologists had a strong research program of their own, and they did not like being told how to do their science. Ross Harrison (1935), Chairman of the section of zoological sciences of the American Association for the Advancement of Science, addressed his colleagues in words evoking the political anxieties of the mid-1930s:

Now that the necessity of relating the data of genetics to embryology is generally recognized and the *"Wanderlust"* of geneticists is beginning to urge them in our direction, it may not be inappropriate to point out a danger of this threatened invasion. The prestige enjoyed by the gene theory might easily become a hindrance to the understanding of development by directing our attention solely to the genom, whereas cell movements, differentiation, and in fact all of developmental processes are actually effected by the cytoplasm.

He dispaired that identical genes could produce the different types of cells seen in the adult organism.

REDEFINING DIFFERENTIATION: THE ADAPTIVE ENZYMOLOGY OF DEVELOPMENT

Genes could not be relevant to embryology until geneticists could explain how identical genes could generate different types of cells. In 1934, T.H. Morgan speculated that every gene, even though present in every cell, might not be active in every cell. Forty year earlier, as a newly minted embryologist, Morgan had worked with Hans Driesch, a leading proponent of epigenesis in his time. Driesch (1894) had just finished writing his *Analytische Theorie de organischen Entwicklung*. Here, he put forth the following view for the mechanism of epigenesis:

Insofar as it contains a nucleus, every cell, during ontogenesis, carries the totality of all primordia; insofar as it contains a specific cytoplasmic cell body, it is specifically enabled by this to respond to specific effects only. . . . When nuclear material is activated, then, under its guidance, the cytoplasm of its cell that had first influenced the nucleus is in turn changed, and thus the basis is established for a new elementary process, which itself is not only the result but also a cause.

In 1934, Morgan dusted off his copy of Driesch's book and referred to it when he updated the account of epigenesis:

The initial differences in the protoplasmic regions may be suppose to affect the activity of the genes. The genes will then in turn affect the protoplasm, which will start a new series of reciprocal reactions. In this way we can picture to ourselves the gradual elaboration and differentiation of the various regions of the embryo.

In other words, cells that contain different types of cytoplasm would be able to activate different batteries of genes.[7]

This may be a pleasant hypothesis, but there was no evidence in its favor. Indeed, there were other models as consistent with the data. Richard Goldschmidt (1939) proposed that timing differences in the genes determined the phenotype of the cell. Ernest Everett Just (1939) proposed that the cytoplasm contained all the developmental determinants and that the nucleus was like a refuse bin for unused determinants. In the late 1930s, even Curt Stern admitted that this idea could not be disproven (see Gilbert, 1988).

The evidence that differentiation is caused by the differential expression of batteries of genes from identical genomes came from the study of unicellular organisms. As Burian pointed out, Hämmerling's experiments on *Acetabularia* checked some of these models and showed that the nucleus produced substances that were essential in constructing the cellular phenotype, at least in unicellular organisms. But an explanation of how *identical* genes could yield *different* cell types came from studies of enzyme synthesis in yeast and *E. coli*.

I would like, therefore, to pick up this story in 1947, for here is where the discourse on development changes from the tissue level to the intracellular level. Here is where differentiation becomes defined as changes in cytoplasmic proteins, and adaptive enzyme synthesis becomes used to model metazoan embryogenesis. The explicit linkage of enzymatic adaptation and cellular differentiation was made by Jacques Monod at the Growth Society meeting of that year. Monod introduced the phenomenon of enzymatic adaptation as a possible solution to the problem of how identical genomes can synthesize different "specific" molecules:

> The widest gap, still to be filled, between two fields of research in biology, is probably the one between genetics and embryology. It is the repeatedly stated – and thus far unsolved problem – of understanding how cells with identical genomes may become differentiated, that of acquiring the property of manufacturing molecules with new or, at least, different specific patterns of configurations.

Monod, however, did not feel that there was enough data from embryonic cells to warrant concluding that the same mechanisms guided both micro-organisms and eukaryotic embryos.

A more systematic mapping of the phenomena of enzymatic adaptation to differentiation was made that year, at the symposium of the Society for Experimental Biology, 'Growth in Relation to Morphogenesis' (Danielli and Brown, 1948). Numerous topics were discussed. There were several papers on the roles that hormones have in regulating cell growth and development in animals and plants. Two other papers, those by Johannes Holtfreter on neural induction in amphibians and by Hörstadius and Gustafson on sea urchins, discussed the biochemistry of animal development. Hans Grüneberg and Ernst Hadorn spoke on mutations that effected the embryos of mice and *Drosophila*, respectively, and three papers dealt specifically with the question of how the nuclear genotype could produce different types of cells. These three papers reached a remarkable consensus and represented the approaches of a microbial geneticist, a vertebrate embryologist, and a plant physiologist.

The most critical of these papers was that of Sol Spiegelman of the Washington University School of Medicine. Expanding on a paper that had been published the preceeding year, Spiegelman (1948) starts his paper by redefining the problem of differentiation in biochemical terms. First, the problem of differentiation must move from being seen as a morphological property of tissues to being seen as a biochemical property of individual cells. Differentiation is to be seen not in terms of tissue structure but "as the controlled production of unique enzyme patterns". This redefinition, he stated, would focus our attention on "the relationship between the genes in the nucleus and the properties of the cytoplasm".

One of two starting points for Spiegelman's synthesis is the Beadle and Tatum studies in *Neurospora* that showed that an altered genotype creates altered enzymatic properties in the cytoplasm. Enzymes were the link between genotype and phenotype. Genes controlled the production of enzymes; enzymes controlled the phenotype of the cell. The second source of Speigelman's theory is the work done by J.P. Greenstein and colleagues that showed that the cells of each tissue had different patterns of enzymes. Indeed, this is what would be expected if differentiation were the regulation of unique constellations of enzymes.

The next question, then, was how could the supposedly identical genome of each cell type be regulated to produce different constellations of enzymes in different types of cells. The answer, claimed Spiegelman, might come from the study of adaptive enzymes in microbes such as yeast. Here are instances where the identical genome produces different enzymes under different environmental conditions. "A population of individual cells placed in contact with a particular substrate acquires, after some lapse of time, the enzymes needed to metabolize the substrate." Moreover, "once the enzyme has been induced, its maintainance requires the continued presence of the substrate". The use of term *induced* here is critical, for it links this effect directly to embryology. While the phenomenon of adaptive enzymes had been known since the turn of the century in a variety of microbes, Spiegelman claimed that the yeast system is the best since the cells can be cultured in nitrogen deficient medium which would prevent their growth. When the enzymes are assayed, one could be certain that the genome has not changed nor were rare variants selected by differential growth on the substrate.

So Spiegelman used the production of adaptive enzymes by yeast cells as a model for metazoan embryos. One could certainly study the biochemistry of cells much better in the clonal micro-organism than in the constantly changing embryo! What did Spiegelman find? Firstly, he showed that the inducer could cause identical genomes to synthesize different proteins. Secondly, he showed that the kinetics of enzyme production approximated the curve expected of an autocatalytic reaction. The amount of β-galactosidase in yeast after induction with lactose was almost exactly what was predicted by equations in which the velocity of enzyme synthesis was a function of the amount of enzyme already present (i.e., an autocatalytic synthesis).

New enzyme production was not merely a consequence of activating an inactive precursor, stabilizing an unstable protein, or creating more enzymes from the nuclear genes. Rather, it appeared that the more enzyme was produced, the faster the reaction proceeded. "On these grounds one is led to propose that enzyme formation is mediated by a mechanism which is inherently autosynthetic . . . Accepting this assumption of autosynthesis possesses definite implications for the problem of gene action."

In addition to the ideas that differentiation was caused by the controlled synthesis of unique enzyme patterns, and that enzyme synthesis was autosynthetic, Spiegelman added a new notion: competition for scarce recourses. The different enzymatic reactions compete for a limited amount of amino acids and energy. As the synthesis of new adaptive enzyme proceeded, the synthesis of certain other enzyme systems declined drastically. The two alternative biochemical pathways could be in competetion with each other. In this model, the "protein molecule was stable, but the enzyme-forming system involved in its formation was a poor competitor for protein material in the absence of substrate. This results in the loss of protein to other synthesizing systems". Spiegelman used the "analogy of predator-prey relationships encountered in the ecology of higher organisms" to make his point. After showing that a trivial explanation, that of enzyme stability being dependent upon substrate concentration, can be excluded, Spiegelman concluded, "Apparently the extent, severity, and kind of competition do, however, determine the types and amounts of enzymes found in the cytoplasm of cells. Consequently, any theories of gene action . . . must provide a mechanism whereby these agents can perform their functions by influencing the outcome of the competitive interactions amongst enzyme-forming systems."

This led him into a plasmagene model of differentiation. Using radioactive phosphate as a tracer, Spiegelman showed the correlation that the synthesis of new protein parallels a transfer of radioactive phosphate out of the nucleus and into the cytoplasm and that this radioactive phosphate binds to proteins other than those being synthesized. These correlations had also been made that year independently by Caspersson and by Brachet. Based on the new research by Lippmann and by the Cori's, Spiegelmann postulated that the nucleoproteins were energy donors that make specific protein synthesis possible. (He notes that H.J. Muller had independently suggested the same conclusion.) These nucleoprotein units were the plasmagenes, and they were the competitors in the ecological drama of differentiation. The genes of the nucleus produced the plasmagenes which migrated into the cytoplasm. Like the genes, these plasmagenes were also self-duplicating entities. Moreover, they were able to synthesize enzymes. Substrates were envisioned to stabilize the plasmagene complex enabling them to duplicate and to make more enzymes. Here, the plasmagene-enzyme complex is labile unless stabilized by the substrate. This hypothesis "would suggest that nucleoprotein is the most active fraction of the cell in stimulating and guiding differentiation". Moreover, "the fate of any given cell during morphogenesis will be determined by the outcome of the competitive interactions amongst its original plasmagenic population. In these terms, one can understand the multipotency of the early embryonic cell as well as the restriction of morphogenic plasticity which accompanies the progress of differentiation."

Spiegelman concluded that one could readily rewrite experimental embryology texts in terms of plasmagene theory, but that it would do little good, since so very little is known. Nevertheless, Spiegelman was confident that expressions such as competence, induction, evocation, and biological fields could be given more precise meanings in this new molecular scheme.

And it seems that at this point, embryologist C.H. Waddington concurred. Waddington had been active in attempting to synthesize genetics and embryology for the past decade. He published one of the first papers linking genes to phenotypic defects observable in embryos, and he championed the concept of competence whereby multiple pathways of reactions would exist in a tissue (Waddington, 1940a,b; see Gilbert, 1992). The inducer merely caused the cell to follow one of these pathways and not others. He had already linked genetics to embryology by showing that certain mutations could cause alterations in cell fate analogous to those caused by inducer tissues.

Waddington (1948) maintained that all the events of morphogenesis were gene-directed processes. Even tissue formation which may result from differential cell stickiness, he claims, depends upon genes that make the substances that cause the cells to become adhesive. However, while Spiegelman stressed different protein constellations as the *end result* of differentiation, Waddington saw cytoplasmic differences as also being the *cause* of differentiation. This view of a reactive genome that could be influenced by the cytoplasm characterized certain embryologists of the 1940s and 1950s, especially Waddington and Paul Weiss. As we will see, the ability to see genes as a library of potencies acted upon by cytoplasmic factors was important in Waddington's later adoption of the operon model.

Differentiation, in Waddington's view, was self-reinforcing, irreversible, and canalyzed. By the last term, he meant that development could only proceed down certain paths. Not all potencies were realized. A cell might become a nerve or a skin cell depending upon whether it is induced by the notochord. It does not become anything that is both. (When it does, it may be malignant.) Given these properties of development, how might differential cytoplasmic synthesis be established and maintained?

Waddington's solution, for which he gives nobody else credit and which he presented as his own, is a competition model for the resources of the cell.

In a cell which has not yet been epigenetically determined as either neural or epidermal, the gene complement must endow it with the potencies for carrying out the synthesis of either of the two types of proteins; there must, in fact, be alternative chains of synthesis, each leading from certain genes to certain cytoplasmic proteins, which are in competition for much the same substrate materials. In such circumstances, comparatively slight changes in the available raw materials might shift the whole dynamic system from one path into another.

The progressive substrate changes would lead to changes in new enzyme synthesis, which would, in turn, create further substrate changes.

But "something further is necessary to account for the self-reinforcing character of differentiation". Waddington reminds us that before overt differentiation, there is a stage called determination where obvious differences are not seen, even though the cells are committed to become one cell type and not another. The cells become progressively different from each other. The simplest explanation of this, he claims, would be if "protein synthesis involves an autocatalytic event". These autocatalytic events, moreover, would be in competition with one another. "If one has a number of autocatalytic synthetic processes competing for the same collection of substrates . . . it seems inevitable that there would be certain sets of syntheses which were compatible in the sense that they could go on simultaneously, while other sets would be incompatible and unrealizable."

This would lead to the canalization of development and would explain the stability and irreversibility of differentiation. This is what Spiegelman had also noted. Waddington noted that others such as Darlington, Lindegren, Spiegelman, and Sonneborn, have proposed such self-duplicating plasmagenes on genetic evidence, but that this type of entity fitted in well with the observed facts of embryology.

The paper by the botanist, K. Mather (1948), gave a different perspective on the plasmagene theory. His plasmagenes were critical for the continuity of cytoplasmic function in the absence of the genome. Like Waddington (who uses snail coiling mutants to the same end), Mather found that there were cases like that of pollen sterility in *Nicotinia* where the effect was seen in the absence of the nuclear genes controlling this effect. However, upon analysis, the cytoplasm was seen to be given these traits by the parental nucleus. Similarly, Mather quoted Hämmerling's experiments on *Acetabularia* to the effect that the nucleus controlled the type of cap that is formed in this alga, but the cytoplasm was a reservoir of morphogenic substances. There were plasmagenes, to be sure; and they were seen as gene products that reproduced themselves in the cytoplasm under certain chemical conditions. These plasmagenes were seen as mediators between the genome and the cell phenotype. But since plant development is not characterized by the irreversibility and stability of differentiation seen in animal development, the competitive interactions postulated by Spiegelman and Waddington were downplayed. The chemical conditions that would enable one plasmagene to multiply whereas another would not are not discussed. So having stated the need for plasmagenes, he did not consider them as playing major roles in cell determination.

Mather saw the the plants' developmental canalization as being influenced more on the nuclear level. Unlike Waddington and Spiegelman, Mather believed that the genome is fluid and that the association of a gene with other genes could influence its activity. Moreover, unlike the other two, Mather thought that whether one gene produced one or many products was still an open question. Different amounts of the cytoplasmic raw materials could influence the nature of these products and the arrangement of the gene with other genes. He postulated, like Waddington, that the genes both reacted to the cytoplasmic molecules in their environment and created new cytoplasmic environments by this action. All three papers saw adaptive enzymes as a fitting model for cell differentiation in multicellular eukaryotes.

ENZYME INDUCTION AND EMBRYONIC INDUCTION

The phenomena of adaptive enzymes was brought closer to embryology by a nomenclature agreement published in *Nature*, 1953. Here, the major researchers in the field, M. Cohn and J. Monod of the Institut Pasteur, M.R. Pollock of the National Institute for Medical Research in London, S. Spiegelman, now at the University of Illinois, and R.Y. Stanier of the University of California at Berkeley agreed to a uniform terminology for enzyme formation (Cohn et al., 1953). No longer would the term adaptive enzyme be used, for the term "adaptive" denoted a change which increased the evolutionary fitness of the organism.[8] Rather, the process would be known as enzyme induction and "any substance thus inducing enzyme synthesis is an enzyme 'inducer'. An enzyme-forming system which can be so activated by an exogenous inducer is 'inducible'." At the end of the letter, the signatories note that

the exposure of an organism to a single inducer which is also a substrate may result in the induction of a sequence of enzymes, since the metabolism of the primary inducer gives rise to the formation of a succession of intermediary metabolites each of which serves as an inducer for the enzyme which converts it to the next member of the metabolic chain. This phenomenon is termed "sequential induction" (simultaneous or successive adaptation).

This letter takes the microbial enzyme formation away from evolutionary and ecological biology and closer to embryology. Induction was, of course, the key concept of vertebrate embryology, and successive (secondary) inductive chains were the mainstay of embryology texts. The notochord induced the neural plate which induced the lens which induced the cornea, etc. Of chief concern was the "primary" inducer, the Organizer, discovered by Spemann and Mangold in 1924. According to Løvtrup and colleagues (1978), "few compounds, other than the philosopher's stone, have been searched for more intensely than the presumed agent of primary embryonic induction in the amphibian embryo", and Harrison (quoted by Twitty, 1966) referred to the amphibian gastrula as a "new Yukon to which eager miner were now rushing to dig for gold around the blastopore". The concept of induction was at the core of both the morphogenesis and differentiation questions and was probably the most powerful principle in embryology during the 1920s through the 1950s. The linkage of microbial enzyme induction to this central concept of embryology was an extremely important devise in asserting the relevance of microbiology for unravelling the problems of development (and perhaps cancer as well[9]), and for asserting the fundamental unity of all living things.

Although the concept of adaptive enzymes had been linked to embryological induction, there had been no great technical or conceptual advances that were immediately made. Curt Stern (1955) reviewed the data on gene synthesis of proteins and found the situation very similar to what it had been in 1947. He agreed with Mather that the identity of the genome in differentiated cells was an unproven speculation and that slight modifications of the genome could indeed have occurred during development. He summarizes the competition model, but leaves out the plasmagenes. The cytoplasm is seen to act directly on the genes and not on their intermediate products. After reviewing enzymatic adaptation (and calling it that), Stern claims that there is no convincing evidence for the existence of the postulated plasmagenes.

However, by 1958, the idea that microbial enzyme synthesis could model eukaryotic development had been digested, processed, and had elicited a reaction. In this year, a symposium on the Chemical Basis of Development was conducted at the Johns Hopkins University. It was an eclectic gathering, and there was much discussion of induction, both embryonic and microbial. By this time, another line of evidence had emerged against the plasmagene/enzymatic adaptation view of differentiation. The inducers of the microorganismal enzymes were small substrate molecules. But evidence from Toivonen, Kuusi, Yamada, and Niu was suggesting that the inducer of the amphibian neural tissue was a protein. Clement Markert concluded that there was very little evidence in embryos for the type of adaptive enzyme synthesis found in microbes (1959, p. 6). Nor was there any particular evidence in favor of plasmagenes. Markert thought that differentiation could be measured by the constellation of proteins in the cell, but his hypothesis of differential gene activation is pretty much the same as Morgan framed it in 1934. However, he is able to buttress the notion of genetic identity of each nucleus with the results of King and Briggs on frog nuclei and Pavan and Beerman on *Drosophila* chromosomes. Markert's model of differentiation concerns cytoplasmic proteins that are able to enter the nuclei and bind to the chromosome in order to activate specific genes.

But there are plenty of speakers who supported the idea that embryonic induction can be modelled by microbial induction, and Bentley Glass (1958) noted that "inasmuch as some of the same enzyme systems significant in vertebrate development can be readily followed in bacteria, the approach holds considerable appeal". Glass criticised Markert for not recognizing that suppressor genes can shift the bacterial phenotype from one state to another. Melvin Cohn (1958) discussed the β-galactosidase inducible system of *E. coli*, and discussed it in terms of embryonic development. "If maintainance in bacteria is to enjoy the dignity of being compared to differentiation in higher organisms, then it should persist over many generations; and in fact it does." Other analogies to metazoan development were also made, such as heterogeneity of response. The ability to respond to an inducer was found to be dependent upon the permease that enables the inducer to enter the cell. This was related to embryonic competence, although not named as such.

Three other papers, by L. Gorini and W.K. Maas, by H.J. Vogel, and by Boris Magasanik, respectively, looked at feedback mechanisms of regulation in bacteria. Vogel (1958) speculated that the inhibition of the synthesis of some product might actually be due to the product's binding to the gene and shutting it off. Here, then, was another example of differential protein synthesis whereby a small molecule (in this case, the product of a series of metabolic reactions) interacted with the genes to alter their activity. Gorini and Maas (1958) speculated that a "change in the enzymatic constitution resulting from either feedback inhibition or enzyme induction could be the step initiating the series of reactions necessary to produce a differentiated cell". Boris Magasanik (1958) concurred, saying that it was "not unlikely" that cellular differentiation could be controlled in a manner similar to that of negative feedback regulation.

But does anything like this ever happen in real embryonic cells? A paper by Richard Stearns and Adele Kostellow of the Albert Einstein College of Medicine (1958) provided evidence that enzyme induction in *E. coli* and enzyme induction in embryonic blastomeres could be one and the same. They purposefully attempted to mimic adaptive enzyme synthesis in embryonic cells. First, they made *Rana pipiens* embryos more like bacteria and yeasts by dissociating them into blastomeres with versene (EDTA). They then centrifuged them according to density to get mixtures enriched for surface cells, presumptive ectoderm, presumptive mesoderm, and presumptive endoderm. In normal embryos, tryptophane peroxidase is found only in the liver, an endodermal organ. Moreover, the levels of the enzyme could be induced in the liver. When they determined if tryptophane peroxidase could be induced by tryptophane in embryonic cells, they find that no cell could be induced before gastrulation, but after gastrulation, the gut precursor cells show induction. Moreover, lactose was found to be able to induce β-galactosidase. It seemed that, despite Markert's objections, embryonic and microbial induction were the same after all.

It is here, too, that we get to hear one of those statements of impatience raised by molecular biologists when they confront embryologists. In the discussion following Dr. Magasanik's talk, Sol Spiegelman (1958) gave a little oration:

I have found it difficult to avoid the conclusion that many of the investigators concerned with morphogenesis are secretly convinced that the problem is insoluble. I get the feeling that many of the intricate phenomena described are greeted with a sort of glee as if to say, "My God, this is wonderful, it is so complicated we will never understand it."

It seems to me that perhaps the time has come to abandon this joyful pessimism and its attendant conviction of incomprehensible complexity. In particular, I should like to make a plea for a more optimistic view based on a belief in simplicity. The phenomena of morphogenesis can hardly be as complicated as implied by the welter of apparently unrelated observations constituting the literature of embryology. . . . It is no longer relevant these days to phrase questions of cell physiology in terms of other than chemically defined entities. It seems to me that the same is true for morphogenetic events.

Here we see the *Kulturkampf* of molecular biology and embryology.

Also by this time, RNA has been discovered and has been found in animal cells as well as in plants. At this symposium, Mirsky and Allfrey reported that while large quantities of RNA were made in the nucleus, only a small fraction was getting into the cytoplasm where the proteins were made. Joe Gall also reported on the visualization by autoradiography of RNA on the lampbrush chromosomes of larval flies and amphibian oocytes. Gall postulated that the locations of the ^{32}P-labeled RNA were genetic loci. By following the tracer, Gall found that the RNA went from the nucleus into the cytoplasm.

In the next three years, evidence for the existence of a messenger RNA would increase. Again, this data would come from *E. coli* and yeast. This RNA, representing less than 5% of the total RNA of the cell would have a short half-life, a rapid incorporation of phosphorus, and the ability to associate with 70S ribosomal particles. Its base composition resembled that of the organism (or in phage-infected bacteria, that of the virus), making it readily distinguishable from the long-lived ribosomal RNAs and from the short soluble RNAs. The ribosome would be established as the site of protein synthesis. The presence of a reusable messenger RNA made the plasmagene hypothesis untenable (Judson, 1979).

These data were synthesized with the ongoing program of Monod's laboratory looking at mutations of the lactose-synthesizing genes of *E. coli*. The story of the operon model has been told many times before (see Schaffner, 1974; Judson, 1979), and I will not repeat it here except to say that it demonstrated that induction worked at the genomic level. In inductive systems, the inducer blocked a gene-encoded repressor (either RNA or protein) from binding at an operator site adjacent to the structural genes. This prevented their activation. If the inducer were present, the gene made its mRNA which bound to the ribosomes to make the proteins. If the inducer were not present, the repressor was able to bind to the gene region and block transcription. In this way, the same genome could give different enzymes depending on whether or not the inducer was present. In their closing statement of a major 1961 review article, Jacob and Monod emphasized that operon-like control mechanisms may be a universal part of gene regulation.

It has been repeatedly pointed out that enzymatic adaptation, as studied in micro-organisms, offers a valuable model for the interpretation of biochemical co-ordination within tissues and between organs of higher organisms. The demonstration that adaptive effects in micro-organisms are primarily negative (repressive), that they are controlled by functionally specialized genes and operate at the genetic level, would seen greatly to widen the possibilities of interpretation. The fundamental problem of chemical embryology is to understand why tissue cells do not express, all the time, all the potencies inherent in their genome.

This fit in perfectly with what Markert had said in 1958 about cytoplasmic proteins binding to regions of the DNA to activate or inactivate specific genes. It also was congruent to the embryologists' view of a reactive genome that would not only produce new cytoplasmic substances but that would take orders from the cytoplasm. In fact, the operon model of development was brought into embryology texts immediately by those people who had been looking for a synthesis of genetics and embryology.

Between 1961 and 1963, at least three major textbooks were published which attempted to synthesize genetics and embryology through molecular biology. Each of them used the operon model, but in different ways. The first book, that of geneticists Ruth Sager and Francis J. Ryan, *Cell Heredity* (1961), included the information that was to become the operon model. However, as the book had been written before Jacob and Monod's integration, the data were left to stand without being extrapolated to other systems. Rather, they were used to model intracellular enzyme synthesis only. Other eukaryotic regulatory systems (such as those found by McClintock in Maize) were represented as being similar. But there was no attempt to model metazoan development by adaptive enzymes.

Waddington's 1962 books, *New Patterns in Genetics and Development*, however, commenced with a chapter relating the Jacob and Monod operon model to neural induction in amphibians. After detailing the genetics of the *lac* operon, he noted two generalizations that are important for its application to embryology. First, there were regulatory genes in addition to structural genes; and second, the gene is the target of the repression or activation, not the protein. Waddington links this to eukaryotic regulatory systems. Like Sager and Ryan, he showed the similarities between the *lac* operon and the control elements discovered in plants. Next, he quoted the results of Mechelke, who in 1961 correlated the puffing of dipteran chromosomes with ecdysone secretion preceeding pupation. Certain regions of the chromosome puff out before the others, suggesting to Waddington "just the kind of intrachromosomal activity which Jacob and Monod's Operator is supposed to carry out".

Waddington then began to extrapolate from the operon model to metazoan embryos.

If a structural gene controlled by an operator in the first system produced a substance which functioned as a repressor in a second system, we would have the possibility of 'cascade repression'; and if there were a number of links of this kind, complex systems might be built up which exhibit some of the tendency towards irreversibility which is commonly found in embryological systems but which is hardly accounted for on the simple Jacob-Monod scheme.

Waddington is sufficiently aware of history to relate all this back to T.H. Morgan's 1934 statement that the initial protoplasmic regions determine which of the genes are active.

If metazoan embryos utilize operon-like systems for development, two predictions can be made. First, in mosaic embryos, the Jacob-Monod regulators of development should be found in the regionalized cytoplasm of oocytes. Moreover, there should be genes controlling the formation of these repressors. Waddington proposed the *deep-orange* locus as a candidate for such a gene even though, he admitted, the data were not totally consistent with this identification.

Second, embryonic inducers might function by blocking a naturally occurring repressor that is already in the competent cell. He claims that he, Needham, and Brachet (and also Holtfreter) were too pessimistic when they found that unnatural molecules could induce neural plate formation. Indeed, they had proposed that the real inducer lay within the competent cell but was bound by an inhibitory molecule. He now modeled induction by two adjacent cells. One cell, the competent cell, could synthesize a molecule that was the functional activator of gene activity. However, this molecule was bound by a repressor formed in the same cell. The inducing cell could synthesize an inducing molecule that acted to remove the repressor from the activator. By modeling embryonic induction by microbial enzyme induction, Waddington was able to harmonize the data showing that the competent cell, itself, had inducing activity and the data that showed that numerous compounds that had no chemical similarity with each other could all cause induction to occur. Some of these molecules would resemble the gene activator, while others would resemble the evocator that separated the repressor from the gene activator.

For Waddington, the operon model vindicated his 1936 hypothesis that the actual inducing molecule was produced by the competent (responding) cell. In his autobiographical notes (Waddington, 1975), he maintained,

We showed that, in these terms, the specificity resides in the cells that react to induction – we called it "the masked evocator". This is very similar to the situation discovered by F. Jacob and J. Monod many years later in bacteria, where again the specific repressor molecules are internal to the cells which react to enzyme-inducing substances.

As Sahotra Sarhar has pointed out (as editor of this paper), Waddington was able to take the quantitative data of Jacob and Monod's operon model and to distill it to a qualitative model that was applicable to differential gene expression between cells.

In 1963, John Moore published his synthesis of embryology and genetics, *Heredity and Development*. This book, like Morgan's 1934 volume, was divided into two sections: the first concerned genetics and the second concerned embryology. In the last chapter of the book, he attempted to fuse the two units together. His bridge was the *lac* operon. "Not only does it satisfactorily account for many genetic facts, but it also provides an obvious way of explaining the role of genes in early development." Embryologists, he wrote, were adamant that non-genetic phenomena could influence what the identical genes actually do, but the geneticists were set against it. "This point of view, which once would have been reasonable to an embryologist but not to a geneticist, now seems reasonable to both." He illustrated this using the example of polarized light and pH (two definitely non-genetic influences) on *Fucus* development. Indeed, reading Moore, one gets the impression that the *lac* operon model was a victory of the embryological view over the genetic.

Moore's book consists largely of the geneticists' fly *Drosophila* and the embryologists' frog *Rana pipiens*. It ends, however, with a story of unification and harmony effected by the molecular biologists' microbe *E. coli*.

A generation ago, few embryologists or geneticists would have predicted that a synthesis of their fields would be made possible by studies on the bacterium *Escherichia coli*. But this microscopic creature, with no embryology of its own, has shown a way. A decade from now it may be difficult to distinguish between a geneticist and an embryologist, as they advance their science beyond what each might independently achieve.

The year 1963 also saw the first experiments using transcription inhibitors to demonstrate the importance of differential gene transcription in the embryo (Nemer et al., 1963; Scott and Bell, 1963). The search for the "eukaryotic promoter" had started, and the molecularization of embryology had begun.

Department of Biology, Swarthmore College,
Swarthmore, PA, USA

NOTES

[1] The globin studies are extremely important for integrating transmission genetics, population genetics, molecular genetics, medical and human genetics, and developmental genetics into a coherent field. (Only behavioral and plant genetics were left out). Both the thalassemias and the globin switching were seen as important analogues to embryonic cell differentiation. These became even more important after the operon concept had introduced the notion of regulator gene sequences into embryology. The expansion of embryology into developmental biology (which permitted the data on adult cells, red blood cell precursors and lymphocytes to model embryonic development) also allowed the globin studies to become used for embryonic differentiation.

[2] The professional dimension depends upon the view of the relationship between genetics and embryology. If genes only put the final touches on an organism's phenotype, genetics would just be part of commercial animal and plant breeding.

[3] Biochemistry and molecular biology are both reductionist in that they both attempt to explain living processes solely in terms of chemical and physical principles. They start, however, from different premises. The biochemists have traditionally maintained that metabolism is the *sine qua non* of life (and that the cell is therefore the simplest organism), whereas molecular biologists have seen replication as the fundamental property of life (and elect the virus as life's simplest form). See Gilbert (1982). Embryology didn't adapt well to either reductionism and got a bad reputation for its intransience. When I was hired at Swarthmore College, the former head of the Biology Department, Robert Enders, asked if I were a mystic like the other embryologists he knew.

[4] Berrill's book is probably the last major embryology book that resisted both molecularization and vertebratization, the highwater mark of that school of embryology where entocodons were more important than anticodons. It had, as we shall see, the misfortune of being published just when the operon story was breaking. In recent years, there has been more criticism of the genetic program model. Embryologists are now experimenting with new versions of the morphogenetic field concept, and the notion that these fields are the fundamental unit of development and evolution is being used by a former student of Berrill, Brian Goodwin (1982), Eddy De Robertis and colleagues (1991) and John Opitz (1993). Indeed, one cannot clone or make an antibody to the genetic program. No such thing exists. The genome is less like a programmed score than it is an orchestra wherein each member plays a single note and has perfect hearing. Upon hearing a certain phrase, a performer plays its note, which becomes part of a

new phrase, *et cetera*. For a further analysis of the aesthetical dimension in embryology and the distinctions between it and genetics/molecular biology, see Gilbert and Faber (in press).

[5] The relationship between molecular biology, molecular genetics, and genetics is complex. I agree with Burian (this volume) in claiming that molecular biology is a set of techniques and assumptions that can span the disciplines. The first discipline with which it fused was genetics. Genetics had not been able to solve two major problems – the nature of mutation and the way in which genetic information/material was copied. The Watson-Crick model allowed these problems to be solved. The result was molecular genetics. Thus there is a continuity between genetics and molecular genetics that is very strong. Molecular biology/molecular genetics could not begin to solve the two major problems of embryology – differentiation and morphogenesis – until the late 1980s. Whether this will be an integration of genetics and embryology occasioned by molecular biology or a hostile takeover of embryology by molecular genetics remains to be determined.

[6] The power of genetics to explain embryological enigmas was probably first shown by Sturtevant in 1923. His two-column paper in *Science* disagreed with the snail coiling inheritance patterns published by embryologists. Using genetic tools, he predicted what the patterns should have been had they done their experiments correctly. Seven years later, he was shown to be correct. The historiography of genetics *vis-à-vis* embryology is being detailed in the report, *Bearing Crosses: The Historiography of Genetics and Embryology*, that is presently to be submitted by this author. As genetics became more and more restricted into the gene mapping program, Morgan portrayed embryology as more and more muddled. Morgan's 1932 essay in *Science* became the centerpiece for the history of American genetics, and its supersessionist view of genetics rising above embryology has been perpetuated until very recently.

[7] Here we also see, in its anlagen, the notion of feedback between nuclear genes and cytoplasmic proteins which will become the hallmark of the operon-models of eukaryotic gene expression. Other biologists, notably Haldane (1932) and Goldschmidt (1939) were also postulating that genes acted at different times during development; but they did not place so much value on the cytoplasmic feedback.

[8] Interestingly, Sapp (1987) has argued that the French microbiologists had originally used such nomenclature precisely for its evolutionary, even its Lamarckian, connotations.

[9] The relevance of adaptive/inducible enzymes to cancer research had been a point that microbial biologists strove to mention. In 1946, Spiegelman and Kamen pointed out, "The problem of cancer involves explaining the appearance of a sudden *hereditable* change in somatic cells analogous in several ways to enzyme adaptation or cellular differentiation." These ways were not listed. The success of this approach is appreciated by this author who was for two years privileged to study *E. coli* ribosome synthesis on an NIH cancer grant.

REFERENCES

Abir-Am, P. (1991). 'The Philosophical Background of Joseph Needham's Work in Chemical Embryology'. In: Gilbert, S.F. (ed.), *A Conceptual History of Modern Embryology*. New York: Plenum Press, pp. 159–80.

Allen, G.E. (1985). 'Thomas Hunt Morgan: Materialism and Experimentalism in the Development of Modern Genetics'. *Trends Genet.* **1**: 151–4, 186–90.

Allen, G.E. (1986). 'T.H. Morgan and the Split Between Embryology and Genetics, 1910–1935'. In: Horder, T.J., Witkowski, J.A. and Wylie, C.C. (eds.), *A History of Embryology*. Cambridge: Cambridge University Press, pp. 363–95.

Bateson, W. (1922). 'Evolutionary Faith and Modern Doubts'. *Science* **40**: 1412–5.

Berrill, N.J. (1941). 'Spatial and Temporal Growth Patterns in Colonial Organisms'. *Growth Symposium* **3**: 89–111.

Berrill, N.J. (1961). *Growth, Development, and Pattern*. San Francisco: Freeman.

Cohn, M. (1958). 'On the Differentiation of a Population of *Escherichia coli* with Respect to β-galactosidase Formation'. In: McElroy, W.D. and Glass, B. (eds.), *A Symposium on the Chemical Basis of Development*. Baltimore, MD: Johns Hopkins Press, p. 459.

Cohn, M., Monod, J., Pollock, M.R., Spiegelman, S. and Stanier, R.Y. (1953). 'Terminology of Enzyme Formation'. *Nature* 172: 1096.

Danielli, J.F. and Brown, R. (eds.) (1948). *Growth in Relation to Differentiation and Morphogenesis*. Cambridge: Cambridge University Press.

Darden, L. (1991). *Theory Change in Science: Strategies from Mendelian Genetics*. New York: Oxford University Press.

De Robertis, E.M., Morita, E.A. and Cho, K.W.Y. (191). 'Gradient Fields and Homeobox Genes'. *Development* 112: 669–78.

Dobzhansky, D. (1937). *Genetics and the Origin of Species*. New York: Columbia University Press.

Driesch, H. (1894). *Analytische Theorie der organschen Entwicklung*. Leipzig: W. Engelmann.

Fischer, J.-L. and Smith, J. (1984). 'French Embryology and the "Mechanics of Development" from 1887 to 1910: L. Chabry, Y. Delage and E. Bataillon'. *Hist. Phil. Life Sci.* 6: 25–39.

Gilbert, S.F. (1978). 'The Embryological Origins of the Gene Theory'. *J. Hist. Biol.* 11: 307–51.

Gilbert, S.F. (1982). 'Intellectual Traditions in the Life Sciences: Molecular Biology and Biochemistry'. *Perspect. Biol. Med.* 26: 151–62.

Gilbert, S.F. (1988). 'Cellular politics: Ernest Everett Just, Richard B. Goldschmidt, and the Attempts to Reconcile Embryology and Genetics'. In: Rainger, R., Benson, K.B. and Maienschein, J. (eds.), *The American Development of Biology*. Philadelphia, PA: University of Pennsylvania Press.

Gilbert, S.F. (1992). 'Induction and the Origins of Developmental Genetics'. In: Gilbert, S.F. (ed.), *A Conceptual History of Modern Embryology*. New York: Plenum Press, pp. 181–206.

Gilbert, S.F. (1992). 'Cytoplasmic Action in Development'. *Quart. Rev. Biol.* 66: 309–16.

Gilbert, S.F. and Faber, M. 'Looking at Embryos: The Visual and Conceptual Aesthetics of Emerging Form'. In: Tauber, A.I. (ed.), *Aesthetics and Science: The Elusive Synthesis* (in press).

Glass, B. (1958). 'A Summary of the McCullom-Pratt Symposium on the Chemical Basis of Development'. In: McElroy, W.D. and Glass, G. (ed.), *A Symposium on the Chemical Basis of Development*. Baltimore, MD: Johns Hopkins Press, pp. 856–8.

Goldschmidt, R.B. (1939). *Physiological Genetics*. New York: McGraw-Hill.

Goldschmidt, R.B. (1955). *Theoretical Genetics*. Berkely, CA: University of California Press, p. 247.

Goodwin, B.C. (1982). 'Development and Evolution'. *J. Theoret. Biol.* 97: 43–55.

Gorini, L. and Maas, W.K. (1958). 'Feed-back Control of the Formation of Biosynthetic Enzymes'. In: McElroy, W.D. and Glass, B. (eds.), *A Symposium on the Chemical Basis of Development*. Baltimore, MD: Johns Hopkins Press, pp. 459–78.

Haldane, J.B.S. (1932). 'Time of Action of Genes and its Bearing on Some Evolutionary Problems'. *Am. Nat.* 66: 5–24.

Haraway, D.J. (1976). *Crystals, Fabrics, and Fields: Metaphors of Organicism in Twentieth Century Developmental Biology*. New Haven: Yale University Press.

Harrison, R.G. (1937). 'Embryology and its Relations'. *Science* 85: 369–74.

Harwood, J. (1993). 'Styles of Scientific Thought: The German Genetics Community 1990–1933'. Chicago: University of Chicago Press.

Huxley, J and De Beer, G.R. (1934). *The Elements of Experimental Embryology*. Cambridge: Cambridge University Press.

Jacob, F. (1988). *The Statue Within*. New York: Basic Books.

Jacob, F. and Monod, J. (1961). 'Genetic Regulatory Mechanisms in the Synthesis of Proteins'. *J. Mol. Biol.* 3: 318–56.

Judson, H.F. (1979). *Eighth Day of Creation*. New York: Simon and Schuster.

Just, E.E. (1939). *The Biology of the Cell Surface*. Philadelphia: Blakiston.
Kohler, R. (1994). *Lords of the Fly*. Chicago: University of Chicago Press.
Lenoir, T. (1982). *The Strategy of Life*. Dordrecht: Reidel.
Lillie, F.R. (1927). 'The Gene and the Ontogenetic Process'. *Science* **66**: 361–8.
Løvtrup, S., Landström, U. and Løvtrup-Rein, H. (1978). 'Polarities, Cell Differentiation, and Induction in the amphibian Embryo'. *Biol. Rev.* **53**: 1–42.
Magasanik, B. (1958). 'The Metabolic Regulation of Purine Interconversions and of Histidine Biosynthesis'. In: McElroy, W.D. and Glass, B. (eds.), *A Symposium on the Chemical Basis of Development*. Baltimore, MD: Johns Hopkins Press, pp. 485–90.
Malacinski, G.M. (1990). *Cytoplasmic Organization Systems*. New York: McGraw.
Markert, C. (1958). 'Chemical Concepts of Cellular Differentiation'. In: McElroy, W.D. and Glass, B. (eds.), *A Symposium on the Chemical Basis of Development*. Baltimore, MD: Johns Hopkins Press, p. 6.
Mather, K. (1948). 'Nucleus and Cytoplasm in Differentiation'. In: Danielli, J.F. and Brown, R. (eds.), *Growth in Relation to Differentiation and Morphogenesis*. Cambridge: Cambridge University Press, pp. 196–216.
Mendel, G. (1865). 'Versuche über Pflanzen-Hybriden'. *Verh. naturforsch. Vereines Brünn* **4**: 1–47 (citation on p. 42). Thanks to J. Opitz for pointing this out to me.
Monod, J. (1947). 'The Phenomenon of Enzymatic Adaptation and its Bearing on Problems of Genetics and Cellular Differentiation'. *Growth Symp.* **11**: 223–89. I thank Dick Burian for pointing out this reference to me.
Moore, J.A. (1963). *Heredity and Development*. New York: Oxford University Press, p. 236.
Morgan, T.H. (1926a). *The Theory of the Gene*. New Haven: Yale University Press.
Morgan, T.H. (1926b). 'The Genetics and Physiology of Development'. *Am. Nat.* **60**: 489–515.
Morgan, T.H. (1932). 'The Rise of Genetics'. *Science* **76**: 261–88.
Morgan, T.H. (1934). *Embryology and Genetics*. New York: Columbia University Press.
Needham, J. (1932). *The Great Amphibian*. London: Student Christian Movement.
Needham, J. (1936). *Order and Life*. New Haven, CT: Yale University Press, p. 99.
Nemer, M. (1963). 'Regulation of Protein Synthesis in the Embryogenesis of the Sea Urchin'. *Metabolic Control Mechanisms. Natl. Cancer Inst. Monogr.* **13**: 141–54.
Nijhout, H.F. (1990). 'Metaphors and the Role of Genes in Development'. *BioEssays* **12**: 441–6.
Opitz, J. (1993). 'Blastogenesis and the Primary Field in Human Development'. *Birth Def. Orig. Art. Series.* **29**(1): 1–34.
Oyama, S. (1985). *The Ontogeny of Information*. Cambridge: Cambridge University Press.
Roll-Hansen, N. (1978). '*Drosophila* Genetics: A Reductionist Research Program'. *J. Hist. Biol.* **11**: 159–210.
Sager, R. and Ryan, F.J. (1961). *Cell Heredity* New York: John Wiley and Sons, pp. 321–38.
Sander, K. (1986). 'The Role of Genes in Ontogenesis: Evolving Concepts from 1883–1983 as Perceived by an Insect Embryologist'. In: Horder, T.J., Witkowski, J.A. and Wylie, C.C. (eds.), *History of Embryology*. Cambridge: Cambridge University Press, pp. 363–95.
Sapp, J. (1987). *Beyond the Gene: Cytoplasmic Inheritance and the Struggle for Authority in Genetics*. New York: Oxford University Press.
Saxén, L. (1973). 'Tissue Interactions and Teratogenesis'. In: Perrin, E.V. and Finegold, M.J. (eds.), *Pathobiology of Development – or Ontogeny Revisited*. Baltimore, MD: William and Wilkins, pp. 31–51.
Schaffner, K. (1974). 'Logic of Discovery and Justification in Regulatory Genetics'. *Stud. Hist. Phil. Sci.* **4**: 349–85.
Scott, R.B. and Bell, E. (1963). 'Protein Synthesis During Development: Control Through messenger RNAs'. *Science* **145**: 711–13.
Spemann, H. and Mangold, H. (1924). 'Induction of Embryonic Primordia by Implantation of Organizers from a Different Species'. In: Willier, B.H. and Oppenheimer, J.M. (eds.), *Foundations of Experimental Embryology*. New York: Hafner, pp. 144–84.

Spiegelman, S. (1948). 'Differentiation as the Controlled Production of Unique Enzymatic Patterns'. In: Danielli, J.F. and Brown, K. (eds.), *Growth in Relation to Differentiation and Morphogenesis*. Cambridge: Cambridge University Press, pp. 286–325.

Spiegelman, S. (1958). 'Discussion Following the Talk of B. Magasanik', In: McElroy, W.D. and Glass, G. (eds.), *A Symposium on the Chemical Basis of Development*. Baltimore, MD: Johns Hopkins Press, Chemical Basis of Development, p. 491.

Spiegelman, S. and Kamen, M.D. (1946). 'Genes and Nucleoprotein in the Synthesis of Enzymes'. *Science* **104**: 581–84.

Stearns, R.N. and Kostellow, A.B. (1958). 'Enzyme Induction in Dissociated Embryonic Cells'. In: McElroy, W.D. and Glass, B. (eds.), *A Symposium on the Chemical Basis of Development*. Baltimore, MD: Johns Hopkins Press, Chemical of Basis of Development, pp. 448–57.

Stern, C. (1955). 'Gene Action'. In: Willier, B.H., Weiss, P.A. and Hamburger, V. (eds.), *Analysis of Development* Philadelphia, PA: Saunders, pp. 151–69.

Sturtevant, A. (1923). 'Inheritance of Direction of Coiling in *Limnaea*'. *Science* **58**: 269–70.

Sturtevant, A. (1932). 'The Developmental Effect of Genes. Int. Congress of Genetics'. Quoted in Keller, E.F. (in press). 'Rethinking the Meaning of Genetic Determinism'. In: *Tanner Lectures 1993*. Salt Lake City: University of Utah Press.

Twitty, V.C. (1966). *Of Scientists and Salamanders*. San Francisco: Freeman, p. 39.

Vogel, H.J. (1958). 'Comment on Possible Roles of Repressors and Inducers of Enzyme Formation in Development'. In: McElroy, W.D. and Glass, B. (eds.), *A Symposium on the Chemical Basis of Development*. Baltimore, MD: Johns Hopkins Press, Chemical of Basis of Development, pp. 479–84.

Waddington, C.H. (1939). *An Introduction to Modern Genetics*. London: Allen and Unwin, p. 135.

Waddington, C.H. (1939b). 'Preliminary Notes on the Development of the Wings in Normal and Mutant Strains of *Drosophila*'. *Proc. Natl. Acad. Sci. USA* **25**: 299–307.

Waddington, C.H. (1940). *Organisers and Genes*. Cambridge: Cambridge University Press.

Waddington, C.H. (1948). 'The Genetic Control of Development'. In: Danielli, J.F. and Brown, R. (eds.), *Growth in Relation to Differentiation and Morphogenesis*. Cambridge: Cambridge University Press, pp. 145–54.

Waddington, C.H. (1962). *New Patterns in Genetics and Development*. New York: Columbia University Press, pp. 14–36.

Waddington, C.H. (1975). 'The Practical Consequences of Metaphysical Beliefs on a Biologist's Work: An Autobiographical Note'. In: *The Evolution of an Evolutionist*. Ithaca, NY: Cornell University Press, p. 3.

Weiss, P. (1939). *Principles of Development*. New York: Holt.

Weiss, P. (1962). 'From Cell to Molecule'. In: Allen, J.H. (ed.), *Molecular Control of Cellular Activity*. New York: McGraw-Hill.

Weiss, P. (1968). *Dynamics of Development: Experiments and Inferences*. New York: Academic Press.

Wilson, E.B. (1925). *The Cell in Development and Inheritance*. Third edition. New York: Macmillan, p. 1112.

Wimsatt, W. (1984). 'Reductionist Explanation: A Functional Account'. In: Sober, E. (ed.), *Conceptual Issues in Evolutionary Biology*, Cambridge: MIT Press, pp. 477–508.

ALFRED I. TAUBER

THE MOLECULARIZATION OF IMMUNOLOGY*

> He scrawled "G.O.D." on the backboard. "That is the question: the generation of diversity."
> – Robert S. Schwartz, 1970

INTRODUCTION

The molecularization of immunology has been an exciting example of the power of applied molecular biology and practice. In fact, one might well argue that one of the earliest, not to say richest, products of the new genetic methodology of the 1970s was dramatically harvested in determining the basis of immunoglobulin assembly from constituent gene components to yield highly diverse, and specific, antibodies (see Appendix). The problem of how specific antibodies are produced against a highly varied universe of antigens[1] dates from the earliest formulation concerning the nature of immunity; thus to appreciate the context of the contemporary findings, it is necessary to delve into the formative arguments of the discipline. For approximately the first 25 years of modern immunology, a debate raged concerning the mechanisms mediating the immune reaction. From our modern perspective, which accepts a highly integrated immune system, the opposing camps artificially chose either cellular or humoral mechanisms.[2] As discussed in detail below, the cellularists were those who advocated immune cells as the mediators of immunity, whereas the humoralists argued for factors found in the fluid phase as the defensive agents. The crucial issues embedded more than an argument over particular mechanisms of immune defense, but arose from different research traditions and metaphysical assumptions, which pertain to the fundamental concept of organismal identity.

Immune mechanisms are generally regarded, first and foremost, to discriminate self from nonself, and in so doing, establish both identity and integrity. These in fact represent distinct functions. The issue of integrity pertains to those activities that defend the host from the foreign, repair the damaged, remove the effete, and destroy the malignant. These integrity-preserving activities depend on immune effector functions. But primary to this behavior is the issue of identity: what is to be protected, preserved, repaired? The mechanisms employed by the immune system to preserve identity logi-

---

* This paper is partially excerpted from *The Immune Self: Theory or Metaphor?* by A.I. Tauber; Cambridge University Press, 1994.

cally must follow its establishment. Self, pertaining to immune identity, is not given, but in fact is a historic and changing entity, altered by each immune encounter, and in a sense continuously recreated. Challenged by constant engagements with the organism's own constituents and the foreign, immune identity must constantly be redefined, reasserted, and redetermined. I address elsewhere the highly complex issue of immune identity as defined as the self; the historical antecedents, philosophical meanings, and current common usage of *self* in our culture-at-large, reveal a problematic metaphor (Tauber, 1994). This is a question that cannot be adequately pursued here, and we will assume immune identity as the *self* with no further analysis of what we presume with such a vocabulary, and simply designate identity functions as representing the afferent arm of the immune reaction, the recognition phase, where matters of immune perception are mediated. It is at this juncture that the immune self emerges, and how this identity perceives the foreign, knows itself, and establishes organismal identity is the subject of immunology. Clearly the afferent and efferent arms of the immune process are locked together in a highly integrated fashion, analogous to the nervous system's sensory perceptive modalities, inexorably connected to the cognitive and motor responses; separate anatomic structures, but functional only as linked processes.

As mentioned, the molecularization of immunology dates from the early cellular-humoral argument. The triumph of discovering the molecular anatomy of immunoglobulin structure and the generation of diversity represents in this historical context an important culmination of the humoral agenda. In fact, the question posed by the humoralists (How is immune integrity preserved?) was quite different from that posed by their cellular opponents (What is immune selfhood, i.e., identity?) whose fundamental query still lingers only partially answered. Following a historical description citing the sources of the cellular/humoral argument and its immediate consequences, the second portion of this article examines the molecularization of an immuno-genetic identity, i.e. the attempt to define a molecular definition of the self. Again, in the elucidation of such a genetic signature of the self in the major histocompatibility complex (MHC), molecular biology has offered crucial insight. But we may ask whether the questions to which molecular biology has so successfully been applied address the deepest issue of immunology – what is the immune self? I will argue that such a question in fact can hardly be answered by a reductionist approach alone and must be fitted into a biological construct at a higher level of complexity. And further, the very question of immune selfhood may no longer usefully serve as an operative construct. Thus, immune identity frames this discussion, and a more detailed discussion of the early history of immunology now follows, recognizing that in the discipline's intellectual foundation, we seek the structure of its questions and the relevance of our current answers.

METCHNIKOFF

The first author of modern immunology is Elie Metchnikoff. The justification for such an assignment is the subject of a recent monograph (Tauber and Chernyak, 1991) and several shorter essays (Tauber and Chernyak, 1989; Chernyak and Tauber, 1990; Tauber, 1990, 1991). There the interested reader will find a thorough examination of how Metchnikoff, a Russian embryologist working in the wake of Darwin's *Origin of Species*, sought to unravel the obscure function of amoeboid mesodermic cells throughout phylogeny. The nutritive function of simple animals was fulfilled by such cells, but in organisms with an intestinal gut, the role of these wandering cells was not apparent. Metchnikoff synthesized diverse data from embryology (the fate of the mesoderm), microbiology (bacterial etiology of infectious diseases), and pathology (inflammation) to erect a novel theory of immunity. The phagocyte, the cell which eats, retains its earliest phylogenetic role not as the simple nutritive organ, but the defender of the host. It continues to eat, but now to protect. In fact this was the simplest interpretation of phagocyte function and the one fastened upon by the pathologists/microbiologists of the time. But Metchnikoff's theory was far more subtle.

The history of immunology as a modern discipline dates from this hypothesis (presented initially in 1883), because Metchnifoff's phagocytosis theory proposed that the organism employs *active* mechanisms to 1) define itself, and 2) defend against pathogens. In both cases this definition of immune processes is based on the host's active, as opposed to passive restitution from insult. This simply infers that the animal responds "purposefully" to maintain its identity, i.e. it employs an adaptive system of complex humoral and cellular factors both for its defense and to establish, as well as maintain, organismal identity. Defensive response to micro-organisms constitute an *evoked* immune challenge. Our review of immunology's history will offer an overview of this hard-won position. But in addition, Metchnikoff's thesis argued that immunity was a defining process, i.e. before integrity (defense), identity must be established. This issue is conceptually more difficult to appraise, for it is generally regarded that there is a signature of selfhood registered in the library of a diverse polymorphic system of alleles, the so-called major histocompatibility complex (MHC). In this view, the self is thus given, not developed, by its MHC genome. This identity issue, as being derived from Metchnikoff's phagocytosis theory, has been largely neglected, but remains a vibrant concept underlying modern immunology. But before discussing this problem in the current context, let us consider how Metchnikoff's ideas were received and what the consequences of the scientific debate that ensued were.

Metchnikoff's central thesis of organismal identity, as a manifestation of his embryological interests, were not problems of immediate interest to the immunological community. To put it simply, immune processes in the first half of the twentieth century were generally regarded not as those which establish identity, but rather those that serve to protect the organism (i.e., integrity). In this view, the immune reaction does not establish what is the self, but at best discriminates between the self and the other. But for Metchnikoff, immunological processes are, first and primarily, those activities which establish (constitute) organismic identity, and only because of secondary phenomena, do they *protect*. The early criticism against Metchnikoff was not "immunologic" in the sense we now accept. Paul Baumgarten, Ernst Ziegler, Carl Weigert and others (Tauber and Chernyak, 1989, 1991) contended that "immunity" (the state of not being infected by the same pathogen twice) was a passive phenomenon; for instance, Louis Pasteur extrapolated from the test tube in assuming that the micro-organism exhausted an essential nutrient during the first infection and was thus unable to survive in a host starved of that factor. Such passive theories were the model of immunity and rested upon an ancient metaphysical understanding of health and disease, the balance of humors, and the organism's ability to restore its wholeness (Tauber and Chernyak, 1991; Chernyak and Tauber, 1990). In tracing the reception of Metchnikoff's phagocytosis hypothesis, it is interesting that the earliest critics fastened on two issues: 1) did the cells in fact kill bacteria or simply participate in "necrophagocytosis", i.e., a scavenging or clean-up function, and 2) the apparent vitalistic nature of the theory. The first question, which truly was an argument over mechanism persisted throughout Metchnikoff's life. When the humoral theory was first advocated, Metchnikoff argued that the phagocytes released soluble "ferments" and never acknowledged that his immune cells were either subordinate or even complementary to humoral factors (Metchnikoff, 1905). In natural immunity, i.e. the pre-sensitized state, cellular mechanisms may play a more crucial role, but even in "special" cases illustrated by Emil von Behring's antitoxin generation (1890) or Richard Pfeiffer's phenomenon (1894), Metchnikoff took an adamantly opposed position. Von Behring demonstrated that animals injected with either tetanus or diptheria toxin (with Shibasaburo Kitasato) generated an antitoxin (antibody) response, which neutralized the toxin and protected animals against the lethal effects of infection with these microbes. This was a dramatic demonstration of humoral immunity, but because it was a "special" case of sensitization, i.e., prechallenge, Metchnikoff viewed the results as not generally applicable. Pfeiffer championed the next experimental assault on Metchnikoff's theory in demonstrating the apparent cell-free killing of pre-sensitized guinea pigs injected with cholera. Metchnikoff argued that the humoral factors were in fact escaped "ferments" from phagocytes (Metchnikoff, 1905).

Metchnikoff was responding to a deliberate attempt to disprove the phagocytosis theory initiated in several German laboratories, which (almost exclusively along nationalistic lines) mounted an aggressive attack on the cellularists at the Pasteur Institute, where he had emigrated in 1888 to escape the political unrest in Russia following the assassination of Alexander II (Tauber and Chernyak, 1991). George Nuttal, an American studying in Carl Fluegge's laboratory, was given the initial assignment. In a series of experiments with a variety of animals, he demonstrated that bacteria were killed independently of phagocytes, and others following his lead could soon proclaim a humoral theory of immunity to replace Metchnikoff's cellular mechanism. With the demonstration by Behring that neutralizing antitoxins (antibody) to diptheria and tetanus toxins might be elaborated and used to passively protect against deleterious effects, we note the germ of what was to become the molecularization of immunology, since these studies served as the foundation of immunochemistry (Tauber and Chernyak, 1991; Silverstein, 1989). As late as the early 1930s, whether antibodies were definite chemical substances was still in debate (Heidelberger, 1932–1933) and was not firmly resolved until their identification as gamma globulin (thus the name immunoglobulin) was characterized by Arne Tiselius and Elvin Kabat in 1938 (Tiselius and Kabat, 1939). But it is clear that the growing ascendency of reductionism in biology as an ideal was evident in the earliest period of immunology's self-definition. Immunochemistry, although not coined as a term until Svante Arrhenius so dubbed his aspirations (Arrhenius, 1907), was the clear alternative to Metchnikoff's organismically-based theory. By this time (1890s) the issue of an active response was well-accepted and the issue had become one of mechanism. And it is here that a critical analysis must begin.

Metchnikoff clearly polarized the young field. He led a school of research governed by a vision of immunology as a dynamic descriptive biology. The ascendency of immunochemistry was based on conflicting reductionist principles, as discussed below. These early molecularists had little interest in Metchnikoff's well-known philosophical orientation that arose from his views of biology, nor the wide-ranging implications he projected for the field (Metchnikoff, 1907). These musings were scarcely mentioned in the files of correspondence endorsing his nomination for a Nobel Prize or the secret critiques offered by the Committee's reviewers (discussed in detail elsewhere (Tauber, 1992, 1994). These documents show apparent ignorance of the underlying intellectual basis of Metchnikoff's theory, although he had published several books on the fundamental philosophy of his science. For Metchnikoff to share the Nobel Prize in 1908 with Paul Ehrlich, a synthesis was apparent: complementary roles for humoral and cellular (viz. phagocytic) mechanisms were recognized, and the recent enthusiasm for humoral augmentation of phagocytosis by opsonization of targeted bacteria was viewed as both of therapeutic promise and also as an important contribution for understanding the respective function of these two limbs of the immune response against pathogens (Wright, 1910). But this view of Metchnikoff's role in the establishment of immunology is to deprive us of understanding the very intellectual foundation of immune theory.

Metchnikoff had focused his conceptual orientation around the most general property of phagocytes, which was not simply a metaphorical gendarme of the organism, but an independent center of activity, representing in that function its own evolutionary origin; this phylogenetic and comparative pathological perspective structured his future immunological thinking. Metchnikoff's evolutionary position turned Darwinian struggle into conflict *within* the individual organism.[3] From this point of view, it is understandable why Metchnikoff stressed so often that the struggle of phagocytes with intruders (that is, the inter-species struggle) was only a secondary effect of their normal activity. Thus the essence of inflammation (and immunity) was seen by Metchnikoff not primarily as an expression of inter-species conflict, but arose from the conflicting (sometimes harmonized) relations between parts (that is cell lineages) of the organism. By the same reasoning, the "host-parasite" relation did not play a paradigmatic role for Metchnikoff's phagocytosis theory, but rather served as a subordinate theme. The phagocytosis theory deviates from the thrust of nineteenth century conceptions concerning the nature of pathology and its dynamic role in defining health. He *began* with disharmony. Health then was actively sought. Restoration of health implies an initial harmony, but for Metchnikoff this state remains an ever elusive ideal. In this scheme, the phagocyte became the mediator of the process that defined host integrity and identity. In his view, the phagocytes determined organismal integrity simply by striving for their aggrandizement. Metchnikoff's theory was viewed as incipiently vitalistic – an alarming indictment to a struggling reductionist science. In his great enthusiasm, he at times presented the phagocyte as almost autonomous, and when charged with vitalistic heresy, argued that the psyche also had an evolutionary history that might be traced to the phagocyte's behavior![4] Their activity, competing with other cell lines was Darwin's struggle of species turned inward into a struggle within the organism. Here then is a profoundly novel concept of health: no longer are the ancient humors in balance, but life's cellular components are in conflict; health is not given, it is actively pursued. The potential disharmonious assembly of evolved constituents must now strive for harmony. For Metchnikoff, this was a dynamic process, attained in conflict. Thus Metchnikoff saw "health" as mediated by an essentially pathological process, the expression of a center of activity, phagocytes, (viz. immunity) that defined the normative.

REDUCTIONISM PREVAILS

The conflict between Metchnikoff and the humoralists must be regarded as a particular case of a more general struggle taking place throughout biology and medicine, namely the ascendency of reductionism as the strategy governing research practice and theory. When reductionism was declared as a specific program by German physiologists in the 1840's (Lenoir, 1982; Galaty, 1974;

Kremer, 1990; Gregory, 1977) it was a deliberate attempt to discredit vitalism in biology.[5] Hermann Helmholtz and his like-minded colleagues wished to establish the exclusive domain of physics and chemistry as the sciences to explain organic processes. In the process, teleology and holism (the latter we would describe as organismic biology) was assaulted as allied, if not intimately linked with the vitalistic doctrine. In the 25 year period when immunology was born, 1883–1908 (the brackets of Metchnikoff's active immunology career), we might look at almost any biological discipline and observe the emergence of new standards and expectations moulded by what might well be regarded as a new scientific ethos. It is not my intent to carefully delineate the intellectual or social forces that gave rise to this conflict, but certain parameters should be declared to place the source of immunology's disciplinary borders in perspective. At *fin-de-siècle* there still remained a persistent balance between the reductionist and descriptive sciences, but the next century would not be so generous. Metchnikoff was an early victim of the newer standards. Descriptive sciences were indiscriminately associated with a former age, disparaged as riddled with romantic subjectivism and susceptibility to observer bias. If the science was not definable in quantitative terms, *ipso facto*, its credence fell immediately into question under the auspices of positivist criteria, which reductionism embraced as its own intellectual apparatus. Suffice to note here that reductionism was buoyed as a prevailing scientific approach because it claimed a more rigorous and "scientific" method of explanation then it competitors. The power of reductionism in twentieth century biology is not to be denied, although its goals are coming under increasingly critical scrutiny (e.g., Tauber and Sarkar, 1992; Sarkar, 1991, 1992); its success as a research program has dominated biology in virtually all its venues of investigation and explanation.[6]

Reductionism may be divided into three broad categories: constitutive, explanatory and theoretical (Mayr, 1982, pp. 59–67; Sarkar, 1992). Constitutive reductionism asserts that the material composition of the organic and the inorganic are the same. Since the purge of vitalism following its last grasps before World War I in the arguments of Hans Driesch and Henri Bergson, this position is non-controversial. Explanatory reductionism argues that to comprehend the whole, its components must be defined. But this approach does not account for interaction of parts, and it is this interdependence that ultimately characterizes the various components in relation to the whole. Thus the lower levels of a hierarchy offer only limited information on the characteristics of higher levels (Polanyi, 1969). Or to view the issue conversely, it is well-recognized that new characteristics of a system emerge from its connected constituents. Emergence was also debated in the nineteenth century, accompanying the drive for the sufficiency of reductive explanation (Tauber, 1991). Finally, reduction may operate if the theories and laws of one field of science are shown to be a special case of another branch of science, or one set of scientific claims are logical consequences of another set. The theory must allow for emergent properties by explicitly invoking "dictionaries" or "correspondence" or "translation" rules to go from one level of organization to the next (Schaffner, 1967). As Ernst Mayr illustrates, although meiosis may be explained as a biophysical process, it is only meaningful as a biological concept. A courtship ritual can only be understood in the context of behavioral or reproductive biology and has no distinction when viewed simply as a musculo-skeletal gyration (Mayr, 1982).

Perhaps the most formal issue is to maintain the epistemological and ontological questions as distinctive. For example, is reduction employed to explain some theory, law, or fact, or is it employed to demonstrate the more fundamental composition of a given entity? (Shimony, 1987). The former is an epistemological question, the latter is an ontological one – although obviously different issues, there may be deep connections between their underlying assumptions (Sarkar, 1992). During the course of scientific investigation, what entities are sought might well influence the type of explanation, and conversely, the explanatory success of a theory might well increase confidence in its ontology (i.e., the entities used to state its propositions). Philosophers today who debate the role of reductive thinking in biology appreciate its limitations, but they have yet to formulate a new theoretical construct for understanding complex biological phenomena beyond the reductionist paradigm developed in the 1840s. Other than to note a pluralistic appreciation that biological processes cannot be simply reduced to more elemental analyses without invoking transformation rules or systematics to explain the complexity of the organism, biologists have as yet to erect the scaffold of a "new biology", where laws governing complex hierarchial systems cannot rely solely on characterizing isolated phenomena. It is from the historical and philosophical perspectives that the restrictions of reductionism are best appreciated. The power of the reductionist program has driven organic phenomena to the molecular and genetic level of investigation, but in the process of establishing its hegemony, the holistic basis of an earlier biology was lost. The limitations of these discarded scientific programs left us bereft of certain strengths of their metaphysical structure. The organism is more than the sum of its measured functions; further definition of particularities without a substantive grasp of their inter-relationship reduces the phenomena to isolated *essentia*. (Polanyi, 1969; Koestler and Smythies, 1971; Salthe, 1985; Yates, 1987; Tauber, 1991; Sarkar, 1992).

In the twentieth century, the reductionism debate was articulated as whether biology was to be based upon mechanistic principles or holistic materialism (Allen, 1978). Explicitly formulated by Alfred North Whitehead, "organic mechanism" (1925) was a *process* biology, where the *interactions* of parts were of more interest than discerning the components themselves. Working at more complex levels of organization (e.g. organs, tissues, systems), the holistic materialists were concerned with integration of function, but this in itself did not distinguish them from such severe mechanists as Jacques Loeb (1912). Loeb argued that not only were chemical interactions the most fundamental, they *determined* all higher levels of organization and thereby were predictive. In short, the whole was the sum of its parts; thus the components, once defined, could be placed together into a coherent whole, and the properties of the whole were derivable from the properties of the individual constituents. The holistic materialists, like Walter Cannon and Lawrence Henderson, accepted the mechanical interpretation to a point, but believed that certain properties belonging to a component could only be described as it interacted with other parts in the context of the intact organism. These properties derived from the interaction itself. Examples might include nervous system function, the buffering capacity of the blood, or even the complex array of stimulating and inhibiting factors on muscle construction.

There can be no doubt that one of the important tasks of general physiology, which differentiates it from the physical sciences is that it has to take account of simultaneous activities of widely different kinds harmoniously interacting.

Such interactions become more conspicuous as structures become more differentiated and finally take the extreme forms of harmonic and nervous integrative actions. But it is evident that in the organism every activity is more or less integrative. Or, more precisely stated, every physiological phenomenon must be studied, not only in isolation, but also in relation to other phenomena with which it will in general be found to be related in a manner usually called adaptive [emphasis added]. (Henderson, 1928, pp. 14–15)

For the holistic materialist, "new characteristics emerge from the interaction of parts", which "are not merely quantitatively more complex", they are qualitatively different (Allen, 1978, p. 106). We will have occasion to apply such an analysis to immune function as currently understood, but suffice here to return to our historical and philosophical outline of the development of immunology.

FROM IMMUNOCHEMISTRY TO MOLECULAR BIOLOGY

Returning to immunology, the fundamental animosity between Metchnikoff and his detractors was more than methodological. Specifically, their debate was concerned with the very nature of immunity. The phagocyte as possessor of its own destiny and mediator of the organism's selfhood was simply viewed as vitalistic, which I view as but a disguised critique of the more basic conflict between reductionist versus organismic biology. With the discovery of an augmented antibody reaction, autoimmunity, and diverse immunopathology by 1910, an organism-centered agenda was set, but we see that the reductionist approach prevailed in this branch of biology, and these other issues were not actively pursued after the first decade of the twentieth century (Silverstein, 1989). By 1920, those questions which were to dominate post-World War II medical immunology (tolerance, transplantation medicine, autoimmunity) were dormant and the question of immune specificity became paramount. The chemical reaction was of central concern. Metchnikoff's highly abstract and nebulous question of how the organism was defined was not of immediate *experimental* interest. In fact few understood his program. The humoralists asked a simpler question: what were the mechanics of recognition? Immunology, in the chemo-reductive mode of the twentieth century, chose for the next 50 years to follow the latter issue. Only after 1945 did biology demand an approach to the Metchnikovian problem, and we see then a turn to such questions as immunological tolerance and transplantation; these issues are the modern expression of Metchnikoff's basic question: How was the self defined? This issue became heuristically valuable in judging the fruits of immunochemistry's labor, for the molecularization of immunology would ultimately have to be assessed as a contribution towards understanding the biology of this problem.

To outline the theories of antibody formation is at the same time to trace the apparent congruent conceptual history of immunological specificity. These problems focus upon the central issue of immunology and have served as the subject of several historical surveys (e.g., Silverstein, 1989; Mazumdar, 1974, 1989; Moulin, 1991). Neils Jerne dissected that history between two competing scientific communities (Jerne, 1967): the biologists, who Jerne called cis-immunologists, were concerned with characterizing the biological response to antigen, whereas the chemists, or trans-immunologists, studied the antibody molecule as its own entity, i.e., structural basis of specificity, size of the antibody repertoire, and quantitative relationships. (He viewed the dividing point between the cis and trans positions at the receptor on the antigen-sensitive lymphocyte, an issue to be discussed below). This history does not in fact easily fit into such neat compartmentalization, which demonstrates a highly complex interplay of competing agendas and purpose, but Jerne's scheme *is* useful in broadly viewing how successfully the molecularization of immunology fulfilled its biological assignment.

Clearly a major theme in deciphering the history of immunology has been to trace the approaches to antibody specificity, a biological problem that followed paths towards chemical solutions. But we must not allow those developments to define our own assessment of how successfully that research program fulfilled the requirements of addressing immunology's paramount issues. The question posed by Paul Ehrlich and the other humoralists only admitted a chemical answer: what constitutes immune specificity and how is it mediated? To that problem we have a comprehensive story, to which Ehrlich's side chain theory for antibody formation (discussed below) was the hypothesis by which later models were to be measured. But was serology, the dominant immunochemical field, immunology? Certainly not to a Metchnikovian, for Ehrlich's chemical approach as a trans-immunologist does not pose the same question presented by Metchnikoff, who asked not so much how is integrity accomplished, but how is identity, i.e. self-definition, established. This was a biological question which the immunochemists did not address until the 1940s and 1950s, when modern biologists, in the guise of medical challenges, again posed the question in terms of immune tolerance, organ transplantation, and autoimmunity. Jerne poignantly expressed the same issue in regards to his distinction of cis- and trans-immunologists:

This dichotomy can be related to a saying attributed to Francis Crick: "if you cannot study function, study structure." I do not know where to place the correct emphasis, whether "if you *cannot* study function, study structure," or "if *you* cannot study function, study structure." Anyway, the trans-immunologists appeared on the scene about ten years ago, when decades of experimentation had demonstrated that immunologists, indeed, were unable to study function (Jerne, 1967).

Today, it would be fair to argue that each field of immunological investigation is oriented, if not committed, to molecular definition. We will outline the major developments of immunology's molecularization and then return to the issue as to how successfully that strategy has addressed the more basic problem of immune identity.

Ehrlich

Paul Ehrlich had entered immunology in 1890 by working in the newly established Institute for Infectious Diseases, led by one of Metchnikoff's arch German antagonists, Robert Koch (Baumler, 1984). Ehrlich came as an accomplished scientist, whose doctoral dissertation on histochemical staining established his central notion of differential chemical affinities of cells, the core concept of his later articulated side chain theory (discussed below). He showed that basic, acid, or neutral dyes differentiated blood cells (1878) and thus established important principles for histochemistry that were soon applied to visual identification of the tubercle bacillus and later modified for other bacteria in the Gram stain. Ehrlich's immunological research began with studies of antibody formation to the plant toxin ricin, demonstrating that the toxin-antitoxin reaction behaved like a chemical reaction in its heat dependence. Further characterization was however stymied because the content of antitoxin in the immune sera was highly variable and no standardization had been made, a problem particularly relevant to Behring's antidiptheritic serum that was viewed as having broad therapeutic potential.

"It is beyond question that, in dealing with the problem of the diptheria curative serum, whether from the practical therapeutic, or the purely scientific point of view, it is necessary to use sera of *accurately known value.*" (Ehrlich, [1897] 1957, p. 107, English edition). So began Paul Ehrlich's seminal paper to quantify antibody and place the humoral theory on a firm immunochemical basis. It is not surprising that Ehrlich successfully applied chemical principles to immunity. As already noted, his earliest research in histology sought the affinity of biological materials for aniline dyes, and he was to apply the same basic principles of variable chemical affinities to each of the diverse areas (e.g., anti-microbial and cancer chemotherapy) of his productive research career. In the case of the immune reaction, Ehrlich understood toxins to exert their effects by exhibiting differential affinity for target tissues, and antitoxins then were to compete for binding in a highly selective fashion. His first experiments of the immune response to ricin were viewed as "toxin and its antitoxin influence one another by a direct chemical interaction" possibly by the "formation of double salts" (ibid., p. 114). He directly applied models from organic chemistry:

> It must be assumed that this ability to combine with antitoxin is attributable to the presence in the toxin complex of a specific group of atoms with a maximum specific affinity to another group of atoms in the antitoxin complex, the first fitting the second easily, as a key does a lock, to quote Emil Fisher's well known simile (ibid.).

From this perspective, Ehrlich left to others to determine the effectiveness of phagocytes in the host response to infection and reserved for himself to establish the chemical principles of the immune response: Ehrlich successfully quantitated diptheria antibody, providing "the universally desired basis for a really scientific test. *With this information, the immunity-unit is no longer an arbitrary concept but is an exactly determinable quantity, and one, therefore, which can be reproduced afresh at any time*" (ibid., p. 123, English edition [emphasis in original]).

To account for the antibody response, Ehrlich postulated that cells projected so-called "side chains" from their surface, which served as the receptors for what we now call antigens, and that upon challenge, the host reacts by shedding excess side chains to combine with the offending substance by chemical affinity. There are several highly evocative and prescient concepts inherent in this theory. First, Ehrlich is credited with the first modern formulation of the receptor, whose destiny in all areas of communicative biology has been extraordinarily rich in twentieth century cell biology. More specifically for the history of immunology, Ehrlich rejected outright the notion of an adaptive response, that by necessitating production of "new groups of atoms as required, would involve a return to the concepts current in the days of [an obsolete] natural philosophy". He chose instead what he called "normal enhancement of a normal cell function", i.e., a pre-formed antibody that would be secreted upon appropriate stimulation. Each of the subsequent theories of antibody generation were juxtaposed against this hypothesis.

Instruction Theories

Ehrlich had originally proposed the side chain theory as an alternative to Hans Buchner's proposal (1893) that antitoxin might be formed directly from toxin. This antigen incorporation theory was soon dismissed (although revived in various forms as late as 1930) when antibody was shown to be continually formed in the absence of antigen or in augmented amounts upon second stimulation with minute quantities of antigen challenge (Silverstein, 1989, pp. 61–4). A different version for antigen participation in forming antibody were the "instruction" theories, which appeared in many guises, and constituted the major countervailing alternative until revival of Ehrlich's basic concept by the clonal selection theory: The rationale was based on the obvious need for a highly varied antibody repertoire, whose information content was so enormous that the basis for antibody generation arising *de novo* to each new challenge appeared a daunting biological task. A seemingly simpler mechanism simply utilized antigen to instruct antibody synthesis. In 1905 Karl Landsteiner explicitly stated that following the stimulus of immunization, differently constituted products would be formed (Landsteiner and Reich, 1905); antibody in this view was not a given, pre-formed cell product (i.e., side chain), but a completely new substance.

The basic concept of antigen serving as a template was proposed by Oscar Bail and co-workers, beginning in 1909 (Bail and Tsuda, 1909), who believed that persistent antigen impresses its specificity on host factors, is released and continues its mating to enhance the titre of specific antibody. (Not surprisingly, such a process was thought potentially capable of being mimicked in vitro, and efforts in that regard were made [Manwaring, 1930a,b; Pauling and Campbell, 1942a,b]). The chemical nature of such a process was left unspecified until the globular protein nature of antibody was determined. In the 1930s, various investigators (William Topley, Friedrich Breinl, Felix Haurowitz, Stuart Mudd, Jerome Alexander) recognizing the protein character of immunoglobulin suggested that antigen, carried to the site of protein synthesis, served as a direct template, around which the antibody would be synthesized, and thus dictate a unique amino acid sequence (Topley, 1930; Breinl and Haurowitz, 1930; Mudd, 1932; Alexander, 1931). As Silverstein observed (Silverstein, 1989, p. 69), this theory fulfilled several criteria: 1) the host was not required to contain explicit information to code for the multitudinous antigen library of the environment; 2) the high ratio of antibody molecules to antigen was explained by this synthetic process; 3) it accounted for immunological specificity.

Upon this stage, Linus Pauling introduced modern chemical theory to immunology in an attempt to rationalize the template theory. There was a high degree of congruence between the interests of protein chemistry and immunochemistry of the 1930s and 1940s (Kay, 1989). The study of the biological specificites of "giant protein molecules" in this period eventuated into one of the major schools of molecular biology. In this pre-DNA era, proteins were generally considered the primary agents of heredity (as well as cellular regulation), and thus were thought to be copied from pre-existing protein prototypes; this template concept was in turn based on complementarity (nothing to do with Niels Bohr's complementarity principle) and weak intermolecular interactions. Viruses, enzymes, and antibodies were each subject to such study, and thus immunochemistry attained a central role in the research program of early molecular biology. Pauling's major innovation was the introduction of stereochemical complementarity as the mechanism of biological specificity, including that of the putative antigen template and the antibody formed on it. It is interesting to note that he convinced Max Delbrück, who came to molecular biology from a radically different background, the validity of this position (Pauling and Delbrück, 1940; Fischer and Lipson, 1988).[7] Pauling's interest in immunology dated from 1936, when he gave a seminar at the Rockefeller Institute, and was queried by Karl Landsteiner on possible mechanisms of antibody formation (Pauling, 1970). It was Landsteiner, who 20 years earlier had conjugated artificial antigen (hapten) to protein carriers, and elicited immune reactions in animals (Landsteiner, 1945). Because under normal circumstances, such an encounter would not occur, Landsteiner discounted pre-existing antibody, à la Ehrlich, but assumed a *de novo* chemical synthesis. Based on the earlier template theorists (Mazumdar, 1989, pp. 13–32), Pauling offered his own sophisticated chemical version (Pauling, 1940): assuming identical polypeptide chains, antibodies might differ only in configuration of the antigen-combining portion at the end of the molecule; antigen, serving as a template, generated a complementary configuration (again analogous to Emil Fisher's lock and key model, originally employed by Ehrlich), allowing antibody synthesis about any antigen natural or artificial. This instructional theory propelled research in serological genetics at Caltech (supported by the Rockefeller Foundation) (Kay, 1989) and attempted to generate antibody *in vitro* (Pauling and Campbell, 1942a,b), which, parenthetically, is a superb case study of early venture capital speculation in biotechnology.

To CST and G.O.D.

MacFarlane Burnet raised fundamental biological concerns that were not addressed by the template model: 1) the general kinetics of antibody production both in the case of initial antigen challenge, and most especially with augmented secondary responses in terms of quantity and type of antibody produced, and 2) the persistence of antibody production as a function not only of the cells originally stimulated but of their descendants (Burnet, 1941). Herein lies a herald to Jerne's distinction of cis- and trans-immunologists. By the late 1940s, the appreciation of protein synthesis as under control of an information-laden genome raised the spectre of the antigen interacting with that genome (Silverstein, 1989, pp. 72–5). The incorporation of genetic processing allowed for maintenance of information in the cell and transfer during proliferation, which would explain both the persistence of antibody formation and the accentuated booster response seen upon repeated antigen challenge. The second focus of biologic concern in this period was the resurgence of interest in natural antibody, i.e., the pre-existing pool of immunoglobulin that was formed independent of overt antigen challenge and thus is not explained by instructional models. Finally, the template theories could not address the problem of tolerance, the failure to launch an immune reaction against a given antigen and the fact that tolerance can be acquired for foreign antigens administered before or at birth (Ada and Nossal, 1987).

Jerne, in 1955 re-stated and revised Ehrlich's original hypothesis that the antibody repertoire pre-existed and that small amounts of antibody was synthesized under normal conditions, and an augmented response was initiated upon antigen challenge (Jerne, 1955). Under "natural selection" conditions, antigen served as "selective carrier" of antibody, and signalled reproduction of the appropriately reactive antibodies by an unspecified mechanism. Despite ignoring the chemical basis for antibody specificity, Jerne's model offered a biological alternative to instruction theories, which was soon pursued by Burnet, David Talmage and Joshua Lederberg, who developed what Burnet called the clonal selection theory (CST) (Burnet, 1959). Both David Talmage and Burnet had suggested that cells were selected for multiplication when the antibody they synthesize matches the invading antigen (Burnet, 1957; Talmage, 1957). Burnet (like Ehrlich) placed a phenotypically restricted "natural antibody" on the surface of the cell, which upon interaction with antigen triggered cellular differentiation and unique antibody synthesis (proved by Gustav Nossal and Lederberg, 1958) through clonal multiplication. Other aspects of the theory suggested mechanisms for tolerance (deletion of self-reacting clones during ontogeny) and auto-immunity (re-appearance of sequestered antigen), but a mechanism to explain antibody diversity was not apparent.

Fittingly, the dialogue between cis- and trans-immunologists leaves the story to now be completed by the molecular biologists (Leder, 1982; Tonegawa, 1985). The (William) Dreyer-(J. Claude) Bennett hypothesis of 1965, deduced from known immunoglobulin structure (light and heavy chains each contained a constant, i.e., invariable, and a variable region) postulated that the protein was encoded by a single constant gene and the variable region was encoded by a discontinuous segment composed of several hundred or thousand separately encoded variable region genes (Dreyer and Bennett, 1965). A mechanism was then required to bring the genetic information together to join two sequences of DNA to form a single sequence in an antibody-producing cell, by so-called somatic recombination. The hypothesis was addressed to account for the power to generate extraordinary diversity at the genomic level. With the discovery of restriction endonucleases, experimental proof was offered by Susumu Tonegawa who demonstrated that the arrangement of light-chain genes was different in antibody-producing tumor cells as compared to embryonic cells. The activation of these genes during development was accompanied by somatic recombination or gene shuffling (Hozumi and Tonegawa, 1976). With cloning techniques, Tonegawa then showed that the constant (C) and variable (V) regions were indeed encoded far apart in non-producing immunoglobulin cells, but in antibody-producing tumor cells, the genes were close together, with an intervening (J, for joining) sequence.

Antibody diversity derives from heterogeneity in the V genes (ca. 150), J genes (ca. 5), and a hypervariable region (adding another order of complexity to the gene product) that actually forms the direct antigen binding site to yield 7500 ($5 \times 150 \times 10$) genetic combinations to form various light chains. The heavy chain has a similar selective strategy between V, J and a hypervariable D region, which yields 2.4 million possible different heavy chains (Figure 1). Thus from approximately 300 separate genetic segments in embry-

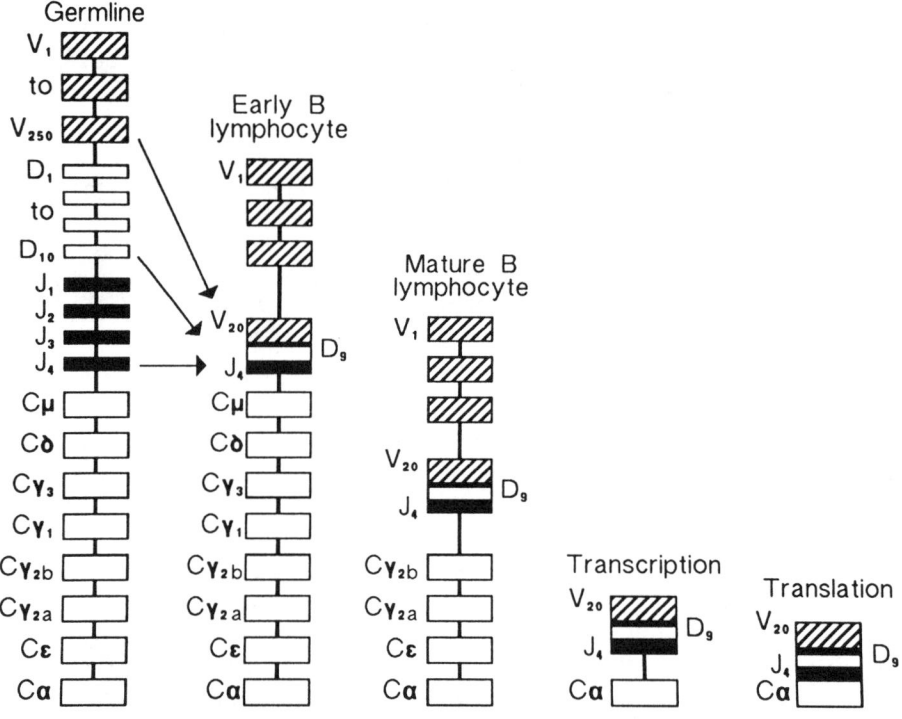

Figure 1. Molecular processing of immunoglobulin heavy chain. A diagrammatic representation of the steps of DNA rearrangement and RNA splicing required to synthesize one of the more than 80,000 possible mouse heavy chains. After the VDJ rearrangement (see text) class switching occurs by deletion of the intervening DNA. The gap between J and C is removed by RNA splicing before translation into the complete heavy chain.

onic DNA, the combinational possibilities yield 18 billion ($7.5 \times 10^3 \times 2.4 \times 10^6$) possible antibodies, which may be supplemented by the extraordinarily high rate of somatic mutation in these genes (1×10^{-4} generation). A molecular basis for clonal selection is now available: antigen (after processing, discussed below) "selects" from the available antibody repertoire and by clonal stimulation drives the proliferation of the appropriate antibody-producing lymphocytes. Our current theory appears to vindicate Ehrlich's original hypothesis offered almost 100 years ago.

THE MOLECULARIZATION OF THE SELF

The Genetic Self

G.O.D. is a mechanistic explanation of a secondary *immune* phenomenon, namely how immune specificity is achieved. What is not addressed is how the self and the other is first differentiated. To that issue a biological model was offered by Burnet, i.e., purging of "self-reactive" clones in ontogeny defines post-partum identity. The theory, first presented in 1949 (Burnet and Fenner, 1949) was more fully developed with CST a decade later (Burnet, 1959). Although the molecularization of that issue has achieved enormous sophistication, its biological meaning is still ambiguous. To explore this issue we must first ask the question, What is the genetic definition of the immunological self?

The issue of immune selfhood was implicit from the first descriptions of auto-reactive antibody and Ehrlich's "horror autotoxicus" at the turn of the century (Tauber and Chernyak, 1991). But molecularization of the self as it is currently understood originated in studies in the 1930s and 1940s that ultimately defined the antigens determining immunological compatibility and which serve as a genetic signature of the self. These proteins are coded in clustered loci termed the major histocompatibility complex (MHC). Obviously the MHC does not exist to frustrate our attempts to transplant, but is thought to serve as a mechanism to introduce the foreign to a thymus-dependent pool of lymphocytes (T-cells); the T cell receptor (generated by similar molecular mechanisms utilized to manufacture immunoglobulin) recognizes antigen complexed (presented) to MHC. Nonself is thus recognized in the context of self (described in the next section and Appendix).

The history elucidating the MHC has been chronicled as emanating from disparate fields of study: tumor transplantation (dependent on the development of inbred mich strains) and blood group typing (Klein, 1986, pp. 1-22; Shreffler, 1988). As Jan Klein notes, mouse fancying spread to Europe and America from Japan and China in the nineteenth century, and using inbred strains developed by American breeders, early twentieth century geneticists interested in the newly rediscovered Mendelian laws used such starting stock and continued to breed brother-sister pairs: Clarence C. Little, as a student of William E. Castle at Harvard, was originally interested in determining the degree to which Mendelian factors could be modified in mammals by selective breeding; Halsey J. Bagg studied the effects of heredity on animal behavior and learning; and Leonell C. Strong, interested in studying the genetic basis of cancer, went from Thomas Hunt Morgan's laboratory to join Little. With successfully inbred mouse strains, the genetic factors affecting tumor biology were then capable of being examined. The basis for such studies were the findings by Carl O. Jensen that tumors originating in a mouse stock strain did not survive when transplanted into wild mice, concluding that there were "race" restrictions on transplantation (Jensen, 1903). The first studies of transplantable tumors were made by Leo Loeb in 1901, and he made similar observations – not knowing Jensen's work – but without understanding the genetic implications (Loeb, 1908). Ernest Tyzzer pursued Loeb's findings and explicitly concluded that although tumor susceptibility was heritable, it did not follow simple Mendelian segregation: after crossing the two mice races, he showed that all the F_1 hybrids grew tumors, but none of the F_2 generation did (Tyzzer, 1909). Little proposed that tumor susceptibility was controlled by several genes, any one of which could cause rejection, so that the number of such tumor-prone individuals would be $(3/4)^n$ (n = number of genes), and if there were 15 genes $(3/4^{15})$ then only 1% of the F_2's would be susceptible (Little, 1914).[8] When the experiment was conducted with many animals, in fact 1.6% of the F_2 mice were tumor transplantable (Little and Tyzzer, 1916). Little extended this observation using normal tissue (spleen) transplanted into two crossed inbred strains (Little and Johnson, 1921).

Before discussing how the theory accounting for the immunological basis for transplantation was proposed in the late 1930s, the second tributary, blood group research, must be described. As Klein notes (Klein, 1986, p. 5) both Karl Landsteiner and Ehrlich (with Julius Morgenroth) described blood groups in 1900. The latter found that hemolysins were generated when erythrocytes of one goat were injected into another, but the immune serum did not react against the cells of the recipient; Ehrlich and Morgenroth defined four blood groups by immunizing four goats and finding antibodies, which they called *isolysins*, that reacted differently against unimmunized animals. At the same time, Landsteiner found that certain human sera clumped red cells of other individuals. A finding he pursued by systematically performing a hemagglutination matrix study of various human sera and erythrocytes to define the first three human blood groups. (Note, Landsteiner used pre-immunized antibodies and his detection system was agglutination, not lysis.) The genetic issue was addressed by Emil von Dungern and Ludwik Hirschfeld who showed, first in dogs and then in humans, that blood group antigens followed strict Mendelian laws (von Dungern and Hirschfeld, 1910), and Felix Bernstein later proposed that the antigens were controlled by allelic genes at a single locus (Bernstein, 1924).

The integration of the genetics of tumor transplantation and blood groups is credited to that most eclectic and profound biologist, J.B.S. Haldane, who in 1933, while visiting Little at the Jackson Laboratory at Bar Harbor, postulated that tumor rejection was due to an immune reaction directed against alloantigens: tumor resistance factors were in this scheme analogous to blood group antigens and tumor rejection was comparable to destruction of incompatible blood cells.[9] Hoping to test the hypothesis, Haldane brought back to University College London Little's C57 and Strong's A and CBA inbred mice strains. Haldane was unable to perform the experiments himself, but successfully encouraged Peter A. Gorer (who had just completed his medical studies) to pursue the project (Klein, 1986).

First Gorer had to define the mouse blood groups, which he accomplished with his own serum, distinguishing the erythrocytes of the three Jackson Laboratory mice strains. Immunizing rabbits, he obtained antisera: I, shared by A and CBA; II, expressed strongly in A, but weakly in CBA; and III, shared by A, CBA and C57 (Gorer, 1936a,b). These anti-sera then enabled him to determine the antigen correlation to tumor rejection: When an A strain tumor was transplanted, the tumor grew in all the A but none of the C57 mice. (The portion of $F_1 \times C57$ backcrosses and the F_2's that were tumor susceptible suggested two loci of control). When the animals were typed with the rabbit anti-II serum, Gorer found that lacking antigen II correlated with tumor rejection, and of those that possessed the antigen, 1/4 of the F_2 and 1/2 of the $F_1 \times C57$ backcross mice resisted the tumor. Further, antibodies against antigen II were found in sera from animals that rejected tumor. Thus one of the blood group gene codes for antigen shared by malignant and normal tissues (Gorer, 1937a,b, 1938).

The major issue in deciphering individual resistance genes was to identify individual loci, which were both numerous and had the same effect. George Snell, a geneticist, undertook the laborious task of backcrossing mice so that strains would differ at only one resistance locus, by selecting tumor-resistant animals (Snell, 1948; Snell et al., 1953). (These so-called congenic strains, which now number over 300, became the single most significant resource for cellular immunology, the term designating in fact lymphocyte immunology.) Gorer and Snell collaborated to show, by linkage studies, that the resistance factor of a congenic line was an allele at the locus coding for the antigen II (Gorer et al., 1948), and they agreed, following Snell's proposal that the tumor resistance factors be designated histocompatibility (H) genes (Snell, 1948; Snell et al., 1953). The corresponding human complex was discovered in 1958 and later designated HLA (human leukocyte antigen); MHC has been identified in the frog, chicken and 15 other mammalian species (Klein, 1986).

As Klein discussed (Klein, 1986), Gorer and Snell utilized two complementary approaches in defining the MHC. Gorer used serologic methods to identify MHC antigens, demonstrating antigenic complexity encoded by several genes. Snell's histogenetic approach, based on cellular immunology, also demonstrated genetic complexity, but this approach measured *in vivo* compatibility of transplanted tissue, and fused with Peter Medawar's re-examination of transplant biology during World War II, when surgical skin grafting gained significant medical urgency (Silverstein, 1989). In Medawar's careful studies, he established three characteristics of rejection: 1) accelerated response to second grafts, 2) specificity, as seen by no acceleration if the donor was different, and 3) the systemic character of sensitization, so that no site was protected. He concluded that allograft reactions were immunologically mediated (Medawar, 1944). But only when Avrion Mitchison, Rupert Billingham, Leslie Brent and Peter Medawar demonstrated in 1954 that transplantation immunity could be adoptively transferred with cells, but not serum, from sensitized donors (Mitchison, 1954; Billingham et al., 1954), were mechanisms of transplantation rejection appreciated as analogous to phenomena dependent on lymphocyte reactivity. Shortly thereafter, the graft-versus host reaction was shown to be caused by immunocompetent cells (Billingham et al., 1955; Trentin, 1956; Simonsen, 1957; Billingham and Brent, 1957; reviewed by Grebe and Streilein, 1976), and the entire issue of regulation of the immune response turned to an analysis of lymphocyte reactivity.

In the 1960s, observations were made in outbred mice (by Hugh McDevitt and Michael Sela) and guinea pigs (by Baruj Benacerref) that indicated marked quantitative differences in their ability to develop humoral and cellular immune responses to certain antigens (McDevitt and Sela, 1965; Benacerraf et al., 1967). Subsequent use of inbred animals and synthetic antigens showed that these responses were under genetic control, a single, autosomal dominant gene located within the MHC, the so-called immune response (Ir) genes (McDevitt and Benacerraf, 1969; Bluestin et al., 1971; McDevitt et al., 1972). In fact a large number of antigens are under Ir control, which exhibit T cell dependency and are required for development of cellular immunity as well as antibody production. Immunogenetics are now highly sophisticated and a confident molecular definition of the immune self, based on the MHC, has emerged. The MHC is 2–3 centimorgans long, containing 2–3 million base pairs, and is functionally divided into two major classes (I and II) and further subdivided into two subclasses (alpha and beta) (and finally into families and subfamilies). In each individual only 2 or 3 gene pairs are functional, (the rest of the DNA serving ostensibly for evolutionary flexibility) and each of the class I and II genes occur as multiple alleles and thus act as alloantigens on transplanted tissue to be recognized by recipient T and B lymphocytes.

The function of MHC is not to obstruct transplantation per se, but is considered as providing the context for the recognition of antigen by T lymphocytes. What function MHC has had in phylogenetic history is still unfathomed and the debate concerning its primitive biological function remains unresolved (Kelsoe and Schulze, 1987). The class I alpha and beta polypeptides associate and appear on all nucleated cells, and the class II alpha and beta proteins similarly associate as heterodimers and are expressed in high quantities on B lymphocytes and activated (or virus-infected) T lymphocytes, and are inducible by interferon in antigen presenting cells (APC) (e.g. macrophage and dendritic cells). When T cells encounter a molecule on the surface of an APC, it does so by "seeing" the antigen complexed to MHC, a process called MHC-restriction. MHC is the genetic foundation for the immune recognition apparatus, for it is on that complex that immunity revolves.

The consensus model may be summarized as follows: antigen digestion by a tissue phagocyte processes the antigen and "presents" its target complexed to its own "self-antigen" system, the MHC. T-cell activation involves three elements – the T cell receptor (TCR), the antigen and the class I or class II MHC molecules (Figure 2). It is as yet unresolved how this interaction occurs from three possibilities: Tcr recognizes either an MHC-antigen complex, an antigen-altered MHC, or antigen alone, (in this case the MHC molecule merely serves as an anchor to present the antigen for recognition). It appears that the first model is the most likely and is generally preferred. In the simplest scenario, a helper T cell binds to the macrophage through its MHC protein (class II) that has bound the processed antigen (the desetope), and as a result, interleukins (a group of molecules involved in signalling between lymphocytes, antigen-presenting cells and other cells) are elaborated to initiate the proliferate phase of the reaction. A clonal selection occurs, as only those lymphocytes from clones that recognize the processed antigen are stimulated to divide. These include not only the helper T cells, but the T cells involved with the actual encounter with the pathogen, the cytotoxic T lymphocyte (CTL). These two lymphocyte classes are but representatives of a complex lymphocyte network; some classes augment immune responses, while others suppress the reaction.

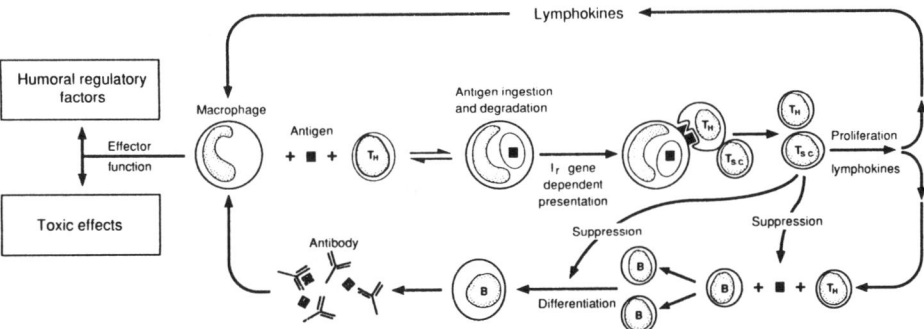

Figure 2. The immune network (see text for explanation).

The immune response is then based on the "selection" of particular lymphocytes, each of which has an immunoglobulin-related receptor of varying affinity, or fitness, for the antigen. To reiterate, the MHC complex serves as a crucial (albeit not exclusive) signature of the self, which when antigens bind to the complex are seen as either "self" or "foreign". This MHC-antigen complex then functions as the docking site for effector immunocytes (cytotoxic or killer lymphocytes) and their ancillary force (helper lymphocytes) (Figure 3). Antigen is thus "processed", lymphocytes are "sensitized", soluble and cellular mediators "stimulate" antibody production from a potentially vast library of "receptors" that constitute a reasonable "fit" of the original antigen. The stimulation and proliferation of a particular lymphocyte that matches its receptor to antigen is referred to as "clonal selection". Current theory holds that heterogenous immunocompetence resides in the unispecificity of surface receptors on antigen-binding cells, which when bound to antigen initiates mitosis and development of a clone that expresses the same receptor specificity. Among the progeny of this clone will be differentiated plasma cells (derived from B lymphocytes) that synthesize and secrete antibody molecules of the same specificity and affinity as the receptor molecules on the original antigen-binding cell (see Figure 4, Appendix).

The Recognition Event

The elucidation of the MHC-associated mechanisms required weaving together diverse research threads centered on the biology of the lymphocyte. Beyond serving as a genetic signature that determined acceptance of transplanted tissue, the MHC during this period quickly became the basis for a molecular definition of antigen presentation to the lymphocyte. In other words, MHC was recognized to encode not only those molecules identified as transplanted tissue signifying "self" versus "non self" (acceptable versus rejectable), but to encode the very molecules on the surfaces of cells *presenting* antigens to (and hence, to some degree, affecting the "immune response" by) lymphocytes themselves. MHC-restricted functional studies led to intensive investigation of the structural and genetic constitution of the histocompatibility locus.

Close on the heels of B cell molecular biology, we thus witness a merger of what began as a biological approach with a molecular definition of selfhood that bears striking parallels with the earlier generation of antibody diversity narrative. The molecular theme of antibody production is seen to be reiterated at the T cell receptor (TCR), and the molecular composition of MHC as the presenting moiety of both self and foreign peptides was discerned as also conforming to the general structure of the immunoglobulin superfamily (albeit without the ability to diversify by recombination). If we focus upon the most important discovery of the 1970s that falls within the Recognition Event, it must be the appreciation of MHC restriction for lymphocyte acti-

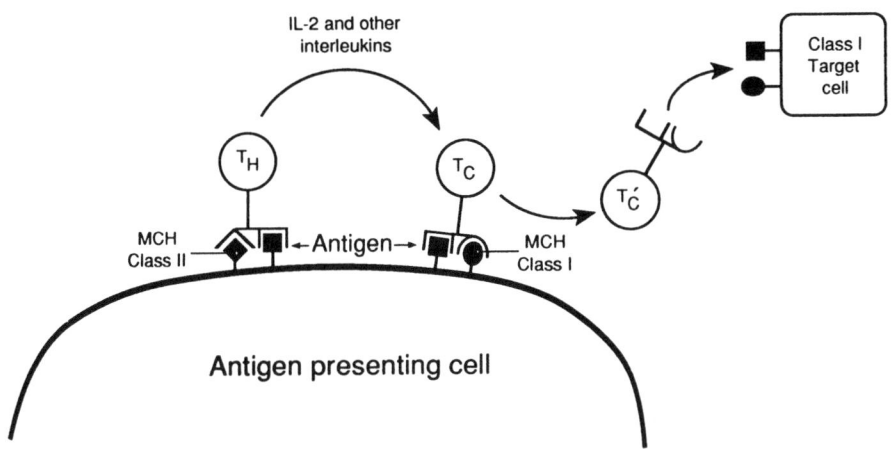

Figure 3. Immune stimulation. MHC class II processing depends on receptor mediated or fluid phase endocytosis of only exogenous antigen, which is degraded in endocytic vesicles that eventually fuse with vesicles containing class II molecules. Cytotoxic T cell (T_C) precursors, specific for an antigen with MHC associated class I complex, are stimulated by helper T cells (T_H) that associate (and in turn are stimulated) through class II antigen presentation. This "helper" function is mediated by various humoral factors (interleukin 2 is the best described of these lymphokines) and is not specific for T_C stimulation by a one to one correspondence with a given epitope. The two major classes of T cells, helper and killer, recognize two different kinds of protein antigen. Broadly, Class I-restricted cytotoxic T lymphocytes recognize cell-associated antigens such as those on virus-infected cells or cells expressing foreign alloantigens, and class II-restricted T cells recognize antigens that enter the system as soluble entities; class I peptides generally derive from endogenously synthesized proteins, while the class II universe derives from proteins taken up by the antigen-presenting cells from the external medium. Degradation and subsequent processing of both soluble and particulate antigens by macrophages, dendritic cells, or B cells is followed by T cell recognition of the associated antigen linked to the class II MHC gene product. We understand this division teleogically as in the first case (class I recognition) pathogens harbored intracellularly should be recognized and attacked by lytic cells, while in the second case (class II recognition), require positive signals from T cells to make antibody or become activated.

Class I molecules bind endogenously processed antigen, which has been committed to a pathway distinct from that used to degrade and present exogenous protein for class II recognition events. As it is currently understood, there are three modalities which lead to antigen degradation and peptide presentation with class I molecules: 1) endogenously synthesized proteins, 2) proteins injected into the cytoplasm or perhaps fuse into the plasma membrane, or 3) phagocytosed cells (i.e., cellular debris) or micro-organisms. Class II presentation required exogenous antigens small enough to be endocytosed. (Reproduced from Chernyak and Tauber (1991) with permission.)

vation (i.e., lymphocytes "see" antigen only in the context of their own MHC; see Appendix). Herein lay a new challenge. Antibody specificity had built on a long history of immunochemical quantitative methods analogous to enzymatic kinetics and affinity profiles. The T cell, despite the highly specific

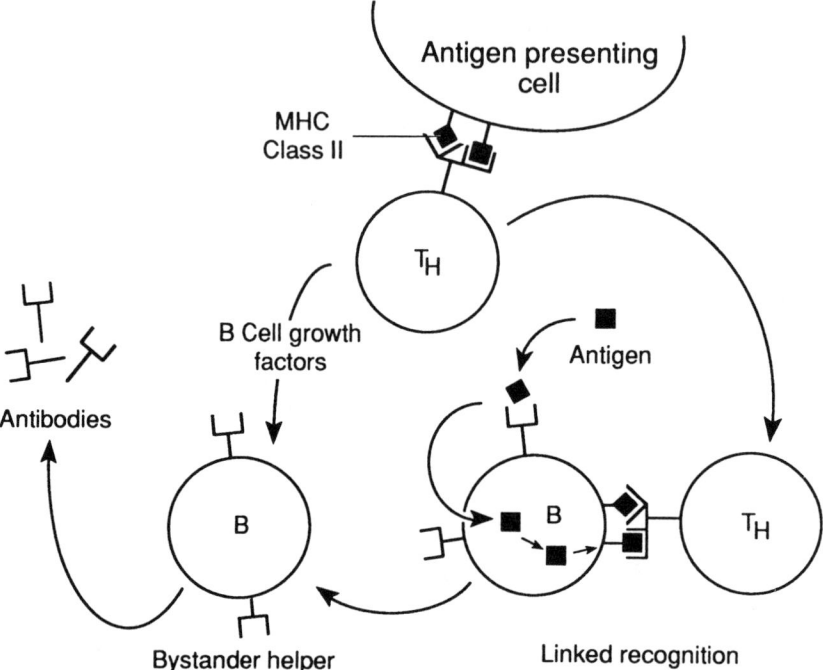

Figure 4. B cell activation. B cell activation, proliferation and differentiation may be initiated by soluble, unprocessed antigen, which is first bound by surface-exposed immunoglobulin, processed (analogously to macrophage digestion) and then presented through MHC Class II receptors to helper T cells. Again interleukins are elaborated and the B cell further differentiates, divides and secretes antibody. Alternatively, B cell response is initiated just as the cytotoxic T cell is stimulated to proliferate, that is through the macrophage recognition limb, with the helper T cell directing the B cell response. In linked recognition, B cells may also serve as an APC, which is antigen specific and MHC restricted. Only those B cells with immunoglobulin receptors for epitopes on the intact antigen can bind it prior to presentation to T_H cells, by the so-called bystander mechanism. T_H cells may stimulate previously primed B lymphocytes to secrete antibodies by elaborating B cell's growth and differentiation factors. B cell stimulation in this case is MHC unrestricted. With regard to the immune response to a given antigen there are common features that are recognized by T cells and antibodies, but the mechanisms are clearly different and the molecular basis for the commonality is not clear. It is possible that B cells may act as antigen presenting cells, and represent, in association with class II, those regions of foreign protein that their antibodies recognize. (Reproduced from Chernyak and Tauber (1991) with permission).

nature of its recognition mechanism, was shrouded in the cruder methods of cellular assays. Beyond the extrapolation of potentially complex cell-cell interactions, the T cell recognition apparatus was now compounded in its complexity with the requirement of MHC mediation for the antigen-lymphocyte encounter.

MHC restriction ultimately addressed two seemingly disparate problems: First, what molecular constituent confers immune identity for tissue transplantation, and second, regarding immune regulation, what is the basis of the two signal mechanism for lymphocyte activation? These questions were at the very nexus of immunology during this period, for in answering the first query, an important hypothesis from which tolerance might be understood was erected, and in answering the second question, a structural and biochemical basis for defining the inner workings of the immune system could be more firmly established. The elucidation of MHC restriction, both in terms of its genetic organization and its role in regulating immune recognition events would serve as the crucial nexus for a theory of lymphocyte activation. Shortly following, and quickly applying the genetic reductionist approach that had proven so efficacious in elucidating B cell genetics, the TCR research saga reflected essentially the same molecular themes.

The biological concerns that defined T cell recognition and the genetic solution to that question highlight the striking parallels between the elucidation of antibody diversity of the B cell and the recognition of antigen in the context of the MHC by the T cell; moreover, it illustrates the quick acceptance of molecular biological applications in immunology. Those advances were obviously important in their own right, but the ascendency of this molecular orientation had profound repercussive effects in other areas of immunology, most obviously in assessing the claims made by biologists whose theoretical models and experimental strategies did not incorporate molecular biological evidence to support their claims. Despite the eventual merger of cis and trans perspectives, proponents of the network hypothesis (see Appendix) would persist in their theoretical musings well into the 1980s (e.g., Bona and Kohler, 1983; Bona, 1987) in marked epistemological contrast to those embracing a molecular approach to immune regulation. If we seek an abbreviation of this contrast, we might regard the molecularization of MHC, giving rise to one kind of immune regulatory model, as most closely reflecting the product of trans immunological research, whereas the network hypothesis remained firmly embedded in the contrasting cis perspective.

The research on MHC initially appeared to offer the most direct link between the trans and cis projects for addressing T cell recognition, for the attempt to forge a synthesis between these competing research efforts underlay this research, much as the solution to antibody synthesis reflected similar aspirations as originally posed by CST. The biophysical triumph of defining the presentation of antigen at the Angstrom level in many respects completes the genetic solution of immune specificity (see Appendix). But this molecular picture left the more encompassing agenda still unresolved: namely, What is the "meaning" of that encounter? There must be a seamless connection between our construction of the recognition event and its regulation. Regulation ultimately must subsume "recognition" in any comprehensive view, and the best-fitting model is still very much in debate. So, following dissection of the MHC-TCR encounter, we are left pondering, What regulates the lymphocyte to respond or ignore the stimulus? What are the dynamics of the reaction? More broadly, What is the basis of immune selfhood and what are the regulatory rules of the immune response in that context.

And now, after molecularization, what is the immune self?

A Critique

When one compares the sophistication of knowledge concerning acquired and natural immunity, there is little doubt that the focus of fascination has been with the "smart weapons". To understand the highly discriminating specificity of immunoglobulin chemistry and the elegance of shuffling gene fragments to custom-produce antibody is true testament to the power of molecular biology. Nevertheless, natural immune mechanisms play an important role in the entire spectrum of infectious disease, cancer surveillance and tissue repair (Nelson, 1989), but the means of identifying targets for immune destruction by antibody-independent means remains problematic.[10] Critics have argued that our pre-occupation with acquired immunity has been misplaced, for it in fact probably represents the more minor mode of defense. Serology is not necessarily a reflection of immunity:

> immunity refers to actual resistance to infection, and serology refers to antigen-antibody reactions. They should never be brought together until they have been thoroughly inspected separately, and rarely then. . . . [K]nowing that he can control serology, man speculated that antibodies neutralize everything right and left, and that hardly anything else is needed to explain the whole of immunity. This simply is not so. (Marshall, 1959)

But even more fundamentally, at the core of both natural and acquired immunity, remains the issue of defining the immunologic self; whether we are concerned with MHC or non-MHC restricted recognition, there is unresolved the issue of what immune cognition truly consists.[11]

Precisely, what is known? Immunology has elegantly succeeded in offering a mechanical explanation of G.O.D., i.e. the basis for understanding how myriad antigens may be countered with a finite number of antibodies, which is no less than the grammar that establishes the immune language. What is not known is how the host decides that an antigen should be countered. How are the logical consequences of the self defined in terms of the other, or negatively, as not other? The question (leaving aside natural immune mechanisms) is thus refined to alert that in explaining G.O.D. or MHC-antigenic complexing, we have only resolved the next immediate question, "How is diversity and immune challenge mechanistically explained?" Solving that problem is in fact an extraordinary accomplishment, no less than culminating a century of immunologic research. So why is such an ungracious, if not gratuitous distinction made?

Let us return to examining the clonal selection theory (CST), which is as close to a paradigm as immunology possesses. First, it accepts the self-nonself discrimination as the central issue of immunology, a perspective derived from the historical development of the field. From Paul Ehrlich's first side-chain theory, which postulated antibody formation in terms not significantly different from our current model, the underlying presumption was that the self was given (i.e. defined, formed, constant) and in need of protection. As already noted, immunity was first postulated as *establishing* organismal integrity and that (immune) defensive functions were but subsidiary to this primary role. We will return to that definition, but in any case, twentieth century immunology followed Ehrlich's lead; to avoid self-destruction, the same notion of immunity as defensive in nature requires "ignorance" of the self, which was provided by Burnet's concept of embryologic purging of those clones that would recognize (and thereby initiate attack of) the self. In this orientation, the object of immune activity is not self, but the other, and consequently antibody production is viewed as the process by which the organism defines non-self. Perhaps paradoxically, this positive definition of the other then defines the self, negatively (Chernyak and Tauber, 1991).[12]

Ehrlich was thus forced to view immune activity directed against the host as "horror autotoxicus", and Burnet postulated the incomplete antibody repertoire to explain the sanctuary of the self against immune attack. Blanks in the immune library represent the negative definition of self.

On the one hand, the other is *essentially* (in it presentation, i.e. in the definitions provided by the immune system) retained and *existentially* eliminated, and on the other hand, the Self is *existentially* retained (protected) and *essentially* (in its immunologic presentation) eliminated. The model . . . proposed positive representations of the other and negative representations of the self (Chernyak and Tauber, 1991).

Ironically perhaps, the self is not defined in *immunological* terms, because immunological self-definition would invoke immune reaction, i.e. *horror autotoxicus*. Is the self defined by its selfness – not according to the theory of embryonic purging of self-reactive clones. From the modern perspective, affinity (i.e., reaction) to the other is the normal expression of immunological phenomenon. Because self is defined negatively, nonself is defined as a tautology: the other is the universe presented by antigen-specific antibodies and lymphocyte receptors. What is this universe? That which is so represented. Immune presentation becomes the definition of the other. The matter becomes increasingly confused when we consider the idiotypic nature of certain antibody specificities (i.e., antibodies reacting to other antibodies) and the general role of either reactive or non-reactive auto-antibodies in immune regulation.

Is self/nonself discrimination truly the organizational nexus of immune function? When posed in this fashion, the theoretical orientation suggests such further question as What is identity? How does the immune system establish selfhood? Is selfhood in fact a useful metaphor? Underlying these issues of immune regulation is the fundamental question of selfhood. If this remains as the fundamental construct, the question persists whether selfhood can be established simply by defining the molecular elements and their direct interactions. Charles Janeway, Jr. voiced a growing skepticism that the G.O.D., TCR, and MHC molecular structures would hardly suffice to explain these problems of immune recognition. In his 1989 Introduction to the Cold Spring Harbor Symposium devoted to immune recognition, he made the keen observation that the simple self/nonself distinction that had formed the basic operative principle of post-Burnetian immunology was collapsing. In what he called the "Landsteinerian fallacy", Janeway exposed the "immunologist's dirty little secret".

> The Landsteinerian fallacy is that all foreign macromolecules are equally able to give rise to an immune response, and, therefore, that the immune response shows no special predilection to respond to infectious agents. I believe that this latter conclusion is wrong, which is why I call it a fallacy. It is becoming clear that immunogenecity, the ability to elicit an immune response in the form of clonal expansion of virgin lymphocytes, requires both the presence of a suitable antigenic determinant to signal the lymphocyte through its antigen receptor *and* distinct signals derived from host cells. (1989, p. 5)

The "dirty secret" refers to use of adjuvant (consisting of killed bacterial debris), which is required to obtain immune response from well-defined proteins or hapten-protein conjugates. The need for adjuvant is unclear, but appears related to the dynamics of obtaining effective antigen uptake by macrophages (thus increasing ligand density), and co-stimulatory activity induced in APCs and/or B cells by the bacterial constituent of the adjuvant. Thus, highly specific immune recognition appear to require the complementary activation of earlier, less specific responses of the macrophage and related non-discriminatory APCs. That immunology would integrate these innate, or natural immune mechanisms with those so carefully elucidated for the clonal selection system of T and B cells was to be pursued by other theorists. Here we need to only note that the concerns raised by Janeway in the late 1980s represent a growing awareness that immune function required a more global perspective than that offered by a simple genetic definition of selfhood.

This circumspective regard for the complexity of immune recognition mechanisms must be set in stark contrast to those who would argue for a genetic definition of the self, buoyed by a growing confidence stemming from the dramatic advances made in the molecular universe of VDJ, TCR, and MHC. As stated in the widely read second edition of James Watson's *Recombinant DNA*:

> We have learned that most of the diversity in antibody and T-cell receptor structure is encoded directly in the genome and further amplified by physical shuffling of these gene segments. Cloning of the MHC genes revealed that, in contrast, self identity is hard-wired, directly encoded in only a handful of genes that differ in their primary DNA sequence from one individual to another. In other words, whereas a newly born individual inherits the potential for the complete diver-

sity of antibody and T-cell receptor specificities, he inherits only his unique identity at the MHC locus. (Watson, Gilman, Witkowski, and Zoller, 1992, p. 305).

The implications of this statement are far-reaching, and we will have occasion to revisit this position from several directions.

I begin with preliminary remarks concerning the "certainty" of immune identity, and in that context explore how the genetic solution might only offer a partial answer to the fundamental quandary of immune selfhood. Specifically, the question of autoimmunity looms over all current notions of differentiating mechanisms that discern host from foreign. Although an understanding of such immune behavior begins with the MHC, its characterization appears to reside at levels of biological organization beyond the gene. In addition, there is the problem of non-MHC-restricted immune reactions, which begs the MHC identity question altogether, and which cannot be ignored (see Appendix and note 12). Since our general concern is to draw a perspective on the effects of the "molecularization" of immunology, it is important to discern where potentially more comprehensive approaches might be required to address complex issues of immune regulation. The authority of molecular biology is not in question. After all, the mechanism of antibody synthesis was at heart solved as a genetic matter, and not in its original immunological guise. Similarly, the T cell recognition solution paralleled those developments. But following these molecular triumphs, immunologists grappling with tolerance and autoimmunity as regulatory problems seek to understand these immunological phenomena as determined by the system as a whole, rather than simply arising from a reduced genetic context. Again a sophisticated structural (viz., genetic) solution is at hand, but the "meaning" of that TCR-antigen encounter is conferred by seemingly complex interactive processes that determine the self/nonself discrimination in time and contexts quite removed from the molecular mating that occurs upon any given local antigen presentation.

So from the classical immunologist's perspective, the immune reaction, whether a cascade of cellular and humoral responses or some sort of silent non-reaction, can hardly be formulated solely as a genetic question. The fundamental issue is immune function, the reactivity or tolerance of immunocytes. Since Burnet, that question has been formulated as a discretionary response to self and nonself elements. Thus immunology for almost half a century has attempted to construct a theory of immune function on a foundation that depends on a heuristic definition of the self. But there is no consensus for such a definition, because none have proven sufficient to account for immune function. The diversity of responses to that question, and their intellectual infrastructure have been discussed elsewhere (Tauber, 1994), but it is useful to briefly review the disparate opinions as to what is a self (Matzinger, 1994, p. 993): The self may be construed as 1) everything encoded in the genome, or perhaps everything minus the immune privileged sites; or 2) only cell surface and soluble molecules (Cohn, 1987, 1989); or 3) a set of bodily proteins existing above a certain concentration (Mitchison, 1993); or 4) a derivative of Jerne's model (discussed in detail later in the Appendix), namely an idiotype-anti-idiotype network (Salaun et al., 1990), or 5) the set of peptides complexed with the MHC, or perhaps only those antigens that win competition for MHC slots (thus reducing the number of self antigens to which tolerance is induced [Waldmann et al., 1988]); or 6) only APCs and thymic epithelium (Zinkernagel et al. [1991]). The genome serves varying roles in these different interpretations of selfhood, and we are forced to consider the reductive strategy in light of how any of these theories command serious attention.

With so much dispute as to what might constitute the self, a growing counter position suggests that the "self" might be better regarded as only a metaphor for the immune system's silence, i.e., its non-reactivity, which in itself is problematic, since this silence might be actively attained through tolerance mechanisms. The critical turning point in assessing CST is the realization that the immune system in fact recognizes selfness as natural autoimmunity, and such host-directed reactivity is physiological. The significance of this orientation has taken some time to sink into the collective consciousness of the discipline, and its ramifications are still not widely appreciated. As Antonio Coutinho wrote with Michael Kazatchkine,

During this century, the evolution of concepts on autoimmunity could be summarized by "never, sometimes, always". Thus from the early "horror autotoxicus" [Ehrlich] to the 1960s, immune autoreactivity was simply not considered. . . . With the first identification of autoreactive antibodies in patients and the subsequent conceptual association with autoaggressive immune behaviors, the "sometimes" phase was entered, necessarily equated with disease. By this time, immunology had laid its foundation on the clonal selection theory, which forbids autoreactive clones in normal individuals. Immunologists thereafter devoted 30 years to discovering ways by which autoreactive lymphocyte clones can be deleted and why they fail to be deleted in autoimmune patients. . . . In the 1970s at least three sets of observations and ideas began to alter this course of events and to herald the "always" period. (1994a, pp. 1–2)

The early observations were the documentation of 1) autoreactive lymphocytes in normal individuals, 2) T cell suppression, and 3) Jerne's idiotypic network and the data supporting it, each of which received a mixed reception. However, in the past ten years, a wealth of data have been accumulated for autoreactive antibodies, B cells and T cells in healthy individuals (reviewed in Coutinho and Kazatchkine, 1994b), so that they concluded, "we are experiencing a major shift in the central paradigm of immunology", (1994a, p. 2), namely that identity must be established positively (unlike the negative self of CST), and defense becomes a secondary function of that identification.

Note, Immune Recognition, the codifying issue that has tied so much of the history of immunology into a coherent fabric, has almost exclusively been defined in terms of the antigen reaction with immunoglobulin and lymphocyte. The molecular successes in these arenas of investigation have overwhelmingly dominated other kinds of studies of kindred problems in natural immunity and have widened the conceptual bridges between these research areas. Aside from the obvious problems of establishing a coordinated theory between these arms of immunology, separated by the vicissitudes of their specific developments, the very character of what constitutes operational definitions of a key concept governing the discipline, namely Recognition, assumes a particular orientation, if not bias, when the scientific fortunes of one school so dominates the others. In a strong sense, "recognition" now carries a particular molecular connotation that prevails, not only within the lymphocyte research community, but well beyond, where other models might be more appropriate for describing the phenomena. Specifically, I am referring to phagocyte recognition mechanisms that are seemingly unrelated to the MHC. The discernment of self/nonself discrimination in a non-MHC context remains enigmatic. (Winchester et al., 1984; Lorenz and Allen, 1988). It is too early to even propose whether immune recognition as understood for the lymphocyte has any direct relevance for phagocyte biology, which raises a cautionary note about the universality, applicability, and ultimately the meaning of MHC-restricted recognition for the immune system as a whole, i.e., in its design of identifying the host and the foreign.

Various critics, perhaps Charles Janeway, Jr. most prominently (1989), have argued for a more integrated approach, recognizing in the phylogenetic and functional inter-relatedness of the "clonal" and "pre-immune" systems, a mutual dependence for vertebrate immunity. But when Melvin Cohn assigns the "Immune System" the domain of T and B cell specificity (and memory) as *immunity*, leaving the "Defense System" the province of other host defense mechanisms against pathogens, tumors, and injuries, he has clearly articulated the deep sentiment of a majority of contemporary immunologists (Cohn, 1995). Immunology, in this guise, has been recast into a very specific mold, and as the discipline has evolved so has its declared aspirations and operative theory narrowed. Perhaps this perspective is part and parcel of a particular orientation concerning the self, or how the self is immunologically defined. If self/nonself discrimination fades as the definitional basis of immune function models (e.g., Irun Cohen [1992], Zvi Grossman [1993], and Polly Matzinger [1994]), then the particular role of the phagocyte, and the more general understanding of immune reactivity will obviously undergo a radical metamorphosis. And well it may if the current meanings of immune identity are indicative of a restrictiveness in the theory that fails to account for such "defense". These proposals have generated interest because in a most general respect, each refers to efforts of defining immunity as *function*, emphasing the process characteristics of the immune reaction. In those discussions, new formulations have arisen to challenge the very notion of the Self as *entity*. A centerless notion, or perhaps a deconstructed one, immune selfhood becomes closely linked to how we might account for the immune system's dynamic properties, that is, how it functions as an ever-changing system. In this view, there is no "final" definition of the self.

From this vantage-point, we discern that the current upheavals in immunology might be less the application of novel molecular biological techniques to problems of immune function, than a significant conceptual turn in the underlying theory of the field. The solution to the generation of antibody diversity might then better be regarded as a finale to a set of problems posed in an older, eclipsed framework. Specifically, we must ponder whether the Burnet-Talmage-Lederberg CST construction, which most would agree "is no longer a theory but a fact" (Klein, 1990, p. 335), is now being seriously challenged. If the paradigm in fact is shifting, then the significance of CST's "proof" takes on a different significance.

This is a theoretical issue, but there is a "political" one as well. The power of the technology that provided VDJ has the potential of directing the research agenda more generally. We must not be misled or awed by methodological innovations that in themselves do not propel novel questions. There is no doubt that the answer to immunological specificity was answered through molecular biological applications. In this case, a powerful new method answered a vexing problem. At the same time, there should be no confusion that the biology framed, and in fact originated, that genetic query. In the enthusiasm for a new technology, we must caution that its frame of reference may set a research agenda more appropriate to its own concerns than to the biological issues at hand. In other words, there is always the danger that because a particular technology has the capability of addressing certain research problems, these problems then become the dominant agenda of the field.

In this regard, the genetic definition of selfhood ("self identity is hard-wired directly encoded in a handful of genes" [Watson et al., 1992, p. 305]) has assumed a certain credibility and therefore evoked a dissenting reaction. Cohen, Grossman, and Matzinger each adopt, albeit to varying extents, what I called a "contextual" theory. Not only are the dynamic elements emphasized, but the way the immune system is structured as developmentally and/or locally self-defining challenges earlier conceptions of immune selfhood as a given entity. This newer view of immune function is fundamentally formulated as a self-seeking, self-organizing activity, whose structure is decentered from any bounded self. Immune identity can only be defined in particular contexts, and from such histories, selfhood emerges.

CONCLUSION

Ambiguity regarding the nature of the immune system remains. On the one hand it is responsible for defending the host, but on a more fundamental level it defines organismal identity. The first function lends itself easily to metaphoric language of defense and related mechanical models. Thus the effectiveness of asking such questions as to how antibody diversity is generated. But the second issue, the definition of the self is far more nebulous and abstract. The problem is to clarify how to view the immune system, i.e. as one that defines the self, or defends it. Although not mutually exclusive, the dual function has, and does, cause confusion in formulating the theoretical foundation of the discipline. What is the Self? It certainly is not a static entity by any organic parameter, least of all by its ever-changing immune history, where each encounter with environmental challenge provokes the *self* to decide acceptance, incorporation or rejection, and finally "immune incorporation" (i.e., antigenic processing and immune response). In each such decision, the self is altered. The point of course is that to know mechanisms of generating antibody diversity is but a subsequent partial solution to a more profound question concerning the boundaries of the self and how it makes the "cognitive" leap to encounter the other. Molecularization of immunology has not answered that question. Here we are not concerned with a readily available genetic definition, but why a given antigen is processed for immune destruction rather than simply eaten (i.e., incorporated). How is the foreign perceived before or even independent of MHC-linkage?

A final note – perception, cognition, response – are the properties of other immune components that do not exclusively use the MHC to identify self/nonself. For instance natural killer lymphocytes, various phagocytes (macrophages, neutrophils) and complement are effective first line mediators of host defense. These, with the proper perspective on MHC-restricted recognition as one, albeit important mechanism of immune recognition, remind us of the hierarchical complexity and emergent character of the immune system. It is this perspective that is addressed by "cognitive" theory which utilizes a scientific structure differing from that of a purely reductive approach. "What is the immunologic self?" remains ambiguous by our current molecular criteria. This is not a question to be resolved at the molecular level, but like other cognitive problems, depends on principles of organizational, systematic, cybernetic, hierarchical or even logical analysis. I am not denigrating the extraordinary accomplishment of molecular biological application to immunology, but only noting that the biological issues demand complementary approaches, where chemical or genetic processes are incorporated into a comprehensive biological perspective and not confused for solutions that in fact only address the phenomena to which they are appropriately applied.

ACKNOWLEDGMENTS

I am most grateful to Eileen Crist, Scott Podolsky, Sahotra Sarkar and Judah Weinberger for their insightful comments and corrections, and to Ann Marie Happnie for her expert assistance in preparing the manuscript. Figures 3 and

4 appeared in Chernyak and Tauber (1991), and are reproduced with permission from Kluwer Academic Publishers, Dordrecht, The Netherlands. This paper was written with support from NIH Grant No. HG 00912-01/02.

Boston University,
Boston, MA, USA

APPENDIX

There are basically two forms of immune recognition – natural and acquired immunity. Natural immunity is characterized by effector cells and soluble factors that do not require specific prolonged induction for their functions, and more specifically do not require opsonization (coating) of their target. Such coating of the pathogen with antibody and other serum proteins offer "handles" for attachment and engulfment by specialized "eating cells" – phagocytes. The encounter with a phagocyte serves as the initial cellular host defense action; it can result in the killing of the micro-organism by a blood or tissue phagocyte, and if the encounter is with an antigen-presenting cell (APC), a more complex interaction occurs, whereby the antigen (that constituent which elicits an immune response) is "processed" so that the immune recognition (i.e. lymphocyte network) process is initiated. A clonal selection occurs, as only those lymphoctyes from clones that recognize the processed antigen are stimulated to divide. These include not only the helper T cells, but the T cells involved with the actual encounter with the pathogen, the cytotoxic T lymphocyte (CTL). These two lymphocyte classes are but representatives of a complex lymphocyte network; some classes augment immune responses, while others suppress the reaction (Figure 2).

This problem of specific recognition is viewed by many as the true province of immunology, whereas natural immunity falls under some other rubric. In any case, immune processing is the basis of lymphocyte and humoral immunity, and serves as the memory system of the immunologic self. The immune reaction formally begins here, at the level of specific recognition, a reaction that reflects the extraordinary ability of vertebrates to recognize, respond, and memorize a myriad number of antigens (to the order of 10^8–10^9). As described, acquired immunity is dependent on generation of specific antibody, which arises from a mixed genome that arrays its few components to make a particular protein.[13] The molecular biology of immune selection is based on the discovery that a receptor for an antigen is generated on the surface of a lymphocyte, by the particular arrangement of a finite number of gene segments in a unique linear array. There are constant regions (i.e., universal, homogenous) of the resultant protein that confer class recognition and function, and variable and hypervariable regions that accommodate a particular "fit" for the antigen. Generation of diversity (G.O.D.) then is explained by the shuffling of these genetic constituents into forming distinct antibodies. Those B lymphocytes with such antibody receptors are stimulated to secrete more antibodies, and the antigen is then neutralized by binding to a limited library each with varying affinities for the antigen. The production of antibody is then a result of antigenic selection for those lymphocytes that will produce proteins that "recognize" that antigen. The selection is from an array of a random population of potential antigenic receptors; fitness, i.e., affinity, of the antibody receptor then initiates the biological response: cell proliferation and antibody synthesis. Thus to the myriad presentation of pathogen, exogenous antigen, and altered self (e.g., effete, cancerous, damaged cells), the host responds by creating an antibody library. A range of antibodies with varying affinities for the target of immune attack are generated; thus a spectrum of antibody-producing lymphocyte cones are stimulated, each generating a single antibody to bind antigen.

For a regulatory system to function in a coordinated and meaningful fashion certain rules for lymphocyte-lymphocyte interactions must be followed: 1) lymphocytes release their message only after an appropriate stimulatory signal, thus resting antigen-specific cells do not constitutively produce lymphokines; 2) activation of cells to produce lymphokines must follow some recognition event, thus a specific activation signal is required; 3) target cells must be able to decipher the message, i.e., the signal initiates an appropriate response; 4) only the target lymphocytes are activated. The data to support these various criteria is in some cases strong and suggests strong cooperation between lymphocyte subtypes. For instance, the T cell must recognize processed antigen on the B cell surface through antigen-specific Ti/T3 receptor, which means that only T cells specific for the processed antigen can be stimulated (Flood, 1988). To look at the reaction from the effective cells vantage, the B cell gathers independent information about the real nature of the antigen. In its interaction with a T helper lymphocyte, Th recognizes a different epitope on the same antigen molecule presented either by an antigen presenting cell, or the B cell itself. The testimony of the T cell (in a yes/no format of stimulation) is crucial in defining the antigen as foreign, because by CST, T cell are "educated" in the thymus where cells reactive to host constituents have been eliminated or silenced (Celada, 1988).

There are common features of antigen recognized by T cells and antibodies, but the mechanisms are clearly different and the molecular basis for the commonality is not clear, for it is possible that B cells may act as antigen presenting cells, and re-present, in association with class II, those regions of the antigen that their antibodies recognize. There are thus multitudinous arrays by which antigen is recognized, presented, and to which several limbs of the immune network must respond: 1) macrophages have diverse function including direct toxic effects (Mogensen 1979), the elaboration of humoral regulatory factors (Nathan, 1987), as well as the antigen-presenting function; 2) lymphocyte cytotoxicity is mediated by both class I (e.g., Mock et al., 1987) and class II associations (e.g., Kane et al., 1989; Chen and Parham, 1989; Eisenlohr and Hackett, 1989; French et al., 1989), and finally, 3) elaboration of antibody for both antibody-dependent cell-mediated cytotoxicity (e.g., Hashimoto, Wright and Karzon, 1983), and both serum and mucosal immunoglobulin proteins generated by activated B cells. Couple these immune sensitized pathways for self-protection with natural (i.e., presensitized) immune mechanisms of the phagocyte, and we discern the extraordinary complexity of the immune response.

The discovery that immune reactions consisted of coordinated sub-populations of lymphocytes led to the designation of an immune system (Moulin, 1991), but this was viewed in two different modalities. The immune system may be modelled as a complex array of antigen-driven interlacing cellular and humoral factors that regulate the immune response by feedback cycles. This model has already been sketched. A complementary hypothesis postulates that in addition to the control mechanisms already described, a self-regulatory *network* is controlled by auto-reactive antibodies and the latter control is the critical element that regulates the immune system. The system is formally called the idiotypic network theory. The theory, first proposed by Jerne (1974), is based on the simple argument that, if the 10^7 (or more) clones of the immune system are capable of recognizing any antigen, they should also be able to recognize each other, and initially focused upon the notion that immunoglobulin idiotypes[14] are organized as a network of complementary shapes. The network hypothesis postulates that the primary encounter with antigen results in the formation of a network of cross-stimulating antibodies and the antibody-producing cells which is maintained even after the initial antigenic insult has been removed. In this model, immune memory is a stable state, independent of the continual presence of antigen.[15] The rudiments of an idiotypic network had already been suggested by Jerne seven years earlier (Jerne, 1967), but the network as a working hypothesis awaited finer definition (Jerne, 1974). Before the network theory was formally proposed, a nomenclature for its components had already emerged (Jerne, 1960): antigen carries several *epitopes* (immunogenic elements) and an antibody, aside from its own potential epitopes, has *paratopes* (combining sites/receptors). Those epitopes that are carried by components of one individual are *idiotypes*, which collectively comprise the *idiom* of the animal, and thus any epitype not represented in the idiom is a *xenotype*. Note that Jerne avoided the use of *self*, and used the more neutral, *idiom*. In the network paper published 14 years later, he adopted a modified definition of idiotype: "a set of epitopes displayed by the variable regions of a set of antibody molecules. Each single idiotypic epitope I shall call *idiotope*. An idiotype then denotes a certain set of idiotopes" (Jerne, 1974, p. 380). The definition reflects the discovery that immunoglobulin possesses unique antigenic determinants (reviewed by Burdette and Schwartz, 1987). Idiotopy arises from the antigenicity of this variable region, which in turn generates anti-variable region antibodies, designated anti-idiotypic antibodies as anti-idiotypes (Burdette and Schwartz, 1987; Nisonoff, 1991).

Jerne's theory postulates that the immune system is in a state of equilibrium, in that the original antibody with a particular paratopic specificity, Ab_1, is held in check by its anti-idiotypic antibody, Ab_2. Ab_2 is directed against the unique protein structure to the variable region, the idiotype, of Ab_1. Introduction of antigen interrupts these interactions in a mutually interactive fashion: Ab_1 binding antigen removes its inhibitory effect on the "internal image" clone as well as its stimulatory effect on Ab_2. As Ab_1 production then ensues, and concentration levels are enhanced, these effects are reversed and equilibrium will be re-established (Jerne, 1974). The immune system may thus be viewed as a network comprised of antibodies and their regulatory anti-idiotypes, in what Jerne (1974) described as "Eigen behavior". On this view, the basis of homeostatic control of the immune reaction, specifically the production of antibody, depends on two mechanisms: 1) antibody may interfere with antigen binding to antigen-specific clones of antigen-binding cells, thus preventing further stimulation, or 2) the suppression of a particular antibody response by another antibody generated specifically to it. This hypothesis predicts that as antigen stimulates synthesis of antibody, the characteristic idiotype becomes more prevalent and in turn induces synthesis of anti-idiotypic antibodies in what is called the Id cascade. As the level of these anti-id antibodies (or cells) increase, they exert a specific suppressive effect on the further production of the original idiotypes. This cycle must continue if the anti-idiotypic antibodies themselves mimic antigens, but eventually the original stimulus is ultimately dampened.[16] A series of Id cascades that are generated to different epitopes associated with a multi-determinant antigen may exhibit the ability to interact with one or another at both the B and T cell level. Thus the model has been extended to T cell regulation through the variable regions of the T cell receptor. The idiotypic network of B and T cells is viewed as possessing regulatory properties (e.g., Masaki and Irimajiri, 1992), and the idiotypic network may then be viewed as a complex array of responder and suppressor lymphocytes, where the introduction of antigen induces two distinct pathways: one leading to immunity, the other to its suppression (Bona, 1987).

The immune network concept has been a fulcrum of wide experimental and theoretical interest and not surprisingly has yielded many modifications (e.g., Stewart and Varela, 1989; Perelson, 1989; Bona and Kohler, 1983; Bona, 1987; Lundkvist et al., 1989; Parisi, 1990; Hebert et al., 1990; Coutinho et al., 1984; Coutinho and Stewart, 1991; Kaufman et al., 1985). But there is a strong, if not dominant school, that regards the matrix of lymphocyte sub-groups and regulatory cytokines as sufficient and that an idiotypic network is simply an unnecessary theoretical appendage that has little direct experimental support (e.g., Cohn, 1986; Langman and Cohn, 1986; Klein, 1990). Although the generation of anti-idiotypic antibodies has been demonstrated, the true functional relevance of such elements is still unclear and much debated.[17] Idiotypy rests on a firm experimental basis, serving research as clonal markers of B cell development and somatic mutation, as a phenotypic marker for germ line V genes, as a probe for receptors on cells, and as therapeutic reagents of various kinds (Nisonoff, 1991). That idiotypy exists is not the issue, but its functional significance remains obscure. Foremost, the idiotypic network theory required that auto-anti-idiotypic antibodies be regulatory, and that under some conditions these antibodies might mimic antigens. Support for both predictions has been forthcoming, but their significance has remained problematic. So it is probably fair to say that twenty years after Jerne published his hypothesis, the idiotypic network hypothesis has had a mixed fate. On the one hand, there is general agreement that there is lymphocyte inner connectedness, but the physiological significance of that network is still debated. The scarcity of quantitative data on the magnitude of the idiotypic contribution to immune regulation and the failure to produce a series of predictions that have directed experimentalists leave Jerne's hypothesis in some sense in limbo. Second, and perhaps more saliently, the hypothesis failed to pose a molecular biological agenda, and on those grounds the network remained tethered to its cellular school. Again, without application of new genetic criteria to the theory, its validation remained outside a rising scientific ethos that clouded research in areas that would not partake of its methods. Finally, Jerne's theory suffered significant critical attacks, but at the same time gave rise to new formulations in answer to its shortcomings. In this sense, the idiotypic network provoked an important challenge to a molecular definition of immune selfhood.

NOTES

[1] An antigen is any substance that invokes an immune reaction by reacting with antibody and/or T lymphocytes. Although definitions are offered in the text, the sophistication of the discussion may well require referral to textbooks (e.g., Klein, 1990; Watson et al., 1992).

[2] Humoral pertains to serum, lymph, and fluids secreted from mucosal surfaces. In this context, the primary immune responses are mediated by antibodies (i.e., immunoglobulin), lectins (proteins that recognize carbohydrates on the surface of micro-organisms) and/or complement. In contrast to these humoral defenses is cell-mediated immunity which historically were those reactions mediated by cells as opposed to those which were originally referred to as "colloid" factors. Today these cellular reactions refer specifically to delayed-type hypersensitivity reactions and cytotoxic T lymphocyte (CTL) reactions (e.g., graft rejection, contract sensitivity, tuberculin response).

[3] How competing cell lines might account for evolution's march towards "individuality" has again been advocated by Leo Buss (1987), whose argument has been critiqued (Falk and Sarkar, 1992; Gilbert, 1992; Chernyak and Tauber, 1992).

[4] "This biological theory has often been considered too vitalistic in its tendency. I need only quote Frankel's outspoken criticism of my theory from this point of view. 'The phagocyte theory presupposed extraordinary powers on the part of the protoplasm of leucocytes, to which are attributed sensations, thoughts and actions, in fact a kind of psychical activity'. The sensibility of the phagocytes is not an hypothesis which can be admitted or rejected at will, but an established fact, which cannot be ignored, as it is by Frankel. Whether they possess powers of thought and volition, as this author accuses me of assuming, is quite beside the question, though we are justified in considering that they possess a germ of these qualities and that their sensibility, like that of various vegetable and animal unicellular organisms represents the lowest stage in the long series of phenomena which culminate in the psychical activities of man. . . . The accusation of vitalism and animism, which is unjustly cast at the phagocyte theory, might really be more appropriately applied to my opponents, who maintain that the psychical acts of the higher animals are fundamentally different in their nature from the more simple phenomena peculiar to the lower organisms" (Metchnikoff [1891], 1968, pp. 192–3).

[5] "Vitalism took several guises in the nineteenth century and *causa vitae* as ascribed to either: 1) a unique ensemble of activities that emerge when ordinary elements are organized in an extraordinary way; 2) certain inexplicable properties of the organism; 3) imposed causal agents present only in living systems, or 4) active life as a collective or summated expression of the immanent life of matter itself (a distinctly minor position)" (Hall, 1969, p. 222).

[6] It is important to note that the mechanical philosophy of the seventeenth century was

reductionist (i.e., physical laws and facts may be explained by, "reduced" to, local interactions of matter), but was ultimately discarded. In the nineteenth century, the success of explaining optics by classical electromagnetism and the laws of thermodynamics by the principles of statistical mechanics, buoyed those enthusiasts who wished to apply the same strategy to other scientific and social endeavors (Sarkar, 1992).

[7] As Fischer and Lipson document (Fischer and Lipson, 1988, pp. 128–9), Pauling and Delbrück wrote their *Science* letter as a rebuttal to Pascual Jordon's hypothesis that macromolecules had a special quantum mechanical resonance attraction for each other, so that replication was accomplished by homologous "synapsis". Synthesis by coupling complementary structures, rather than identical replication, was in fact first suggested by J.B.S. Haldane (Haldane, 1937).

[8] Derivation of the formula may be found in Klein, 1975, chapter 1.

[9] It is of interest to compare the fate of the respective views of Ehrlich and Metchnikoff regarding the relevance of immunity to tumor biology. As detailed elsewhere (Tauber and Chernyak, 1991; Tauber, 1991) Ehrlich viewed tumors as competitive for crucial nutritive or growth factors, and because of higher avidity might grow at the expense of normal tissue, or conversely starved if tumor receptors could not competitively obtain those factors. Metchnikoff, in contrast to this passive, non-immunological theory, viewed tumors as nonself intruders that were relatively autonomous and were subject to phagocyte (i.e., immune) surveillance. Like any struggle between host and pathogen, outcome was determined by the ability of host defense mechanisms to prevail, which in his terms were active immune responses.

[10] Immunology obviously encompasses a broad spectrum of research, but as an investigator of the phagocyte, I have on occasion been challenged as to my credentials as an immunologist by what I call the S and M enthusiasts. S and M refer to specificity and memory, those criteria that lymphocyte biologists have pre-empted as the basis of immune function. Two trivial examples: 1) When I was welcomed as an editor of the *Journal of Immunology*, I was queried and asked to defend why phagocyte papers were in fact addressing a legitimate area of immunological interest. Incidentally, they are still included, and in fair proportion. 2) When visiting Niels Jerne in 1978, I told him of my work on the eosinophil, a phagocyte whose apparent immune function is to attack helminths, protozoa and other parasites too large to ingest and pathologically is found to mediate allergic reactions. After an hour of discussion, Jerne mused that he doubted that the eosinophil in fact was an "immune cell" at all. Thus ended my visit to the Basel Institute for Immunology.

[11] The issue of cognition in immunology is an extended metaphor from neural processes and philosophy of mind. To envision "full knowledge" as a definition of cognition is obviously inappropriate, but at the same time, there are end points in the immune reaction where the foreign is perceived as "other" and the "self" responds. But a theory to account for cognition in immunity is most tentative and here the term is used solely in a weak sense – to perceive information as information (This issue is more fully discussed in Tauber, 1994).

[12] Note that APC's do not appear to have the capacity to differentiate self from nonself (Winchester et al., 1984), yet there are mechanisms by which phagocytes engulf organisms by relatively non-specific recognition mechanisms involving carbohydrate binding by proteins with a high degree of specificity for such interactions. These "primitive" recognition proteins are called lectins and the process is termed "lectinophagocytosis" (Ofek and Sharon, 1988). Although this is an interesting question, and unrestricted MHC recognition processes are undoubtedly important in phagocyte surveillance functions (Nelson, 1989), this issue has not been carefully examined and cannot be discussed theoretically with the same molecular sophistication as afforded to T cell MHC-mediated recognition processes.

[13] Antibody comprises several classes of immunoglobulin, which have a basic structure: heavy and light polypeptide chains, with constant and variable regions linked together; the variable region constitutes the antigen-binding site, and the heavy chain serves as the link to cellular receptors.

[14] According to Jerne's theory, antibody may recognize antigenic sites on another antibody. This immunogenic region (idiotope) in the variable region of the "recognized" antibody then serves as the linkage site for the "recognizing" antibody. The idiotope collective is designated the "idiotype".

[15] There is evidence that persistent antigen stimulation is required for memory (Gray and Skarvall, 1988) and models have been proposed to account for such findings. (Antra et al., 1991) This is however a matter we need not delve into.

[16] "In a network arrangement, the idiotype of antibody I can bind to the variable region of the immunoglobulin on the surface of B-cell 2, thereby stimulating the cell to secrete antibody 2. In this way, an idiotype produced during the immune response to a particular antigen can stimulate the production of a corresponding anti-idiotype. Antibody 2, a kind of auto-antibody, is called an auto-anti-idiotype. Antibody 3 (an anti-anti-idiotype) has also been found during the immune response, but there the circuit usually ends because of the resemblance of antibody 3 to antibody 1. In a self-contained immune network, the designation 'idiotype' or 'anti-idiotype' is arbitrary; and idiotype is just as likely to be an anti-idiotype" (Burdette and Schwartz, 1987, p. 220).

[17] "For reasons that are not entirely clear, some immunologists have felt that antigens and lymphokines do not suffice in the regulation of the lymphocyte response. They have therefore come up with elaborate regulatory schemes and are now seeking experimental support for their ideas. The best known among these schemes is the *network hypothesis*, which was inspired by the fact that the elements in the nervous system are known to interconnect. This fact was then *artificially* [emphasis added] imposed on the immune system, first on antibodies and then on lymphocytes" (Klein, 1990, pp. 386–7).

REFERENCES

Ada, G.L. and Nossal, G. (1987). 'The Clonal-Selection Theory'. *Sci. Am.* **257**: 62–9.
Alexander, J. (1931). 'Some Intracellular Aaspects of Life and Disease'. *Protoplasma* **14**: 296–306.
Allen, G. (1978). *Life Sciences in the Twentieth Century*. Cambridge: Cambridge University Press.
Anita, R., Levin, B. and Williamson P. (1991). 'A Quantitative Model Suggests Immune Memory Involves the Colocalization of B and Th Cells'. *J. Theor. Biol.* **153**: 371–84.
Arrhenius, A. (1907). *Immunochemistry: The Application of Physical Chemistry to the Study of Biological Antibodies*. New York: Macmillan.
Bail, O. and Tsuda, K. (1909). 'Versuche über bakteriolytische Immunkörper mit besonder Berücksichtigung des normalen Rinderserums'. *Z. Immunitätsforsch.* **1**: 546–612.
Baumler, E. (1984). *Paul Ehrlich. Scientist for Life*. New York: Holmes and Meier.
Benacerraf, B., Green, I. and Paul, W.E. (1967). 'The Immune Response of Guinea Pigs to Hapten-Poly-L-Lysine Conjugates as an Example of the Genetic Control of the Recognition of Antigenicity'. *Cold Spring Harbor Symp. Quant. Biol.* **32**: 569–75.
Bernstein F. (1924). 'Ergebnisse einer biostatistischen zusammenfassenden Betrachtung über die erblichen Blutstrukturen des Menschen'. *Klin. Wochenschr.* **3**: 1495–7.
Billingham, R.E. and Brent, T. (1957). 'Tolerance. A Simple Method for Inducing Tolerance of Skin Homografts in Mice'. *Transplant. Bull.* **4**: 67–71.
Billingham, R.E., Brent, L. and Medawar, P.B. (1954). 'Quantitative Studies on Tissue Transplantation Immunity. I. The Survival Times of Skin Homografts Exchanged Between Members of Different Inbred Strains of Mice'. *Proc. R. Soc. London, Ser. B.* **143**: 43–80.
Billingham, R.E., Brent, L. and Medawar, P.B. (1955). 'Acquired Tolerance of Skin Homografts'. *Ann. N.Y. Acad. Sci.* **59**: 409–16.
Bluestin, H.G., Green, I. and Benacerraf, B. (1971). 'Specific Immune Response Genes of the Guinea Pig. II. Relationship Between the Poly-L-Iysine Gene and the Genes Controlling

Immune Responsiveness to Co-polymers of L-glutamic Acid and l-tyrosine in Random-Bred Hartley Guinea Pigs'. *J. Exp. Med.* **134**: 471–81.

Bona, C.A. (1987). *Regulatory Idiotypes*. New York: J. Wiley and Sons.

Bona, C.A. and Kohler, H. (1983). *Immune Networks*, vol. 418. New York: Ann. N.Y. Acad. Sci.

Breinl, F. and Haurowitz, F. (1930). 'Chemische Untersuchung des Präzipitates aus Hämoglobin und Anti-Hämoglobinserum und Bemerkungen über die Natur des Antikörper'. *Z. Physiol. Chem.* **192**: 45–57.

Burdette S. and Schwartz, R.S. (1987). 'Idiotypes and Idiotypic Networks'. *N. Engl. J. Med.* **317**: 219–24.

Burnet, F.M. (1941). *The Production of Antibodies*. New York: Macmillan.

Burnet, F.M. (1957). 'A Modification of Jerne's Theory of Antibody Production Using the Concept of Clonal Selection'. *Austr. J. Sci.* **20**: 67–9.

Burnet, F.M. (1959). *The Clonal Theory of Acquired Immunity*. London: Cambridge University Press.

Burnet, F.M. and Fenner, F. (1949). *The Production of Antibodies*, 2nd edition. London: Macmillan and Company.

Buss, L.W. (1987). *The Evolution of Individuality*. Princeton, Princeton University Press.

Celada, F. (1988). 'Does the Human Mind Use a Logic of Signs Developed by Lymphocytes 10^8 Years Ago?'. In: Sercarz, E.E., Celada, F., Mitchison, N.A. and Tada, T. (eds.), *The Semiotics of Cellular Communication in the Immune System*. NATO ASI Series, vol. H23. Berlin: Springer-Verlag, pp. 71–9.

Chen, P.P. and Parham, P. (1989). 'Direct Binding of Influenza Peptides to Class I HLA Molecules'. *Nature* **337**: 743–5.

Chernyak, L. and Tauber, A.I. (1990). 'The Idea of Immunity: Metchnikoff's Metaphysics and Science'. *J. Hist. Biol.* **23**: 187–249.

Chernyak, L. and Tauber, A.I. (1991). 'The Dialectical Self: Immunology's Contribution'. In: Tauber, A.I. (ed.), *Organism and the Origins of Self*. Dordrecht: Kluwer Academic Publishers, pp. 109–56.

Chernyak, L. and Tauber, A.I. (1992). 'Concerning Individuality'. *Phil. Biol.* **7**: 489–99.

Cohen, I.R. (1992). 'The Cognitive Principle Challenges Clonal Selection'. *Immunol. Today* **13**: 441–4.

Cohn, M. (1986). 'The Concept of Functional Idiotype Network for Immune Regulation Mocks All and Comforts None'. *Annl. Inst. Pasteur Immunol.* **137**: 64–76.

Cohn, M. (1987). 'The Ground Rules Determining any Solution to the Problem of the Self/Nonself Discrimination'. In: Matzinger, P., Flajnik, M., Rammensee, H.-G., Stockinger, G., Rolink, T. and Nicklin, L. (eds.) *The Tolerance Workshop*, Basle Editions, pp. 3–35.

Cohn, M. (1989). 'The Apriori Principles Which Governs Immune Responsiveness'. In: *Cellular Basis of Immune Modulation*. New York: Liss, pp. 11–4.

Cohn, M. (1995). Role of the Top Down and Bottom Up Thinking About the Immune System. Unpublished lecture delivered at 'Symposium on the Theory and Science of immunology', The Wistar Institute, University of Pennsylvania, September 12, 1995.

Coutinho, A., Forni, L., Holmberg, D., Ivars, F. and Vaz, N. (1984). 'From an Antigen-Centered, Clonal Perspective of Immune Responses to an Organism-Centered, Network Perspective of Autonomous Activity in a Self-Referential Immune System'. *Immunol. Rev.* **79**: 151–68.

Coutinho, A. and Kazatchkine, M. (1994a). 'Autoimmunity Today'. In: Counho, A. and Kazatchkine M. (eds.), *Autoimmunity: Physiology and Disease*. New York: Willey Liss, pp. 3–6.

Coutinho, A. and Kazatchkine, M. (eds.) (1994b). *Autoimmunity: Physiology and Disease*. New York: Wiley-Liss.

Coutinho, A. and Stewart, J. (1991). 'A Hundred Years of Immunology: Paradigms, Paradoxes and Perspectives'. In: Cazenave, P.A. and Talwar, G.P. (eds.), *Immunology: Pasteur's Heritage*. New Delhi: Wiley Eastern Ltd., pp. 175–99.

Dreyer, W.J. and Bennett, J.C. (1965). 'The Molecular Basis of Antibody-Formation: A Paradox'. *Proc. Natl. Acad. Sci. (U.S.A.)* **54**: 864–8.

Ehlrlich, P. (1897). 'Die Wertbemessung des Diphtherielheilserums and deren theoretische Grundlagen'. *Klin. Jahrb.* **60**: 299–326. English translation (1957), 'The Assay of the Activity of Diptheria-Curative Serum and its Theoretical Basis'. In: *The Collected Papers of Paul Ehrlich, Vol. 2*, compiled and edited by F. Himmelweit, M. Marquardt and H. Dale. London: Pergamon Press, pp. 107–25.

Eisenlohr, I.C. and Hackett, C.J. (1989). 'Class II Major Histocompatibility Complex-Restricted T Cells Specific for a Virion Structural Protein That Do Not Recognize Exogenous Influenza Virus'. *J. Exp. Med.* **169**: 921–31.

Falk, R. and Sarkar, S. (1992). 'Harmony from Discord'. *Phil. Biol.* **7**: 463–72.

Fischer, E.P. and Lipson, L. (1988). *Thinking About Science Max Delbrück and the Origins of Molecular Biology*. New York: W.W. Norton.

Flood, P.M. (1988). 'Modes of Communication Within the Immune System. Action or Reaction?'. In: Sercarz, E.E., Celada, F., Mitchison, N.A. and Tada, T. (eds.), *The Semiotics of Cellular Communication in the Immune System*. NATO ASI Series, Vol. H23. Berlin: Springer-Verlag, pp. 120.

French, R.A., Tang, X-Lin, Anders, F.M., Jackson, D.C., White, D.O., Drummer, H., Wade, J.D., Tregear, G.W. and Brown, L.E. (1989). 'Class II-Restricted T-cell Clones to a Synthetic Peptide of Influenza Virus Hemagglutinin Differ in Their Fine Specificities and in the Ability to Respond to Virus'. *J. Virol.* **63**: 3087–94.

Galaty, D.H. (1974). 'The Philosophical Basis of Mid-Nineteenth Century German Reductionism'. *J. Hist. Med. Allied Sci.* **29**: 295–316.

Gilbert, S.F. (1992). 'Cells in Search of Community: Critiques of Weismannism and Selectable Units in Ontogeny'. *Biol. Phil.* **7**: 473–87.

Gorer, P.A. (1936a). 'Detection of Antigenic Differences in Mouse Erythrocytes by Employment of Immune Sera'. *Br. J. Exp. Path.* **17**: 42–50.

Gorer, P.A. (1936b). 'Detection of Hereditary Antigentic Difference in Blood of Mice by Means of Human Group A. Serum'. *J. Genetics* **32**: 17–31.

Gorer, P.A. (1937a). 'The Genetic and Antigenic Basis of Tumor Transplantation'. *J. Pathol. Bacteriol.* **44**: 691–7.

Gorer, P.A. (1937b). 'Further Studies on Antigenic Differences in Mouse Erythrocytes'. *Br. J. Exp. Pathol.* **18**: 31–6.

Gorer, P.A. (1938). 'The Antigenic Basis of Tumor Transplantation'. *J. Pathol. Bacteriol.* **47**: 231–52.

Gorer, P., Lyman, S. and Snell, G.D. (1948). 'Studies on the Genetic and Antigenetic Basis of Tumour Transplantation. Linkage Between a Histocompatibility Gene and "Fused" in Mice'. *Proc. R. Soc. London, Ser. B* **135**: 499–505.

Gray, D. and Skarvall, H. (1988). 'B-Cell Memory is Short Lived in the Absence of Antigen'. *Nature* **336**: 70–3.

Grebe, S.C. and Streilein, J.W. (1976). 'Graft-Versus-Host Reactions: A Review'. *Adv. Immunol.* **22**: 119–221.

Gregory, F. (1977). *Scientific Materialism in Nineteenth Century German*. Dordrecht: D. Reidel Publishing Co.

Grossman, Z. (1993). 'Cellular Tolerance as a Dynamic State of the Adaptable Lymphocyte'. *Immunol. Rev.* **133**: 45–73.

Haldane, J.B.S. (1937). 'Biochemistry of the Individual'. In: Needham, J. and Green, D.E. (eds.), *Perspectives in Biochemistry*. Cambridge: Cambridge University Press, pp. 1–10.

Hall, T.S. (1969). *History of General Physiology*, vol. 2. Chicago: The University of Chicago Press.
Hashimoto, G., Wright, P.F., and Karzon, D.T. (1983). 'Antibody-Dependent Cell-Mediated Cytotoxicity Against Influenza Virus-Infected Cells'. *J. Inf. Dis.* **148**: 785–94.
Hebert, I., Bernier, D., Boutin, Y., Jobin, M. and Mourad, W. (1990). 'Generation of Anti-Idiotypic and Anti-Anti-Idiotypic Monoclonal Antibodies in the Same Fushion'. *J. Immunol.* **144**: 4256–61.
Heidelberger, M. (1932–1933). 'Contributions of Chemistry to the Knowledge of Immune Processes'. *The Harvey Lectures*. Baltimore: Lippincott, pp. 184–201.
Henderson, L.J. (1928). *Blood. A Study in General Physiology*. New Haven: Yale University Press.
Hozumi, N. and Tonegawa, S. (1976). 'Evidence for Somatic Rearrangement of Immunoglobulin Genes Coding for Variable and Constant Regions'. *Proc. Natl. Acad. Sci. (U.S.A.)* **73**: 3628–32.
Janeway, C.A. Jr. (1989). 'Approaching the Asymptote? Evolution and Revolution in Immunology'. *Cold Spring Harbor Symp. Quant. Biol.* **54**: 1–13.
Jensen, C.O. (1903). 'Experimentelle Untersuchungen über Krebs bei Mäusen'. *Centralbl. Bakt. Parasitenk. Infektionskranth.* **34**: 28–122.
Jerne, N.K. (1955). 'The Natural-Selection Theory of Antibody Formation'. *Proc. Natl. Acad. Sci. (U.S.A.)* **41**: 849–57.
Jerne, N.K. (1960). 'Immunological Speculations'. *Ann. Rev. Microbiol.* **14**: 341–58.
Jerne, N.K. (1967). 'Summary: Waiting for the End'. *Cold Spring Harbor Symp. Quant. Biol.* **32**: 591–603.
Jerne, N.K. (1974). 'Towards a Network Theory of the Immune System'. *Ann. Immunol. (Paris)* **125C**: 373–89.
Kane, K.P., Vitiello, A., Sherman, L.A. and Mescher, M.F. (1989). 'Cytolytic T-Lymphocyte Response to Isolated Class I H-2 Proteins and Influenza Peptides'. *Nature* **340**: 157–59.
Kauffmann, M., Urbain, J. and Thomas, R. (1985). 'Towards a Logical Analysis of the Immune Response'. *J. Theor. Biol.* **114**: 527–61.
Kay, L.E. (1989). 'Molecular Biology and Pauling's Immunochemistry: A Neglected Dimension'. *Hist. Phil. Life Sci.* **11**: 211–19.
Kelsoe, G. and Schulze, D.H. (1987). *Evolution and Vertebrate Immunity. The Antigen-Receptor and MHC Gene Families*. Austin: University of Texas Press.
Klein, J. (1975). *Biology of the Mouse Histocompatibility-2 Complex*. New York: Springer-Verlag.
Klein, J. (1986). *Natural History of the Major Histocompatibility Complex*. New York: John Wiley.
Klein, J. (1990). *Immunology*. Boston and Oxford: Blackwell Scientific Publications.
Koestler, A. and Smythies, J.R. (1971). *Beyond Reductionism*. Boston: Beacon Press.
Kremer, R.L. (1990). *The Thermodynamics of Life and Experimental Physiology, 1770–1880*. New York, Garland Publishing, Inc.
Landsteiner, K. *The Specificity of Serological Reaction*, 2nd edition. Cambridge: Harvard University Press (1945) and New York: Dover (1962).
Landsteiner, K. and Reich, M. (1905). 'Über Unterschiede zwischen normalen und durch Immuniesierung enstandenen Stoffen des Blutserums'. *Centralbl. Bakteriol.* **39**: 712–17.
Langman, R.E. and Cohn, M. (1986). 'The "Complete" Idiotype Network is an Absurd Immune System'. *Immunol. Today* **7**: 100–101.
Leder, P. (1982). 'The Genetics of Antibody Diversity'. *Sci. Am.* **246**: 102–115.
Lenoir, T. (1982). *The Strategy of Life: Teleology and Mechanism in Nineteenth Century German Biology*. Dordrecht: D. Reidel and reissued by Chicago: The University of Chicago Press.
Little, C.C. (1914). 'A Possible Mendelian Explanation for a Type of Inheritance Apparently Non-Medelian in Nature'. *Science* **40**: 904–6.

Little, C.C. and Tyzzer, E.E. (1916). 'Further Studies on Inheritance of Susceptibility to a Transplantable Tumor of Japanese Waltzing Mice'. *J. Med. Res.* **33**: 395–453.
Little, C.C. and Johnson, B.W. (1921). 'The Inheritance of Susceptibility to Implants of Splenic Tissue in Mice. I. Japanese Waltzing Mice Albinos, and Their FI Generation Hybrids'. *Proc. Soc. Exp. Bio. Med.* **19**: 163–7.
Loeb, J. (1964) [1912]. *The Mechanistic Conception of Life.* Cambridge: Harvard University Press.
Loeb, L. (1908). 'Über Entstehung einer Sarkoms nach Transplantation eines Adenocarcinoms eines japanischen Maus'. *Z. Krebsforsch. Berl.* **7**: 80–110.
Lorenz, R.G. and Allen, P.M. (1988). 'Direct Evidence for Functional Self Protein/Ia-Molecule Complexes In Vivo'. *Proc. Natl. Acad. (U.S.A.)* **85**: 5220–3.
Lundkvist, I., Coutinho, A., Varela, F. and Holmberg, D. (1989). 'Evidence for a Functional Idiotypic Network Among Natural Antibodies in Normal Mice'. *Proc. Natl. Acad. Sci. (U.S.A.)* **86**: 5074–8.
Manwaring, W.H. (1930a). 'Renaissance of Pre-Ehrlich Immunology'. *J. Immunol.* **19**: 155–63.
Manwaring, W.H. (1930b). 'Biochemical Relativity'. *Science* **72**: 23–7.
Marshall, M.S. (1959). 'The Concept of Immunity'. *Cent. Rev.* **3**: 95–113.
Masaki, H. and Irimajiri, K. (1992). 'Generation of Helper T Cells that Recognize a Cross-Reactive Idiotype Through a Network Mechanism'. *Microbiol. Immunol.* **36**: 279–95.
Matzinger, P. (1994). 'Tolerance, Danger, and the Extended Family'. *Annu. Rev. Immunol.* **12**: 991–1045.
Mayr, E. (1982). *The Growth of Biological Thought. Diversity, Evolution, and Inheritance.* Cambridge: Harvard University Press.
Mazumdar, P.H.M. (1974). 'The Antigen-Antibody Reaction and the Physics and Chemistry of Life'. *Bull. Hist. Med.* **48**: 1–21.
Mazumdar, P.M.H. (ed.) (1989). *Immunology 1930–1980. Essays on the History of Immunology.* Toronto: Wall and Thompson, Inc.
McDevitt, H.O. and Benacerraf, B. (1969). 'Genetic Control of Specific Immune Responses'. *Adv. Immunol.* **11**: 31–74.
McDevitt, H.O. and Sela, M. (1965). 'Genetic Control of the Antibody Response. I. Demonstration of Determinant-Specific Differences in Response to Synthetic Polypetide Antigens in Two Strains of Inbred Mice'. *J. Exp. Med.* **122**: 517.
McDevitt, H.O., Deak, B.D., Shreffler, D.L., Klein, J., Stimpfling, J.H. and Snell, G.D. (1972). 'Genetic Control of the Immune Response. Mapping of the Ir-1 Locus'. *J. Exp. Med.* **135**: 1259–78.
Medawar, P.B. (1944). 'The Behaviour and Fate of Skin Autografts and Skin Homografts in Rabbits'. *J. Anal.* **78**: 176–99.
Metchnikoff, E. (1968). *Lectures on the Comparative Pathology of Inflammation,* 1891. New York: F.A. Starling and E.H. Starling. Trans. New York: Dover.
Metchnikoff, E. (1905). *Immunity in Infective Diseases* (translated by F.G. Binnie). Cambridge: Cambridge University Press.
Metchnikoff, E. (1907). *The Prolongation of Life: Optimistic Studies.* English translation by P.C. Mitchell. London: Heinemann.
Mitchison, N.A. (1954). 'Passive Transfer of Transplantation Immunity'. *Proc. R. Soc. London, Ser.* B **142**: 72–87.
Mitchison, N.A. (1993). 'A Walk Around the Edges of Self Tolerance'. *Ann. Rheumatol. Dis.* **52** (Suppl. 1): S3–S5.
Mock, D.J., Domurat, F., Roberts Jr, N.J., Walsh, E.E., Licht, M.R. and Keng P. (1987). 'Macrophages are Required for Influenza Virus Infection of Human Lymphocytes'. *J. Clin. Invest.* **79**: 620–4.

Mogensen, S.C. (1979). 'Role of Macrophages in Natural Resistance to Virus Infections'. *Microbiol. Rev.* **43**: 1–26.

Moulin, A.M. (1991). *Le dernier langage de la médicine. Histoire de l'immunologie de Pasteur au Sida.* Paris: Presses Universitaires de France.

Mudd, S. (1932). 'A Hypothetical Mechanism of Antibody Formation'. *J. Immunol.* **23**: 423–7.

Nelson, D.S. (ed.) (1989). *Natural Immunity.* San Diego: Academic Press.

Nisonoff, A. (1991). 'Idiotypes: Concepts and Applications'. *J. Immunol.* **147**: 2429–38.

Nossal, G.J.V. and Lederberg, J. (1958). 'Antibody Production by Single Cells'. *Nature* **181**: 1419–20.

Ofek, I. and Sharon, N. (1988). 'Lectinophagocytosis: A Molecular Mechanism of Recognition Between Cell Surface Sugars and Lectins in the Phagocytosis of Bacteria'. *Infect. Immun.* **56**: 539–47.

Parisi, G. (1990). 'A Simple Model for the Immune Network'. *Proc. Natl. Acad. Sci. (U.S.A.)* **87**: 429–33.

Pauling, L. (1940). 'A Theory of the Structure and Process of Formation of Antibodies'. *J. Am. Chem. Soc.* **62**: 2643–57.

Pauling, L. (1970). 'Fifty Years of Progress in Structural Chemistry and Molecular Biology'. *Daedalus* **99**: 909–10.

Pauling, L. and Campbell, D.H. (1942a). 'Manufacture of Antibodies in Vitro'. *J. Exp. Med.* **76**: 211–20.

Pauling, L. and Campbell, D.H. (1942b). 'The Production of Antibodies In Vitro'. *Science* **95**: 440–1.

Pauling, L. and Delbruck, M. (1940). 'The Nature of the Intermolecular Forces Operative in Biological Processes'. *Science* **92**: 77–9.

Perelson, A.S. (1989). 'Immune Network Theory'. *Immunol. Rev.* **110**: 5–36.

Polanyi, M. (1969). 'Life's Irreducible Structure'. In: Greene, M. (ed.), *Knowing and Being, Essays by Michael Polanyi.* Chicago: University of Chicago Press, pp. 225–39.

Salaun J., Bandeira, A., Khazaal, I., Calman, F., Coltey, M., Coutinho, A. and Le Dorian, N.M. (1990). ' Thymus Epithelium Tolerizes for Histocompatibility Antigens'. *Science* **247**: 1471–4.

Salthe, S.N. (1985). *Evolving Hierarchical Systems.* New York: Columbia University Press.

Sarkar, S. (1991). 'Reductionism and Functional Explanation in Molecular Biology'. *Uroboros* **1**: 67–94.

Sarkar, S. (1992). 'Models of Reduction and Categories of Reductionism'. *Synthese* **9**: 167–94.

Schaffner, K.F. (1967). 'Approaches to Reduction'. *Phil. Sci.* **34**: 137–47.

Shimony, A. (1987). 'The Methodology of Synthesis: Part and Wholes in Low Energy Physics'. In: Kargo, R. and Achinstein, P. (eds.), *Kelvin's Baltimore Lectures and Modern Theoretical Physics.* Cambridge: MIT Press, pp. 373–95.

Shreffler, D.C. (1988). 'Seventy-five Years of Immunology: The View from the MHC'. *J. Immunol.* **141**: 1791–8.

Silverstein, A.M. (1989). *A History of Immunology.* San Diego: Academic Press.

Simonsen, M. (1957). 'The Impact on the Developing Embryo and Newborn Animal of Adult Homologous Cells'. *Acta Pathol. Microbiol. Scand.* **40**: 480–500.

Snell, G.D. (1948). 'Methods for the Study of Histocompatibility Genes'. *J. Genet.* **49**: 87–108.

Snell, G.D., Smith, P. and Gabrielson, F. (1953). 'Analysis of the Histocompatibility-2 Locus in the Mouse'. *J. Nat. Can. Inst.* **14**: 457–80.

Stewart, J. and Varela, F.J. (1989). 'Exploring the Meaning of Connectivity in the Immune Network'. *Immunol. Rev.* **110**: 37–61.

Talmage, D.W. (1957). 'Allergy and Immunology'. *Ann. Rev. Med.* **8**: 239–56.

Tauber, A.I. (1990). 'Metchnikoff, the Modern Immunologist'. *J. Leuk. Biol.* **47**: 560–6.

Tauber, A.I. (1991). 'The Immunological Self: A Centenary Perspective'. *Perspect. Biol. Med.* **35**: 74–86.

Tauber, A.I. (1992). 'The Birth of Immunology. III. The Fate of the Phagocytosis Theory'. *Cell. Immunol.* **139**: 505–30.
Tauber, A.I. (1994). *The Immune Self. Theory or Metaphor?* Cambridge: Cambridge University Press.
Tauber, A.I. and Chernyak, L. (1989). 'The Birth of Immunology: II. Metchnikoff and his Critics'. *Cell. Immunol.* **121**: 447–73.
Tauber, A.I. and Chernyak, L. (1991). *Metchnikoff and the Origins of Immunology: From Metaphor to Theory.* New York and Oxford: Oxford University Press.
Tauber, A.I. and Sarkar, S. (1992). 'The Human Genome Project and the Limitations of Reductionism'. *Perspect. Biol. Med.* **35**: 220–35.
Tiselius, A. and Kabat, E. (1939). 'An Electrophoretic Study of Immune Sera and Purified Antibodies Preparation'. *J. Exp. Med.* **69**: 119–31.
Tonegawa, S. (1985). 'The Molecules of the Immune System'. *Sci. Am.* **253**: 122–31.
Topley, W.W.C. (1930). 'Role of Spleen in Production of Antibodies'. *J. Pathol. Bacteriol.* **33**: 339–51.
Trentin, J.J. (1956). 'Mortality and Skin Transplantability in X-irradiated Mice Receiving Isologous, Homologous or Heterologous Bone Marrow'. *Proc. Soc. Exp. Biol. Med.* **92**: 688–93.
Tyzzer, E.E. (1909). 'A Study of Inheritance in Mice with Reference to their Susceptibility Transplantable Tumors'. *J. Med. Res.* **21**: 519–73.
Varela, F.J. (1979). *Principles of Biological Autonomy.* New York: Elsevier-North Holland.
von Dungern, E. and Hirschfeld, Z. (1910). 'Über Verebung gruppenspezifischer Struckturen des Blutes. *Z. Immunitätsforsch. exp. Therap.* **6**: 284–92.
Waldmann, H., Cobbold, S.P., Benjamin, R. and Qin, S. (1988). 'A Theoretical Framework for Self-tolerance as its Relevance to Therapy of Autoimmune Disease'. *J. Autoimmun.* **1**: 623–9.
Watson, J.D., Gilman, M., Witkowski, J. and Zoller, M. (1992). *Recombinant DNA*, 2nd edition. New York: Scientific American Books.
Whitehead, A.N. (1925). *Science and the Modern World.* New York: Macmillan Co.
Winchester, G., Sunshine, G.H., Nardi, N. and Mitchison, N.A. (1984). 'Antigen-Presenting Cells Do Not Discriminate Between Self and Non Self'. *Immunogenetics* **19**: 487–91.
Witkowski, J.A. (1988). 'The Discovery of "Split" Genes: A Scientific Revolution'. *Trends Biochem. Sci.* **13**: 110–13.
Wright, A.E. (1910). *Studies on Immunisation. Diagnosis and Treatment of Bacterial Infections.* New York: William Wood and Company.
Yates, F.E. (1987). *Self-Organizing Systems. The Emergence of Order.* New York: Plenum Press.
Zingernagel, R.M., Pircher, H.P., Ohashi, S., Odermatt, B., Mak, T., Arnheiter, H, Burki, K. and Hengartner, H. (1991). 'T and B Cell Tolerance and Response to Viral Antigens in Transgenic Mice: Implications for the Pathogenesis of Autoimmune Versus Immuno-pathological Disease'. *Immunol. Rev.* **122**: 133–71.

JON BECKWITH

THE HEGEMONY OF THE GENE: REDUCTIONISM IN MOLECULAR BIOLOGY

The successes of genetics in recent years have strengthened a recurrent tendency in biology towards extreme reductionism. There are many uses of the term reductionism, but the one I find most helpful for the purposes of this discussion is that proposed by philosopher Helen Longino (Longino, 1990).

> Reductionism is both a methodological practice and a metaphysical view. Methodologically, reductionism is the practice of characterizing a system or process in terms of its smallest functional units. Metaphysical or ontological reductionism argues that those smallest functional units are what is real and that all causal processes can ultimately be understood as a function of interactions among these least bits. Methodological reductionism is often very useful in guiding researchers to the mechanisms or material constituents of a process. The biochemical analysis of metabolic processes is certainly a positive results of methodological reductionism. Metaphysical reductionism, however, conflates the pragmatic successes of local applications of methodological reductionism with both a guarantor of truth and the promise of universal reducibility.

In this paper I will address the two points raised by Longino as they relate to genetics. I will analyze the stance that the success of genetic reductionism as a research approach points to the necessary hegemony of genetics over other approaches to biological problems. At its extreme, this point of view includes the claim that the study of biological phenomena is of necessity the study of genes – that genes are the starting point. I will also question the attempts to convert the successes of the reductionist genetic research approach into a view of the human condition. Such attempts have been referred to more generally by Levins and Lewontin (1985) as "the confusion of reduction as a tactic with reductionism as an ontological stance". In this realm, the extreme position suggests that all or most human disease and human behavior can be ascribed to genes.

These two facets of genetic reductionism have been intimately connected both historically and today. That is, in periods when genetics has been particularly successful as a research strategy, the attention it has attracted has allowed its use to make statements about important social problems. That use is not a necessary consequence of the achievements of genetics and the resultant societal attention paid to it. Rather, this attention in combination with particular social conditions has provided a powerful source for doctrine.

I am particularly struck by parallels between conditions today and in the early part of this century in the United States. In both periods there are dramatic breakthroughs that generate highly productive periods in genetics research. In 1900, the rediscovery of Mendel's laws opened up the field of genetics as we now know it. Today, we see an era in which technical developments ranging from recombinant DNA, to DNA sequencing, to the polymerase chain reaction have produced a revolution in the ease with which genetic problems can be addressed. In the early 1900s, the new era in genetics was accompanied by a powerful eugenics movement that influenced social policy. Today, the increasing focus on genetics, both within biology and in the media, is beginning to shift public attention to genetic explanations and genetic solutions to health problems and social problems.

THE EUGENICS MOVEMENT IN THE UNITED STATES

In the early 1900s the burgeoning field of genetics was quickly incorporated into the eugenics movement (Ludmerer, 1972; Chase, 1977; Kevles, 1985). The origins of this movement are complex, evolving in part from a cattle breeding association and led by a number of men from the upper social classes. Not only did individuals such as Madison Grant and Robert DeCourcy Ward use the new concepts of genetics to support their claims for the inferiority of certain ethnic groups and of the lower social classes, but also many of the prominent geneticists of the day supported the eugenicists or even became active in the enterprise. According to Kenneth Ludmerer (Ludmerer, 1972), most of the leading geneticists were seduced by, or promoted eugenic theory. For instance, every member of the first editorial board of the journal *Genetics* (1916) gave support to the eugenics movement.

It is of some interest to ask why so many of these scientists supported the new "science" of eugenics, which, in retrospect at least, was based on primitive scientific analysis. Perhaps if we can understand this phenomenon, we will be better prepared to anticipate and avoid such trends today. The early days of genetics were a series of successes where one after the other mendelian trait was shown to follow Mendel's laws of inheritance. This unquestionably powerful new analytic tool may have generated an overweening confidence among geneticists which led them to imagine that the same approaches could be used to explain more complex human traits. Eugenicists argued that such social phenomena as criminality, poverty, intelligence and even seafaring could be attributed to single genes (Allen, 1975), extending the successes of the reductionist approach in an unwarranted fashion. However, it is unlikely that the nature of the science alone explains the prevalence of eugenics ideology among geneticists. Those doing research in genetics in this period came mainly from the upper social classes. At a time of considerable social turmoil, labor strife and major immigration movements, explanations for social phenomena that took away the responsibility for problems from those governing the society, and attributed them to the genetic defects of individuals or groups must have been very soothing to those in the upper echelons of society. Rather than having to surrender any privilege, this class could look on eugenics as a solution.

However, this class analysis cannot explain all the various strands of the eugenics and anti-immigration movements. At various points, these movements included major figures in the labor movement, and socialists such as Margaret Sanger. Furthermore, eugenics ideology was strong among socialists in both Germany and the Soviet Union (the latter until Lysenkoism took power partly in reaction against eugenic theories) (Graham, 1977).

At any rate, this combination of a social movement and an apparently scientific base allowed the eugenicists to have significant social impact (Ludmerer, 1972; Chase, 1977; Kevles, 1985; Allen, 1975). The push for eugenics programs played a role in both state and federal legislation. A majority of states passed laws that allowed sterilization for low intelligence, certain kinds of criminality and other characteristics. There laws were based on the claims of eugenicists that these traits were genetically determined. Many states also passed miscegenation laws forbidding marriage between individuals of different races, based on flawed scientific theories of the inferiority of hybrid species. Finally, the United States Congress passed the Immigration Restriction Act of 1924, which dramatically reduced the number of people allowed in the country from Southern and Eastern Europe, and from other cultures considered inferior. While the factors leading to the passage of this bill are many, eugenicists played a significant role in mustering support for it.

A less tangible thread in generating the atmosphere in which such legislation became possible was the development of popular attitudes towards the issues eugenicists were promoting. These attitudes are often fostered by the popular press and by other societal institutions. The eugenics movement presented its views to the public in many ways. From the presentation of eugenic displays at county and state fairs to the teaching of eugenics courses in most of the major universities and colleges in the country, various sectors of society became exposed to eugenics science and theories (Allen, 1975). The communication of these theories was spread even more widely by its appearance in the popular magazines of the day. Articles with titles such as 'Decadence of Human Heredity' (Atlantic Magazine – 1914), 'Plain Remarks on Immigration for Plain Americans' (Saturday Evening Post – 1921), and 'Danger That World Scum Will Demoralize America' (The Boston Herald – 1921) helped to strengthen eugenic attitudes among the public.

An examination of Popular Science Magazine from the years 1913 and 1915 gives an indication of the influence of eugenicists on popular culture. The following articles all reflected the position of the eugenics movement of that time:
- 1913 – 'Going through Ellis Island'; 'A Study in Jewish Psychopathology'; 'Heredity and the Hall of Fame'; 'The Biological Status and Social Worth of the Mulatto'; 'Heredity, Culpability, Praiseworthiness and Reward'; 'Eugenics with Special Reference to Intellect and Character'; 'Immigration and the Public Health'; 'A Problem in Educational Eugenics'; 'Economic Factors in Eugenics'.

- 1915 – 'The Racial Element in National Vitality'; 'Eugenics and War: The Dysgenic Effects of War'; 'Families of American Men of Science'.

Thus, the evolution of the eugenics movement went from academic theorizing and academic pronouncements, to their translation to the public via the media and other institutions and finally, with the appropriate public attitudes generated, to the formulation of social policy.

By the time the eugenics movement had reached its peak, many of the geneticists had withdrawn their backing. This falling off of scientific support, however, had little effect on the implementation of eugenics policies. In general, the recently disaffected geneticists rarely spoke out against these policies, and, by the time they did, it was essentially too late (Ludmerer, 1972; Allen, 1975).

The studies and proclamations of the US eugenicists were also closely followed in Germany in the 1920s and 1930s. Perhaps the most widely used text in human genetics during this period was by the German geneticists Fritz Lenz and Erwin Baur and German anthropologist Eugen Fischer (Baur et al., 1931). This text, which relied heavily on data and conclusions from the United States was a blatantly eugenicist and biological determinist text. It was full of characterization of races and ethnic groups as exhibiting certain genetically-based personality traits. Members of the German "Racial Hygiene" movement pointed to the laws and influence of the US eugenics movement as support for their positions (Waldinger, 1973; Proctor, 1988; Muller-Hill, 1988). As in the United States, many scientists were ardent supporters of these policies.

With the extreme misuse of genetics by German scientists and finally the Nazi government, some English and American geneticists began to speak out more openly. At the seventh International Congress of Genetics in 1939, a number of geneticists issued a "manifesto" criticizing eugenic programs (Crew et al., 1939). Among the signers were J.B.S. Haldane, J.S. Huxley, H.J. Muller, Th. Dobzhansky and A.G. Steinberg, several of whom had eugenic views themselves, but were appalled by its implementation in Germany. For the most part, the influence of geneticists on the misapplication of their field was too little and too late and thus had a minimal effect.

RECENT HISTORY

The universal revulsion at the Nazi eugenics policies after World War II led to a rejection of many of the general claims of the eugenics movement. In particular, the position that human behavioral traits and social problems had their origins in genetics was replaced by the position that environment was the determining factor in such issues. Some of these positions re reflected in two statements issued by UNESCO in the early 1950s. One of these, prepared by leading physical anthropologists and geneticists (several of them from the group that wrote the 1939 statement), criticized the concept of race and argued that differences in culture, intellectual achievement and behavior between ethnic groups were not genetic in origin (Montagu, 1963).

However, beginning in the late 1960s, scientific arguments for a genetic basis for various behavioral traits began to attract increasing attention. One of the earliest and most dramatic of such claims was the proposal that males with an extra Y chromosome (XYY males) were more aggressive than the average male (Jacobs et al., 1965) and exhibited a susceptibility to lead criminal lives (Price and Whatmore, 1967). Despite the weakness of the initial evidence, the myth of the "criminal chromosome" took hold of the public imagination (*Newsweek*, 1968; Pyeritz et al., 1977). Within a few years of the first findings, it became clear that XYY males were neither hyper-aggressive nor doomed to lives of criminality (Borgaonkar and Shah, 1974; Witkin et al., 1976; Theilgaard, 1983; Bender et al., 1984). But by the time these conclusions were reached, the XYY myth was already being presented as fact in newspapers, popular magazines, high school biology texts and college and medical school psychology and psychiatry texts.

The publicity received by the XYY male story was more than matched by the media interest in the proposals of Arthur Jensen in 1969, that blacks were "Born Dumb" as *Newsweek* put it (*Newsweek*, 1969). Jensen, an educational psychologist, published a lengthy article in the *Harvard Educational Review* (Jensen, 1969), suggesting that evidence from genetics and from IQ tests indicated that blacks were genetically inferior to whites in intelligence. As with the XYY story, the scientific criticisms of this paper led to a recognition within a few years that the complex issue of intelligence could not be approached in the simplistic fashion that Jensen had employed (Kamin, 1974; Block and Dworkin, 1976).

In 1975, entomologist E. O. Wilson generated enormous public interest in the field of sociobiology, when he suggested that there was much new evidence indicating that human social behavior was heavily influenced by genetics (Wilson, 1975). Many sociobiologists argue that animal social behavior, including that of humans, can be explained as a consequence of evolution, and, therefore, is genetically based. The scientific controversy that erupted around sociobiology (Caplan, 1978) still exists today and its course may, as some have suggested, be influenced by new developments in human genetics (see below).

In 1980, two psychologists, Julian Stanley and Camilla Benbow reported results that suggested to them that girls were genetically inferior in math ability to boys (Benbow and Stanley, 1980). The publication of this work in *Science Magazine* was greeted with headlines such as "The Male Math Gene" (Williams and King, 1980), as the media showed great interest in the study (Beckwith, 1983). Finally, over the last 10 years, psychologist Thomas Bouchard and his colleagues have presented to the public amazing stories on the similar behaviors of identical twins (Dusek, 1987) and have published work suggesting that such behaviors as religiosity show strong heritability (Bouchard et al.,

1990). These more recent examples of studies in the field of human behavioral genetics have also generated intense debate among scientists (Fausto-Sterling, 1985; Bazell, 1987; Dudley et al., 1991).

Except for the XYY studies, most of the work described above was not that of geneticists. Thus, the increasing interest in this area cannot be ascribed directly to reductionist attitudes among geneticists. However, the elucidation of the structure of DNA in 1953 (Watson and Crick, 1953), the birth of molecular biology in the late 1950s (Judson, 1979) and the successes in the field of genetics that they engendered were causing increased attention to genetics, both in academic settings and among the public. The rapid expansion in the understanding of how genes work may have given renewed confidence to those who saw human behavior in genetic terms. At the same time, the media and the public, having witnessed the dramatic advances in basic research in genetics may have been more prepared to accept genetic arguments for behavior.

The resurgence of these particular examples of human behavioral genetics may also have been related to social conditions in the United States. The 1960s were in a time of increasing challenges to established social structures and social arrangements. Arguments from genetics were often used to suggest that dramatic social changes were not possible given the hereditary nature of much of human behavior (Beckwith, 1976). According to this reasoning, if intelligence is genetic, then compensatory education programs are doomed to failure. If sex roles as they exist today are a result of our evolutionary heritage, then expecting women to succeed in the same societal realms as men is biologically unwarranted.

THE RECOMBINANT DNA ERA, THE HUMAN GENOME PROJECT AND REDUCTIONISM

While it is possible that the molecular biology of the 1950s and 1960s generated an environment in which reductionist approaches to a wide range of problems seemed the appropriate one, it is much clearer that this is the case for the breakthroughs in genetics in the 1970s. The improvements in DNA sequencing techniques, the development of recombinant DNA approaches to gene cloning and manipulation and a host of advances since have made genetic approaches to biological problems infinitely simpler in any organism, including humans. The success in biology based on this progress have been impressive. A partial list of such achievements includes 1) the mapping and characterization of genes involved in numerous genetic diseases, 2) the working out of developmental pathways at the genetic level in several organisms, 3) the refinement of evolutionary trees based on DNA sequence homology, and 4) an extraordinary increase in the understanding of the development and functioning of the immune system. Increasingly, it is becoming child's play to locate and sequence genes for a host of phenotypic characteristics or developmental pathways. "Child's play" is not a totally inappropriate term. Much of the biological literature today exhibits a sameness in terms of the techniques and approaches used. One can easily train a scientist to use a more or less standard sequence of steps to find a gene which may have profound importance for the study of biological processes or a disease state.

As a result of these technological breakthroughs, the focus of biology has shifted even more dramatically than in the 1950s and 1960s to the analysis of genes. Clearly, the shift has been tremendously productive and exciting. The reductionist approach of focusing on genes has worked for a host of previously intractable biological problems. However, accompanying this transformation of biology has been a strengthening of the extreme reductionist position both toward the science itself and its social applications. As with the period that initiated genetics at the turn of the century, the success of the science have been translated into a world-view. First, some molecular biologists have implied that essentially all biological problems are best approached by studying genes. For instance, according to Walter Gilbert (Gilbert, 1991) "To identify a relevant region of DNA, a gene, and then to clone and sequence it is now the underpinning of all biological science." "The new paradigm – the starting point of a biological investigation will be theoretical [based on gene sequence]."

Second, many leaders of the revolution in molecular biology have publicly claimed a nearly all-explanatory role for genetics. Many of these claims have been associated with the initiation of the Human Genome Project. Here are some of these statements:

James Watson in *Time Magazine*: "We used to think our fate was in our stars. Now we know, in large measure, our fate is in our genes" (Jaroff, 1989).

Norton Zinder calls the human genome sequence "Rosetta Stone" (Hall, 1990) while Walter Gilbert termed it the "Holy Grail of genetics" (Hall, 1990). The latter also stated that, from the sequence "we can have the ultimate explanation for a human being" (DelGuercio, 1987).

Robert Sinsheimer says that the sequence is what "defines a human being" (Hall, 1988).

Charles DeLisi entitled a subsection of his article on the Human Genome Project "The Blueprint for Life" (DeLisi, 1988).

Paul Berg at a recent Stanford conference stated: "Many if not most human diseases are clearly the result of inherited mutations" (Berg, 1991).

Frances Collins suggests that "[HGI] will likely transform medicine in the 21st century into a preventive mode, where genetic predispositions are identified and treated before the onset of illness rather than after illness is under way" (Collins, 1991).

And, in an editorial in *Science Magazine*, Daniel Koshland implied that the Human Genome Project will provide solutions to many of our social problems, including homelessness (Koshland, 1989). (The rationale being that much homelessness is due to mental illness – that mental illness has a genetic basis – and that finding the postulated genes for mental illness will allow cures to be developed.)

These attitudes toward the future of biology and its relevance to social problems reflects again "the confusion of reduction as a tactic with reductionism as an ontological stance" lamented by Levins and Lewontin. The movement within biology to concentrate on genes as the basis of all biological studies is a myopic view of the field. The questions we study in biology arise in many different ways. Some come from the discovery of new genes, but many others are only made possible because of years of descriptive work. There are many complex basic and applied problems that require approaches other than genetics. While the improvement in genetic techniques has occurred at an incredibly rapid pace, comparable improvement in techniques of cell biology have been neglected. The greatest rewards in biology today come for those working in the areas of DNA and gene manipulation. The devaluing of descriptive biological work and of technical innovation in such areas as electron microscopy could ultimately lead to a drying up of the source of that very information that is needed to make sense of genetic studies, or even to stimulate new areas of genetic research. The training of students in the latest technological developments to the detriment of broader biological training can also contribute to an impoverishment of the field. Molecular biologists should not be blinded by the dazzling successes of genetics to the balance in approaches that are required for future progress (Maddox, 1992).

The translation of the reductionist approach to an analysis of everything from human health to the human condition is also problematic. The arguments about health are based, in part, on the finding that some instances of susceptibility to common diseases, such as heart disease or cancer, are correlated with the inheritance of an altered gene. It is likely that more such instances will be found. However, this does not mean that most cancer or most heart disease is related to such susceptibility genes. Further, in those cases where there is a susceptibility, it is usually *only* a susceptibility. The actual development of cancer will be due to many factors, including other genes and the environment. It is not at all clear, that the best way to approach cures or prevention of cancer is a study of a cancer gene as opposed to systematic analysis of environmental factors, and the many other approaches that are currently employed in studying this problem. This is an area full of uncertainty. There are few examples that would give us confidence that gene characterization will lead to solutions to health problems. For instance, researchers have understood the molecular basis of sickle-cell anemia in terms of the amino acid change in the hemoglobin protein for over three decades (Ingram, 1957); but it has been continuing medical studies on the progress of the disease rather than genetic knowledge that have contributed to the significant improvements there have been in survival and health of those suffering from the condition (Kolata, 1987).

Of even more concern are the claims concerning genetics and social problems such as homelessness (Koshland, 1989). It is useful to analyze the content of such claims. First, a social problem is relegated to the realm of medicine or biology, when the roots are often in failings of the society itself. While clearly some of the homeless do have severe mental problems, much homelessness has its roots in economic deprivation. Second, the reliance on genetics to account for mental disorders exaggerates and distorts what we know. That there is relatively convincing evidence that some cases of, e.g. manic depressive illness (MDI) have a heritable component (although a gene has yet to be discovered), does not mean that all MDI can be traced to a gene, and certainly not that all depression has a genetic basis. Furthermore, even in those cases where there is substantial evidence for MDI having a genetic correlate in certain families, it is clear that not everyone who inherits the MDI gene develops the disorder. It seems likely that environmental factors are also important and should be explored in considering how to deal with the disorder. Third, the fact that a gene plays a role in a particular disease does not necessarily imply that genetics will provide solutions. As discussed above, while finding a gene for a particular condition will certainly promote better understanding of that condition, there is no certainty that it will generate cures or treatments.

CONSEQUENCES OF EXTREME RESURGENT REDUCTIONISM

I have already discussed the impact of reductionist thinking on the field of biology itself. However, the consequences are much broader.

The last 15 years have not only witnessed an explosion of genetic information but also the public has been deluged with reports of the discovery of genes for everything from cystic fibrosis to alcoholism. These are exciting times and the publicity for the achievements of genetics is warranted. At the same time, the impact of this publicity has been to promote the conception that genetics is all-explanatory. Reductionist statements from scientists of the sort quoted above only reinforce a distorted perception of the basis of the human condition. Genes are used in the popular media more and more to explain social phenomena. Everything from the attitudes of TV critics (Stewart, 1991), to the basis of violence among soccer fans in Great Britain (Lehmann-Haupt, 1992), to presidential candidate Ross Perot's frugality (Wright, 1992) are ascribed to genes. In the early part of this century the media served as a means of transmission of the perspective of scientists and, thus, helped form public opinion that influenced social policy.

What kind of environment is now being generated by the publicity that genetics has achieved with the grandiose claims that accompany them? In 1988, Melvin Konner (Konner, 1988), one of the most thoughtful proponents of human sociobiology, referred to the then recent mapping of genes for

Huntington's disease (Gusella et al., 1983) and manic depressive illness (Egeland et al., 1987) in the context of the sociobiology debate. (The report of the mapping of the MDI gene was later retracted (Kelsoe et al., 1989).) These discoveries were seen by Konner as a refutation of the critics of sociobiology and, by inference, provided greater support for this theory of human behavior. Yet most critics had never denied that certain human behavioral conditions such as Huntington's or forms of manic depression were correlated with altered genetic material. And the verification of these links says nothing about the likelihood of a genetic basis for male dominance, the tendency to commit rape, or xenophobia.

Thus, it may be that the increased attention to genetics in society will give greater courage to those who argue that our social problems and social inequities are genetic in origin. In recent years, there has been a resurgence of academic controversy over arguments that blacks are genetically inferior to whites in intelligence (Holden, 1991; Selvin, 1991a,b; Palca, 1989; Anderson, 1990; Allen, 1992; Maddox, 1992; Kaufman, 1992). Professors from the University of Denver, the University of California at Berkeley, City College of New York, Harvard University, the University of Delaware and The University of Western Ontario have been embroiled in controversy over this issue. Academics seem to have acquired a new-found boldness in tackling this sensitive topic. (At CCNY, one Professor also propounds the opposite theory, that blacks are genetically superior to whites (Finder, 1991)). I have strong suspicions that the reappearance of the issue at this time is facilitated by the climate in which genetics is made to appear more and more important. Of course, racism is not generated by genetics, and an important source of the renewed interest in these issues arises out of the political climate, including the debate over such issues as affirmative action. But historically, as we have seen, arguments from the scientific community have provided important support for racist ideology and racist political action.

Overall then, the impact of an overextension of the applications of genetics can have profound effects on society. In general, the focus on genetics alone as explanatory of disease and of social problems tends to direct society's attention away from other means of dealing with such problems. At its extreme, a false hope of cures for disease distorts the distribution of resources. Genetic explanations for intelligence, for sex role differences and for aggression, lead to an absolving of society of any responsibility for its inequities.

THE FUTURE

In the eugenics movement of the early part of the century, geneticists played a significant role. Furthermore, even after they became disaffected from the science and politics of the movement, they did little to blunt its effects. Today's geneticists, caught up in the enthusiasm of the successes of the new molecular biology, are contributing to an unbalanced view of the role of genetics and environment. A climate is being created in which social policy and individual attitudes may be formulated on the basis of incomplete or incorrect views of the human condition. A knowledge of history and a little less hubris are called for.

Harvard Medical School,
Boston, MA, USA

REFERENCES

(1968). 'Born Bad?' *Newsweek* **71**: 87.
(1969). 'Born Dumb?' *Newsweek* March 31: 84.
Allen, C. (1992). 'Gray Matter, Black-and-White Controversy'. *Insight* **13**(January): 4–9, 32–6.
Allen, G. (1975). 'Genetics, Eugenics and Class Struggle'. *Genetics* **79**: 29–45.
Anderson, C. (1990). 'Sex, Racism and Videotape'. *Nature* **347**: 6.
Baur, E., Fischer, E. and Lenz, F. (1931). *Human Heredity*. New York: MacMillan.
Bazell, R. (1987). 'Sins and Twins'. *The New Republic* **21**(Dec.): 17–8.
Beckwith, J. (1983). 'Gender and Math Performance: Does Biology Have Implications for Educational Policy?' *J Education (Boston Univ.)* **165**: 158–74.
Beckwith, J. (1976). 'Social and Political Uses of Genetics in the United States: Past and Present'. In: Lappe, M. and Morrison, R.S. (eds.), *Ethical and Scientific Issues Posed by Human Uses of Molecular Genetics*. Annual N.Y. Academy of Sciences, pp. 45–56.
Benbow, C.P. and Stanley, J. (1980). 'Sex Differences in Mathematical Ability: Fact or Artifact'. *Science* **210**: 1262–4.
Bender, B.G., Puck, M.H., Salbenblatt, J.A. and Robinson, A. (1984). 'The Development of Four Unselected 47,XYY Boys'. *Clin. Genet.* **25**: 435–45.
Berg, P. (1991). Talk presented at the Stanford Centennial Symposium 'The Human Genome Project: Biological Nature and Social Opportunities'. *Stanford Centennial Symp.* January 11.
Block, N.J. and Dworkin, G. (1976). *The IQ Controversy: Critical Readings*. New York: Pantheon.
Borgaonkar, D.S. and Shah, S.A. (1974). 'The XYY Chromosome Male – or Syndrome'. *Prog. Med. Genet.* **10**: 135–222.
Bouchard, T.J. Jr., Lykken, D.T., McGue, M., Segal, N.L. and Tellegen, A. (1990). 'Sources of Human Psychological Differences: The Minnesota Study of Twins Reared Apart'. *Science* **250**: 223–8.
Caplan, A.L. (1978). *The Sociobiology Debate*. New York: Harper and Row.
Chase, A. (1977). *The Legacy of Malthus: The Social Costs of Scientific Racism*. New York: Knopf.
Collins, F.S. (1991). 'The Genome Project and Human Health'. *FASEB J.* **5**: 77.
Crew, F.A.E., Haldane J.B.S., Harland, S.C., Hogben, L.T., Huxley, J.S., Muller, H.J. and Needham, J. (1939). 'Men and Mice at Edinburgh'. *J. Hered.* **30**: 371–3.
DelGuercio, G. (1987). 'Designer Genes'. *Boston Mag.* Aug: 79.
DeLisi, C. (1988). 'The Human Genome Project'. *Am. Sci.* **76**: 488–93.
Dudley, R.M., Beckwith, J., Geller, L., Sarkar, S. Jr., Bouchard, T.J., Lykken, D.T., McGue, M., Segal, N.L. and Tellegen, A. (1991). 'IQ and Heredity'. *Science* **252**: 191–2.
Dusek, V. (1987). 'Bewitching Science'. *Science for the People* **19**(6): 19–22.
Egeland, J.A., Gerhard, D.S., Pauls, D.L., Susex, J.N., Kidd, K.K., Allen, C.R., Hostetter, A.M.

and Housman, D.E. (1987). 'Bipolar Affective Disorder Linked to DNA Markers on Chromosome 11'. *Nature* **325**: 783–7.

Fausto-Sterling, A. (1985). *Myths of Gender*. New York: Basic Books, Inc.

Finder, A. (1991). 'Panel Asks That No Action Be Taken Against Jeffries'. *New York Times* Aug. 24: 27–8.

Gilbert, W. (1991). 'Towards a Paradigm Shift in Biology'. *Nature* **349**: 99.

Graham, L.R. (1977). 'Science and Values: The Eugenics Movement in Germany and Russia in the 1920s'. *Am. Hist. Rev.* **82**: 1133–64.

Gusella, J.F., Wexler, N.S., Conneally, M. et al. (1983). 'A Polymorphic DNA Marker Genetically Linked to Huntington's Disease'. *Nature* **306**: 234–8.

Hall, S.S. (1988). 'Genesis: The Sequel'. *Cal. Mag.* July: 62–9.

Hall, S.S. (1990). 'James Watson and the Search for Biology's Holy Grail'. *Smithsonian Mag.* Feb.: 42.

Holden, C. (1991). 'Politics in the Class Room (cont.)'. *Science* **251**: 622.

Ingram, V.M. (1957). 'Gene Mutations in Human Hemoglobin: The Chemical Difference between Normal and Sickle-cell Hemoglobin'. *Nature*: **180**: 326–8.

Jacobs, P.A., Brunton, M., Melville, M.M., Brittain, R.P. and McClemont, W.F. (1965). 'Aggressive Behavior, Mental Subnormality, and the XYY Male'. *Nature* **208**: 1351–2.

Jaroff, L. (1989). 'The Gene Hunt'. *Time* March 20: 67.

Jensen, A.R. (1969). 'How Much Can We Boost IQ in Scholastic Achievement?'. *Harv. Ed. Rev.* **33**: 1–123.

Judson, H.F. (1979). *The Eighth Day of Creation*. New York: Simon and Schuster.

Kamin, L. (1974). *The Science and Politics of I.Q.* Potomac: Earlbaum Associates.

Kaufman, R. (1992). 'U. Delaware Reaches Accord on Race Studies'. *The Scientist* **6**(14):1, 6, 13.

Kelsoe, J.R., Ginns, E.I., Egeland, J.A., Gerhard, D.S., Goldstein, A.M., Bale, S.H., Pauls, D.L., Long, R.T., Kidd, K.K., Conte, G., Housman, D.E. and Paul, S.M. (1989). 'Re-evaluation of the Linkage Relationship Between Chromosome llp Loci and the Gene for Bipolar Affective Disorder in the Old Order Amish'. *Nature* **342**: 238–43.

Kevles, D. (1985). *In the Name of Eugenics: Genetics and the Uses of Human Heredity*. Berkeley: University of California Press.

Kolata, G. (1987). 'Panel Urges Newborn Sickle Cell Screening'. *Science* **236**: 259–60.

Konner, M. (1988). 'New Keys to the Mind'. *New York Times Magazine* July 17: 49–50.

Koshland, D. (1989). 'Sequences and Consequences of the Human Genome'. *Science* **246**: 189.

Lehmann-Haupt, C. (1992). 'Studying Soccer Violence by the Civilized British'. *New York Times* June 25: C17.

Levins, R. and Lewontin, R. (1985). *The Dialectical Biologist*. Cambridge, MA: Harvard University Press.

Longino, H.E. (1990). *Science as Social Knowledge*. Princeton: Princeton University Press.

Ludmerer, K. (1972). *Genetics and American Society*. Baltimore, MD: Johns Hopkins University Press.

Maddox, J. (1992). 'How to Publish the Unpalatable?'. *Nature* **358**: 187.

Maddox, J. (1992). 'Is Molecular Biology Yet a Science?'. *Nature* **355**: 201.

Montagu, A. (1963). 'The UNESCO Statements on Race'. In: *Race, Science and Humanity*. Princeton, NJ: D. Van Nostrand Company, Inc., pp 178–83.

Muller-Hill, B. (1988). *Murderous Science: Elimination by Scientific Selection of Jews, Gypsies and Others, Germany 1933–1945*. Oxford: Oxoford University Press.

Palca, J. (1989). 'AAAS Annual Meeting Draws Largest Crowd of Decade'. *Nature* **337**: 297.

Price, W.H. and Whatmore, P.B. (1967). 'Criminal Behavior and the XYY Male'. *Nature* **213**: 815.

Proctor, R. (1988). *Racial Hygiene: Medicine Under the Nazis*. Cambridge: Harvard University Press.

Pyeritz, R., Schreier, H., Madansky, C. et al. (1977). 'The XYY Male: The Making of a Myth'. In: Ann Arbor Science for the People (eds.), *Biology as a Social Weapon*. Minneapolis: Burgess.
Selvin, P. (1991b). 'Is Vincent Sarich Part of a National Trend?'. *Science* **251**: 369.
Selvin, P. (1991a). 'The Raging Bull of Berkeley'. *Science* **251**: 368–71.
Stewart, S. (1991). '"Dallas" is Dead, but Who Cares?'. *Albuquerque J.* May **2**: B10.
Theilgaard, A. (1983). 'Aggression and the XYY Personality'. *Int. J. Law Psychiatry* **6**: 413–21.
Waldinger, R.J. (1973). *The High Priests of Nature: Medicine in Germany, 1883–1933*. B.A. Thesis, Harvard University.
Watson, J.D. and Crick, F.H.C. (1953). 'A Structure for Desoxyribose Nucleic Acids'. *Nature (London)* **171**: 737–8.
Williams, D.A. and King, P. (1980). 'Do Males Have a Math Gene?' *Newsweek* December **15**: 73.
Wilson, E.O. (1975). *Sociobiology: The New Synthesis*. Cambridge: Harvard University Press.
Witkin, H.A., Mednick, S.A., Schulsinger, F., Bakkestrom, E., Christiansen, K.O., Goodenough, D.R., Rubin, K. and Stocking, M. (1976). 'Criminality in XYY and XXY Men'. *Science* **193**: 547–55.
Wright, L. (1992). 'The Man from Texarkana'. *New York Times Magazine* June **28**: 34.

SAHOTRA SARKAR AND DAVID S. THALER

INTRODUCTORY NOTE TO THE CONTRIBUTIONS BY SARKAR AND THALER

The two papers that follow had their common origin in a session of the conference on 'Methods in Philosophy and the Sciences', New York City, on January 16, 1993. At that meeting Sarkar presented a paper on non-reductionist models in molecular biology, and Thaler was the commentator. Since then, these two papers have developed in divergent ways and many of the original points of disagreement, which were usually matters of emphasis, are no longer present in the contributions being published here. We will, therefore, take this opportunity to describe these disagreements since they help motivate some of the discussions that *are* present in the contributions being published in this volume.

However, first let us note a major point of agreement on which we arrived independently, and which has, as far as we know, gone unnoticed so far in philosophical writing on molecular biology. Sarkar's account of strong reduction in molecular biology, which is an attempt to explicate the substantive assumptions of reductionist explanations (rather than only the formal characteristics on which philosophical attention has almost always been focused) relies on three criteria: (i) *physicalism*, or the demand that the explanatory force in molecular biology comes from physics and chemistry alone; (ii) *hierarchical organization*, with the explanatory force coming from relatively lower levels of the hierarchy; and (iii) *spatialization of the hierarchy*, that is, the hierarchy be determined by the relation "consists of" in physical space. This is an extension of an account of reduction originally given by Sarkar (1989, 1992). The extension is due to continuing discussions with Abner Shimony (see Shimony, 1987).

We both agree that *explanations in classical genetics and much of molecular genetics do not satisfy, and need not satisfy, the strictures of strong reduction*. We also agree that strong reduction did not motivate the research programs of classical and much of molecular geneticists. Of the three criteria just mentioned, genetics only clearly satisfies (ii). It also clearly does *not* satisfy (iii), and largely ignores (i). These points are elaborated with both historical and contemporary examples in Thaler's contribution; they are also alluded to in Sarkar's.

One original point of disagreement in emphasis is still present in the published contributions. Thaler argues that classical genetics, *rather than* enthusiasm about the new disciplines of computer science, cybernetics or information theory, was the source of arguments about coding and information that occurred in the early history of molecular biology. Sarkar does not deny that classical genetics was a critical, if not *the* critical, source for molecular biology. Nevertheless, he continues to argue that the language of "information", "coding", etc., and the types of explanations that were framed in this language, had their source in the disciplines mentioned above. This is a disagreement about the history of molecular biology that will merit further work. Here we have only tried to state the nature and extent of the disagreement as clearly as we can.

At the New York session, there was a further source of disagreement that is no longer apparent in the published contributions. There, Sarkar provided an unequivocally negative assessment of Bohr's hypothesis of complementarity in biology. He was also critical of Delbrück's pursuit of this goal, especially because this hypothesis could not be directly translated into experimental questions. He has dropped this point – and it *was* relevant to the original paper which tried to give a fairly comprehensive survey of non-reductionist explanation in molecular biology – because he has previously argued this point at length (Sarkar, 1989). However, Thaler pointed out that the search for complementarity (and paradoxes in general) can have a beneficial influence on science. This point is developed in the published version of Thaler's contribution and, when stated as generally as he does there, it is independent of the particular context of Sarkar's claims.

Finally, in New York, Sarkar also analyzed network models of immune regulation and argued that they violate criteria (i) of strong reduction, and present an entirely new (and interesting) type of explanation which he called "topological explanation". Since Thaler did not discuss these models, that analysis is not being published here.

REFERENCES

Sarkar, S. (1989). 'Reductionism and Molecular Biology: A Reappraisal'. Ph.D. Dissertation, University of Chicago.
Sarkar, S. (1992). 'Models of Reduction and Categories of Reductionism'. *Synthese* **91**: 167–94.
Shimony, A. (1987). 'The Methodology of Synthesis: Parts and Wholes in Low-Energy Physics'. In: Kargon, R. and Achinstein, P. (eds.), *Kelvin's Baltimore Lectures and Modern Theoretical Physics*. Cambridge: MIT Press, pp. 399–423.

SAHOTRA SARKAR

BIOLOGICAL INFORMATION: A SKEPTICAL LOOK AT SOME CENTRAL DOGMAS OF MOLECULAR BIOLOGY[1]

INTRODUCTION.

Biologists would probably be bored rather than surprised if the most important part of the conceptual structure of contemporary molecular biology is encapsulated in the following three precepts; (i) all hereditary information resides in the DNA sequence of organisms; (ii) this information is transferred from DNA to RNA through the process of transcription, and from RNA to protein through translation; (iii) this information is never transferred from protein to nucleic acid sequence. All these claims are so universally accepted that they usually find their way into introductory biology texts and may well be taken to form the theoretical core of molecular biology (to the extent – and this is a matter of controversy – that molecular biology has any theory).

Most biologists would probably find the following two additional, and somewhat more interpretive, claims only very slightly more controversial: (iv) precept (iii), which is usually called the "central dogma of molecular biology" is an explication, at the molecular level, of the well-known biological fact that acquired characteristics cannot be inherited; and (v) most, if not all, of the behavior of organisms is ultimately determined by their DNA sequences. These last two claims have long had influential critics, and though what will be said here will argue explicitly against claim (iv), and implicitly against claim (v), I will largely leave them aside for some other occasion.

Meanwhile, what I will try to do here, is suggest that the first three precepts (i–iii) are at best misleading and, more likely, simply vacuous in the sense that they do not perform any significant explanatory or predictive role at all. The reason for this is that there is no clear technical notion of "information" in molecular biology. It is little more than a metaphor that masquerades as a theoretical concept and, as I shall argue in detail, leads to a misleading picture of the nature of possible explanations in molecular biology. I will do so by describing how "information" came to be introduced in molecular biology, look at three *different* ways in which it has been construed, examine the biological theories in which these construals are supposed to be embedded, and argue that these theories are either not theories about "information" as it is customarily used in molecular biology (as for instance in the precepts (i), (ii) and (iii) above), or are of little predictive or explanatory value.[2] I will also point out that this failure of information-based reasoning is philosophically interesting because had it not been so, that is, had "information" managed to play a significant explanatory role, then that would have provided a striking example of non-reductionist explanation in biology.

Since several of these arguments will hinge on what constitutes a "significant explanation", it should be incumbent upon me to indicate what I mean by that term. I will not attempt to provide an explication of "explanation" here. Suffice it to note that there is no philosophical consensus at present about the nature of scientific explanation and there is no good reason even to believe that a single explication will capture the various modes of explanation – statistical or deductive, historical or predictive, functional or mechanistic, and so on – that have come into vogue. All I will assume are two "adequacy criteria", one whose violation would preclude a purported explanation from being an *explanation* and another whose violation would preclude an otherwise adequate explanation from being *significant*. Thus, these criteria are intended as necessary but not sufficient conditions, and I am not even assuming that they constitute all such necessary conditions that may be formulated. Since I will be arguing that these criteria will be violated in certain putative significant explanations in molecular biology that I shall analyze, it will suffice for my present purposes not to pursue "significant explanation" any further.

The first of the two criteria is easy enough to formulate: the factors invoked in any explanation must help codify a body of knowledge so that it may be viewed as conforming to some general pattern. The intuition behind this criterion is that new explanations, whatever else they do, must at least add some new structure into the framework of what we already understand. If an explanation is anything like a deductive-nomological one, this criterion would be easily satisfied provided that the universal statement assumed in such an explanation is truly nomological. If an explanation is statistical, the relevant statistical factors must have some broad applicability. Otherwise these factors cannot aid the process of codification that I am assuming to be a necessary feature of an explanation. The emphasis, here, is on that *substantive* issue rather than on the formal structure of the explanation. The point, here, is that no matter how "explanation" is formally explicated, this adequacy criterion must be respected.

The second criterion is also simple: a significant explanation must answer new questions that are recognized to be important. Once again, this is a substantive rather than a formal criterion. Moreover, it is context-dependent. The stage of scientific enquiry determines what is important. The questions that were important to molecular biology in the 1950s are not those that are important now. However, loss of significance of a pattern of explanation is not always simply a function of age: Newtonian or Darwinian explanations are often significant even today. Nevertheless, should the putative scope of a pattern of explanation become circumscribed to smaller and smaller domains over time, its significance is likely to decrease along the way. I will argue, later, that this is precisely what happened to information-based explanations in molecular biology, though this kind of development is by no means the only mechanism for the loss of significance.

These two criteria are undoubtedly very weak. They do not exclude the possibility that there might be more than one explanation, or even more than one significant explanation, of the same phenomenon. They do not demand physicalism, that is, that explanations ultimately be based on physical principle.[3] They do not demand prediction. Nevertheless, I will argue, that explanations in contemporary biology which invoke "information" fail to meet even these criteria.

Similarly, since I will be making claims about reductionist explanations, it is also incumbent upon me to explain what I mean by "reduction" especially since this concept has been construed and explicated in a wide variety of ways by philosophers. I will attempt to do so succinctly but fully in this paper, but not here. Rather, I will take it up when I get to that stage of my argument (section 'Reduction and the Physicalist Alternative'). I will there distinguish five different senses of "reduction", all of which are sometimes relevant in scientific contexts, and argue that whereas a large and interesting class of explanations in molecular biology satisfy the strictures of the strongest of those senses, those that would invoke "information" do not.

BACKGROUND: INFORMATION ENTERS MOLECULAR BIOLOGY

Let me turn to the introduction of "information" into molecular biology. That term, apparently had not been used until 1953, though the concept was certainly implicitly used in earlier discussions, particularly in Schrödinger's (1944) *What is Life?* By 1957 it became part of the standard conceptual apparatus of molecular biologists. Arguably – though I will not pursue this argument here because my purpose is to criticize current uses of "information" rather than to justify its initial introduction – the explicit introduction and systematic use of that concept was the most important part of the reconceptualization involved in the emergence of molecular biology as a conceptual enterprise that was identifiably distinct from its immediate ancestors, notably biochemistry.

This is not the place to attempt to give an adequate history of these conceptual changes though, given the importance of the "molecularization of biology", it is particularly unfortunate that we do not have an adequate sketch of these developments.[4] I will simply list, here, the steps that are most important for the elaboration of the argument of this paper. I hope that this will at least indicate the conceptual background against which the post-1953 developments took place:

(i) During the first three decades of this century it became clear that the molecular interactions which occurred in living organisms were highly "specific" in the sense that particular molecules interacted with exactly one, or at most a very few, reagents. Enzymes acted specifically on their substrates. Living organisms produced antibodies that were highly specific not only to naturally occurring antigens but were equally specific to artificial antigens to which neither they nor their ancestors had ever been exposed (see, e.g., the comprehensive review in Landsteiner (1936)). From the mid-1920s, even the action of genes began to be described using "specificity" (Timoféeff-Ressovsky and Timoféeff-Ressovsky, 1926). In genetics, the ultimate exemplar of specificity became the gene–enzyme relationship in the 1940s: "one gene one enzyme" was perhaps the most important organizing hypothesis of early molecular biology;

(ii) By the end of the 1930s, a highly successful theory of specificity, one which remains central to molecular biology today, had also emerged. Due primarily to Pauling (e.g., 1940) and his collaborators, though with many antecedents, this theory claimed (1) that the behavior of a biological macromolecule was determined by its conformation, and (2) what mediated biological interactions was a precise "lock-and-key" fit between the shapes of the molecules. Implicit in the second assumption is the rather striking idea that biological interactions are mediated by very weak interactions such as hydrophobic bonds rather than the stronger covalent and ionic interactions which are the staples of inorganic reactions. In the 1940s, when the three-dimensional structure of not even a single protein had been experimentally determined, the conformational theory of specificity was still speculative. The demonstration that it was at least approximately true for a wide variety of biological macromolecules in the late 1950s and 1960s has been one of molecular biology's most significant triumphs;

(iii) In 1944, meanwhile, in *What is Life?*, Schrödinger introduced a conceptual scheme that raised the possibility of a startlingly different source of specificity. Schrödinger asked himself how so tiny an object as the nucleus of a fertilized cell could contain all the specifications necessary for the normal development of an individual organism. His answer was " an elaborate code-script" which he compared to the Morse code (1944, p. 61). Though he was willing to countenance codes in more than one dimension, even a linear code based on a 5-letter alphabet and up to 25-letter words could generate over 10^{17} patterns. Thus the arrangement of the units rather than their physical shape became the source of specificity in Schrödinger's model. In the post-war era, *What is Life?* was remarkably influential in orienting a generation of physically minded researchers to biology (though the extent of its influence is usually over-stated (see Olby, 1971; Yoxen, 1979));

(iv) The 1940s also saw significant growth in our knowledge of microbial genetics, starting with Luria and Delbrück's (1943) demonstration of spontaneous mutagenesis in bacteria and continuing, especially, with Avery et al.'s (1944) demonstration that DNA is the likely genetic material and culminating, in a sense, with Lederberg's discovery of genetic recombination in bacteria (Lederberg and Tatum, 1946a,b). "Transformation", and "transduction" were just some of the new terms introduced to describe this phenomena (Ephrussi et al., 1953). In an attempt to navigate through this terminological morass, Ephrussi et al. suggested that "the term 'inter-bacterial information' ... replace those above" (1953, p. 701). It was the first modern use of "information" in genetics, and they went on to emphasize that it "does not necessarily imply the transfer of material substances, and recognize the possible future importance of cybernetics at the bacterial level" (p. 701);

(v) Immediately afterwards – in fact, in the next issue of *Nature* – came the double helix model of DNA (Watson and Crick, 1953a). It showed exactly how DNA was a linear molecule. By base pairing – A:T and C:G –, it also showed a possible way in which the specificities between the two helices could be involved in the formation of exact replicas. Moreover, in their second paper on the model, Watson and Crick (1953b) went on to use "information" explicitly, and implicitly defined it as what the "code" carried:

The phosphate-sugar backbone in our model is completely regular but any sequence of the pairs of bases can fit into the structure. It follows that in a long molecule many different permutations are possible, and it therefore seems likely that the precise sequence of bases is the code which carries the genetic information. (1953b, p. 964)

"Information", however, would not be explicitly defined for another five years, until Crick (1958) identified it with the specification of a protein sequence. Crick thus clearly distinguished two different types of specificity: (i) the specificity of each DNA sequence for its complementary strand, as modulated through base-pairing; and (ii) the specificity of the relation between DNA and protein. The latter was modulated by "genetic information". This notion of information was combinatorial since, as will be described below, by this point, all that was required of the code for it to perform its function was a precise sequence of bases. Since Crick also assumed that the relation between DNA and protein strands was sequential, the amount of information carried by a DNA sequence was directly proportional to its length. Schrödinger's (1944) "arrangement" thus came to encode "information" and a new theory of specificity, distinct from Pauling's conformational theory, resulted.[5]

CODES, TEMPLATES, AND THE CENTRAL DOGMA

What molecular biologists customarily mean by "information" emerged from these developments. The route, however, was somewhat circuitous or, at least, it involved what in retrospect appears to be a very curious digression. That digression was the work of Gamow who was the first to identify a "coding problem". Slowly, but ultimately unequivocally, he proceeded to separate its formal (and, with hindsight, "informational") part from its physical basis and, perhaps most importantly, enticed actual molecular biologists, including Delbrück and Crick, to play the game of solving the coding problem *formally* using only very general facts about the DNA–protein relationship as experimental constraints. Gamow's program was an unmitigated failure, if success is judged by the satisfaction of explicit goals. Nevertheless, in it lies the origin of the standard information-based picture of contemporary molecular biology. I will outline – and criticize – this program next. Other possibilities for "information" will be taken up in subsequent sections.

Template Codes

As just noted, attempts to crack what came to be called the "genetic code" were systematically initiated by Gamow (1954a,b) who developed several detailed (and hinted at many more) general models of the code in the 1950s. Though he explicitly envisioned the formation of a linear amino acid residue sequence of a protein from a DNA sequence as a process of "translation", Gamow's initial attempts to decipher the genetic code were based on the standard stereochemical model of specificity. In Gamow's first scheme, an amino acid residue fit exactly into the diamond-shaped "hole" formed by four DNA bases, two on each helix, with complementary bases forming the lateral diagonal of the diamond. The resultant code was triplet since only three independent bases specified the residue: the two on the top and bottom of the diamond and either one of the complementary base pair along the middle. By an ingenious argument, which already went beyond stereochemistry, Gamow showed that there were only 20 possible diamonds, provided that lateral symmetry (i.e., parity) was irrelevant.[7]

This experimental constraint that there are exactly 20 amino acid residue types, rapidly became what all early models of the code tried to predict. In 1954, Gamow's list of 20 residues that were universal to proteins was faulty, though Crick and Watson soon guessed what has since become the canonical list. Moreover, Gamow (1954b) was willing to countenance an additional five residues that, he thought, also naturally occurred in proteins. His model for the code could easily be extended to these (and, if necessary, up to 32 residue types) provided that the parity of the middle pair was made relevant. Gamow's stereochemical assumptions were unjustifiable as Crick, Watson and others soon realized. Somewhat strangely, Gamow never constructed a physical model to test the putative lock-and-key fit between the residues and the DNA double helix though he routinely referred to such a strategy for testing his model. Moreover, he seems to have been unaware of the possibilities that RNA mediated the translation of DNA to protein, and that protein synthesis took place in the cytoplasm, not at the DNA, both facts which were gradually established during that period.

Nevertheless, even after casting stereochemistry temporarily aside, there were additional empirical problems with the diamond code. As a code, it was overlapping, that is, the middle and top vertices of one diamond formed the lower and middle vertices of the next.[8] This property of the code puts restrictions on the possible adjacencies of different diamonds. For example, six of the diamonds could never follow themselves. Twelve could occur at most in repeats of two, while only the remaining two could repeat themselves indefinitely. These restrictions provided quite stringent experimental tests for the code, especially as the amino acid residue sequence of insulin was beginning to become available from the sequencing work of Sanger and his associates (Sanger and Tuppy, 1951a,b; Sanger and Thompson, 1953a,b). In 1954 Gamow had two sequences available, one 21, and the other 30 residues long. He proceeded to attempt to attribute particular diamonds to residues and immediately found out that the adjacency restrictions on the diamonds were inconsistent with the known sequences.

Gamow (1954b) tried to resolve this problem with *ad hoc* suggestions. Perhaps the adjacency restrictions on the diamonds could occasionally be violated. Perhaps insulin was not a good test since, according to Gamow, it was not a "hereditary protein", that is, one that is coded for by an inherited gene. The first of these possibilities could be tested on the basis of Gamow's stereochemical mechanism, using an atomic model kit but, once again, Gamow made no effort in that direction. That would have doomed the diamond code right from the start. Instead, he turned to statistical studies, essentially abandoning the diamond code. Gamow and Metropolis (1954) tried to correlate residue pairs known from sequencing to those that were predicted from various coding schemes. Their initial analyses found no evidence of adjacency restrictions.

By 1955, Gamow's enthusiasm for the coding problem had sparked efforts towards its solution by many others. By this point Gamow had clearly distinguished the abstract coding problem, "that of translating a four letter code to a twenty letter code" (Gamow et al., 1955, p. 24), from that of finding the mechanism of translation. The latter problem, he continued to insist, had to be solved for a full solution of the coding problem. In practice, however, Gamow and his collaborators altogether ignored the latter problem. In 1955, when Gamow, Rich and Ycas published a comprehensive review of attempts to decipher the code theoretically, it had become clear that the original diamond code was experimentally inadmissible. In its place Gamow et al. (1955) proposed a "triangular code" which came in two versions: compact and loose. Showing increased biological sophistication, this code was based on RNA rather than DNA as the template. RNA was assumed to be helical, which was not unreasonable at that time. Depending on the pitch of the helix, any three adjoining nucleotides on an RNA chain formed either an isosceles or an equilateral triangle, and any four formed two such triangles with two shared vertices. In the compact triangular code each of these triangles corresponded to an amino acid residue. In the loose triangular code, only the last nucleotide was shared as a vertex by two successive triangles. Thus, the compact triangular code was triplet with an overlap of two, whereas the loose was triplet with an overlap of one. Both presented adjacency restrictions. Those posed by the compact version were even stronger than those of the diamond code and could be promptly ruled out using the available amino acid residue sequences. The loose version presented weaker restrictions, but these were so weak that Gamow et al. gave up trying to correlate such codes with sequences. Instead, they turned to statistical considerations. Once again, their analyses and those by Gamow and Ycas (1955) found no such restrictions.

These later statistical arguments of Gamow were probably not fully convincing to those few who considered them. They were almost universally ignored primarily because Brenner (1957) soon demonstrated that the available amino acid residue sequences ruled out all overlapping codes. Brenner assumed that the code was triplet, that the overlap was two, and that the code could be degenerate. Since any two adjacent triplets shared two nucleotides, he pointed out, any given triplet can be preceded or succeeded by at most four different triplets. Let the former be called its "N-neighbor" and the latter its "C-neighbor". Now, assume that each triplet codes for one amino acid residue. Then, given a residue, for every different set of four of its N-neighbors and C-neighbors, one triplet must be assigned to it. Note that more than one triplet can be assigned to a residue in this fashion – the code was allowed to be degenerate. However, using the best-known amino acid residue sequences in 1957, Brenner found that at least 70 triplets were necessary, more than the 64 that were combinatorially possible from a triplet code. Already in 1957, between Gamow's statistics and Brenner's *reductio ad absurdum* demonstration, overlapping triplet codes were dead, and the possibility of deducing the code on formal principles alone must have seemed hopeless to many of its former proponents.

From Comma-Free Codes to the Central Dogma

Crick et al. (1957) thought otherwise. They introduced somewhat more sophisticated ideas implicitly based on assumptions about the efficiency of information storage and transmission. In sharp contrast to Gamow, for them, the only problem to be solved was the "formal" coding problem. The desideratum for success was the ability to obtain the "magic number", 20. The stereochemical mechanism of coding was, no doubt interesting, but only peripheral to the formal problem of coding. Rejecting overlapping codes on experimental grounds, Crick et al. also chose to ignore partially overlapping ones as being improbable. Moreover, they argued that it was "natural" to restrict attention to triplet codes since doublet ones would only allow 16 coding units whereas triplet ones, allowing 64, were clearly sufficient.

From their point of view there were two problems that had to be resolved. The first was that of the potential degeneracy. If 64 triplets only coded for

20 residues, in some cases, several different triplets would have to code for a single residue. There was no experimental reason that precluded a degenerate code, but Crick et al. felt that it was undesirable. The second problem was that of synchronization: if A, C, G, and T represent the four nucleotide base types, is the sequence ACCGTAGT read as ACC, GTA, . . . or CCG, TAG, . . . or CGT, AGT, . . . ? Their solution to both problems was ingenious. By explicitly attempting only to avoid the problem of synchronization, they also removed degeneracy and obtained the magic number, 20.

They assumed that only some triplets had "sense", that is, could code for residues, while others could not. Of the 64 possible triplets, the four with one base type, such as AAA, had to be immediately rejected: otherwise a sequence such as AAAACGA could potentially be ambiguously read as AAA, ACG, . . . or AAA, CGA, . . . This left 60 possibly meaningful triplets. These segregate into 20 sets of three, each set consisting of a triplet and its two cyclic permutations (e.g., ACG, CGA and GAC). Now, from each such set, only one triplet can be meaningful if the possibility of ambiguity is to be avoided. For instance, if ACG and CGA are both meaningful, ACGAGGT could potentially be read as ACG, ACG, . . . or CGA, GCT, . . . Thus, *at most* 20 meaningful triplets were possible. Crick et al. (1957) went on to show that a solution with *exactly* 20 triplets was possible by exhibiting such a set. That solution was not unique. They managed to find 288 different solutions; there can be as many as 408 (Golomb, 1962).

They went on to discuss a possible physical interpretation of the code which is historically important because it suggested that intermediate molecular complexes mediated the translation of a nucleic acid sequence to a polypeptide one. It was one of the first statements of the "adaptor hypothesis", the adaptors eventually being identified with transfer RNA (tRNA). However, these physical considerations did not form the basis for their coding scheme. That basis was provided by the two desiderata: (i) that there should be no degeneracy; and (ii) the synchronization problem should be avoided altogether by the very nature of the code. There was no experimental justification for the desiderata; the only experimental constraint in their analysis was the "magic number" 20. Moreover, these desiderata were not based on any physical considerations. Instead, what was implicitly being used were two claims about the nature of biological information. One was a claim of a certain kind of *simplicity*: a degenerate code was not as simple as a non-degenerate one. The other was, at least approximately, an assumption about *efficiency*: if synchronization was not automatically determined by the nature of the code, ambiguous translation could occur by a shift in the position at which reading began, that is, from a shift in reading frame. Errors could occur and would have, somehow, to be corrected. The comma-free code solved these problems.

Crick et al. had truly freed the coding problem from Gamow's persistent stereochemical worries. But once stereochemistry was cast aside, what remained? In practice, Gamow and his collaborators had never actually built any physical models. They had simply claimed that they had stereochemistry in mind while the data they used – adjacency restrictions, the statistics of adjacencies – refuted their assumptions without ever being based on any physical (or chemical) argument. Once they had failed, in retrospect, it appears that they had no other theoretical framework to fall back upon. Brenner (1957), too, had only negative results to offer but had used the concept of "information" systematically. Crick et al. had not used "information" explicitly but, presumably because they were moving towards it, they had implicitly attributed properties such as simplicity and efficiency to the code. But simplicity and efficiency in doing what? Could it have been "simplicity and efficiency of specificity"? Perhaps, but only at the cost of radically altering the concept of specificity.

Instead, explicitly acknowledging an emergent conceptual framework, no longer centered on physical models of specificity, Crick (1958) turned to "information". His immediate concern was the synthesis of proteins. There were three separate factors involved, he argued: "the flow of energy, the flow of matter, and the flow of information" (p. 144). The former two exhausted anything that physics considered; "information" had finally been liberated for a life of its own. Crick defined it with more care than ever before in a biological context: "By information I mean the specification of the amino acid sequence of the protein" (p. 144). He took it for granted that the genetic information was encoded in a DNA sequence. The physics of the folding of a protein, Crick hypothesized, was taken care of by its amino acid sequence. This became the "sequence hypothesis". Finally, this formalized notion of information was put to additional use:

The Central Dogma

This states that once 'information' has passed into protein *it cannot get out again*. In more detail, the transfer of information from nucleic acid to nucleic acid, or from nucleic acid to protein may be possible, but transfer from protein to protein, or from protein to nucleic acid is impossible. Information means here the *precise* determination of sequence, either of bases in the nucleic acid or on amino acid residues in the protein. (1958, p. 153; italics in the original)

This assumption about information transfer did not arise from physical considerations. They were Crick's way to give a molecular characterization of traditional neo-Darwinism: "it can be argued", he explicitly observed, "that [the protein] sequences are the most delicate expression possible of the phenotype of an organism" (p. 142). Presumably, the assumptions about simplicity and efficiency of information storage, transmission and retrieval that were implicit in the comma-free code were also based on a belief that they were obviously adaptive. However, this point was never made explicit.

Over the next few years, the implicit assumptions of Crick et al. (1957) were made explicit, and other plausible ones added, especially as mathematicians began to get systematically involved in the attempts to decipher the code. In particular, why settle for synchronization alone? In 1958, Delbrück pointed out that, after all, genetic information ultimately resides in DNA which is double-stranded (Golomb et al., 1958). This led to an additional constraint: synchronization must hold for both strands, taking base pair complementarily into account. The "dictionary", Delbrück insisted, would be better off if this additional constraint was also satisfied – this was, as he quite correctly pointed out, a natural extension of the idea of comma-freedom. Delbrück's additional constraint came to be called "transposability". Once it was imposed, it turned out that triplet codes could no longer give 20 residue types; they gave no more than 16. Attention shifted to quadruplet codes and in 1962 Golomb reported, on the basis of computer searches, that there could be at least 57 viable coding units for transposable comma-free quadruplet codes. Twenty had disappeared as a magic number. It was now only a minimal lower constraint, but transposability was gained.

The coding problem, at this point, had not only forgotten physics, but had even abandoned attention to biological specifics in preference to formal arguments about the efficiency in storage and transmission of information. In that spirit, Golomb (1962), also introduced another scheme, that of "biorthogonal codes", based on Hadamard matrices, which had 6 nucleotides coded for each amino acid residue. Its biological motivation, let alone basis, remained rather mysterious. But Golomb was sufficiently convinced of the biological potential of these new formal schemes to observe: "It will be interesting to see how much of the final solution [of the coding problem] will be proposed by the mathematicians before the experimentalists find it, and how much the experimenters will be ahead of the mathematicians (1962, p. 100)". Subsequent developments did not show much respect for the mathematicians.

The Experimental Denouement

What Golomb apparently was unaware of was that, by 1961, it had become clear that the code was triplet (Crick et al., 1961) showing that his speculations – and those of Delbrück –, for all their analytic sophistication, had little relevance for biology. Moreover, that same year, the first codon was experimentally deciphered by Matthaei and Nirenberg (1961a,b) using a cell-free system and RNA sequences. They determined that UUU, a triplet not permitted to be meaningful by the comma-free code, coded for phenylananine. As this result was verified, and other results began to come in, it became clear that the genetic code was not remotely comma-free. It was highly degenerate, but the degeneracy had no pattern that was consistent with any of the schemes predicted by Gamow and his collaborators (see Lanni (1964) for a comprehensive contemporary review). Colinearity of the code was demonstrated in

1964 (Yanofsky et al., 1964) and, by 1966, the entire genetic code was established (Woese, 1967; Ycas, 1969). Synchronization has turned out to be controlled by a variety of mechanisms, none based on considerations of constraints on the flow of information. Failures in synchronization, as exemplified by the expression of frame-shifted sequences, are possible – what these mechanisms routinely do is prevent this kind of polypeptide formation, though this did not become clear until the 1980s (Atkins et al., 1992). What had gone wrong with the comma-free code was that none of the elegant properties that were imposed on the code on the basis of "information" were revered in living organisms. The attempts to decipher the code in the 1950s were recognized to be an unmitigated failure. Even the ideas that the code was fully synchronized and fully sequential eventually came to be modified through the discovery of frameshift mutations and non-coding regions of the genome (see below). The only idea to survive, besides the uniformity of codon length, was Schrödinger's original one – merely that of the existence of a genetic code.

More recent, and presumably more sophisticated, attempts to use information-based reasoning to analyze the genetic code have fared little better (see Yockey, 1992). But, perhaps, in the worst insult to comma-freedom to date, it has been shown that synchronizability does not require comma-freedom (see Neveln, 1990). Where Crick et al. (1957) were too restrictive was in denying the possibility of any cyclic permutation of a triplet. While this was sufficient to ensure synchronizability, the only experimental constraint that motivated this assumption of comma-freedom was the requirement of exactly 20 meaningful words. However, synchronizability can be satisfied by merely requiring, for instance, that a sequence ACGCTATGC be meaningful and not all five possibilities – ACG, CTA, TGC, CGC and TAT – also be meaningful. A code which prevents such sets of five words from being simultaneously meaningful is said to have a "stagger bound" of five. A non-overlapping code with triplet codons and four nucleotide types can have a stagger bound up to nine. These codes automatically ensure synchronizability. Even if the added restriction of at least 20 meaningful words is imposed on them, a larger number of codes are possible.

Most of these codes are degenerate. However, degeneracy does not truly present any significant biological problem for a coding scheme: an organism with a degenerate code remains viable. In contrast, synchronization is biologically important: an organism could be faced with serious difficulties if its code was not synchronized *and* it did not have mechanisms preventing incorrect expression, for instance, through inadvertent shifts of reading frame. In fact, what these points really drive home is that the assumption made in the 1950s, that degeneracy is a problem, had no biological or physical basis. It was an assumption about what should be desired of "information". It reflected a hope, and probably an expectation, that there were principles other than the ordinary ones of physics and chemistry that could be used to explain biological behavior. What the failure of these attempts shows, more than anything else, is the empirical failure of that assumption.

Is the Code Irrelevant?

What went wrong? In 1962 Chargaff attempted to provide an answer. The fault, he argued, lay in the very introduction of the idea of "biological information". At some point between 1937 and 1944, that concept "raised its head and began to sport a multicolored beard which has become ever more luxurious despite numerous applications of Occam's razor" (1963, p. 163). Biological information might explain the highly specific relations between nucleic acid and protein but Chargaff was skeptical that it gave any insight into the equally specific relations between cells and multicellular communities. If there was no continuous "chain of information" from the lowest level to the highest, he argued, there was no justification in claiming that "DNA is the repository of biological information" (1963, p. 165). That argument was provocative though Chargaff's reasons for rejecting any genetic code at all were well known to be poor by 1963.

However, now that the failures of the early systematic attempts to codify the code are well-known, it is worth asking again, what does the use of the concept of a code carrying information still do? At the very most, it provides a succinct look-up table on the basis of which one can predict the sequence of the polypeptide chain that would be determined by a particularly DNA chain provided at least five conditions, discovered since 1966, are fulfilled.[9] Unfortunately, if prediction is the goal, these conditions are quite debilitating:

(i) It must be known whether the code to be used is the usual one or one of the variants. Though the non-universality of the code is well-known, this still does not, at present appear to be a particularly severe constraint by itself because the amount of known variation is not great (see Fox (1987) for a review). At present the most extensive variations have been found in mitochondrial DNA in which, for instance, across all the major kingdoms UGA codes for tryptophan rather than terminate translation, as specified by the usual code. Mitochondrial DNA is special since mitochondria probably arose as independent organisms in biological prehistory that were subsequently symbiotically incorporated into prokaryotes to form eukaryotic cells. Moreover, if prediction is all that is at stake, one could simply limit the use of the standard look-up table to nuclear DNA in eukaryotes. However, in at least four species of protozoa, UAA and UAG can code for glutamine rather than terminate translation even for nuclear DNA. Well across the various species that have been studied at this level so far, UGA is also known to code for amino acid residues that do not belong to the standard set of 20. In the case of some viral DNA sequences, moreover, the UGA and UAG codons are sometimes but not always "read through", that is, ignored as termination signals. This happens within the same system, that is, in the same RNA sequence these codons sometimes result in termination, and are sometimes read through.[10] The genetic code is not nearly as universal as was thought in the 1960s. However, at present, these exceptions are rare enough and though they present a problem, it is not a severe one for the use of the genetic code for predictive purposes in most contexts. The other conditions are far more serious;

(ii) The exact point of initiation of transcription must be known. In practice this, too, is not a severe constraint by itself even though the discovery of frameshift mutations has destroyed any residual belief in a "natural synchronization" of the genetic code and has made the prospect for DNA sequence-based protein sequence prediction even harder than it would have been otherwise. Though a few examples are known, the extent to which frame shifts are present in organisms is at present largely a matter of conjecture (Atkins et al., 1991). Sometimes, frame shifts are used to begin a segment of DNA that codes for a different protein (Fox, 1987);

(iii) All intron-exon boundaries, that is boundaries between coding and non-coding regions of a segment of DNA responsible for a single polypeptide must be known because, in almost all eukaryotes, after transcription, portions of the RNA, corresponding to the introns are spliced out. Moreover, alternative splicing (the production of different RNA segments for translation from the same original transcript) has also been found (Smith et al., 1989). There is no reason to believe at present that the distinction between introns and exons can be made on the basis of sequence information alone thought that is not implausible.[11] Certainly, at present, there is simply not enough data to suggest that enough is known about these boundaries to predict sequences of amino acid residues from DNA sequences alone. It is, of course, possible to use statistical regularities to attempt such prediction for many species. However, these techniques, do not only use "information" in the sense of "coding" – they make systematic use of auxiliary facts about the relation between DNA and protein. When auxiliary facts of this sort are admitted, including codon usage, it is even possible to attempt to predict DNA sequences from protein sequences but all this has little to do with the "information" carried by the code;

(iv) What has just been said about intron-exon boundaries applies with no major qualification to the boundaries between segments of DNA coding for proteins (informally, the "genes") and those that do not. These boundaries are important because a huge fraction of the DNA in the genomes of higher eukaryotes apparently is non-coding. Moreover, in eukaryotes, after transcription, bases are added as "tails" and "caps" to the mRNA;

(v) Besides splicing, several types of mRNA "editing" are also now known, and it must be known whether any of these processes are occurring before the DNA sequence can be used to read off a protein sequence (Cattaneo, 1991).

DNA segments producing transcripts that are subsequently edited are sometimes called "cryptic genes".[12] Moreover, even more unusual behaviors have been observed with mitochondrial RNAs. Bases can be deleted and inserted. The latter, especially, leads to a situation which can be interpreted as the formation of proteins for which there are no genes.[13]

Of course, it would be unreasonable to criticize molecular biologists for not predicting these complexities in the 1960s, long before any experimental evidence for them was discovered. Nevertheless, the point that the code – or the "information" contained in it – is of little predictive value in novel contexts seems unassailable.

Notice that these problems arise in what is still a *static* context. There is no concern yet about *dynamics*: the temporal progress of gene expression, that is, its control and regulation. But, ultimately, no matter what the relationship between gene and protein is, what matters in biology, for predictive and explanatory purposes, is the temporal sequence of events, which characterizes the behavior of an organism. The theories of coding and information, even if they were universal, and even if the speculative models of the 1950s had turned out to be successful, say nothing about the dynamics. This does not mean that they are incorrect. Rather, it means that they are, in a sense, incomplete but, as long as whatever dynamical account that is developed remains consistent with these theories, they can withstand the additional concern for dynamics. However, what this limitation does underscore is that, if the actual prediction of biological behavior through an interval of time is a serious desideratum to be considered, the code by itself, even if it did its part, would stand in need of supplementation by some theory of dynamics. Can considerations about information provide such a theory? The next section will look at one such attempt: the cybernetic models of gene regulation, such as the operon model, which have been around since the 1960s.

Meanwhile, restricting attention only to the static context, the code is at present of little value for predictive purposes unless much else about he nature of the system is also specified. But set prediction aside. Perhaps the concepts of code and information permit something weaker: explanation. At one level, indeed, they do permit some explanation. Given a polypeptide sequence and the DNA sequence responsible for it, one can quite often say that the DNA sequence gives rise to that polypeptide sequence *because of* the genetic code. The trouble is that this is not a particularly revealing explanation, and cannot be pursued too far, once again because of the conditions that have to be satisfied for a successful translation from the DNA to the protein sequences. But what if such exceptions did not exist? Then the explanatory value of the code would be of some significance (and, at least, one-way prediction, from DNA to protein sequences would be possible). Moreover, note that in such explanations, even though the concept of information is not explicitly invoked, its use is implicit because, the use of the concept of a code in this way presupposes an underlying stratum of linear, symbolic information. Unfortunately, because of the absence of any pattern to the code that can be deduced from a putative "theory of information", the explanatory role of information, even in that world without exceptions, would be very limited: the code would still not be of much help in codifying a body of knowledge (which was my adequacy condition for an explanation).

CYBERNETICS AND GENE REGULATION

Recall that when Ephrussi et al. (1953) initiated the use of "information" in molecular biology, they left open the possibility that it would require a cybernetic interpretation. In the excitement following the pursuit of the possibility that information resided in DNA sequence, that other alternative was ignored. However, since the standard sequence-based model has run into trouble, it is probably reasonable, and certainly does no harm, to explore the potential of that forgotten alternative, that is, the possibility that an entirely different notion of "information", derived in some way from cybernetics, can be used to recapture the value of that concept for molecular biology.

The trouble here is that nobody seems to be sure what "cybernetics" actually encompasses (see, e.g., Pierce, 1962). The term "cybernetics" was popularized with almost messianic fervor by Wiener (e.g., 1948). All that is uncontested about the discipline is that it arose out of what was called "servomechanics" during World War II. This was the study of systems such as guided missiles whose trajectories had to be corrected in flight in order to hit a target. This problem arose because the initial data about the target could only be known to a certain level of precision, and subsequent measurements had to be incorporated into a missile's path in order to correct for any errors arising from the initial imprecision. Wiener, in particular, developed mathematical procedures to make such corrections in a wide variety of contexts. This type of correction of behavior was termed "regulation" or even "self-regulation" when the correction procedure was incorporated into the system from the start. The additional data that was responsible for that correction was called "feedback". Presumably what feedback provided to the system was "information" – the role of information in cybernetics never became more explicit.

Under Wiener's influence, in the early 1950s, Haldane prepared a manuscript on the application of cybernetics to genetics. It remained unpublished.[14] However, in Haldane's department at University College London, Kalmus (1950) attempted a somewhat speculative application of cybernetics to genetics. The gene was interpreted as a message and mutations as errors in copying of that message. Kalmus argued that the perpetuation of genes was similar to "one of the devices in electronic calculating machines, which stores sequences of signals by perpetually recreating them". However, he rejected speculations about the linear arrangement of the gene and the possibility that the message is discrete. Moreover, since the action of a particular gene was sometimes felt in a distant cell, Kalmus argued that genes acted more like "broadcasting systems" than "wired telecommunication". That paper had no discernible influence on thinking about molecular genetics in the 1950s.

Operons

Meanwhile cybernetic terminology, particularly "feedback" regulation, control and inhibition crept into biology primarily through their incorporation into discussions of the dynamics of enzyme systems, apparently beginning with Umbarger (1956) and Yates and Pardee (1956). However, the real impetus for explicitly invoking cybernetics within molecular biology came from the operon model for the regulation of bacterial gene expression. The detailed study of gene regulation had begun in the late 1940s with the attempts of Monod and his collaborators to elucidate the control of enzyme formation in bacterial cells (Schaffner, 1974a,b). Since about 1900 it had been known that some organisms such as yeast displayed an ability to ferment a particular type of sugar only when exposed to it for a few hours. This kind of phenomenon came to be called "enzyme induction" in the early 1950s. In the mid-1940s Monod and his collaborators had already established that the type of enzyme induction involved in the lactose metabolism of *Escherichia coli* was under genetic control. One of the enzymes involved was β-galactosidase which hydrolyses β-galactosides such as lactose. Monod et al. (1951) showed that some inducers do not interact directly or indirectly with β-galactosidase. Moreover, some substrates of β-galactosidase did not act as inducers, as Lederberg (1951) had established. Monod et al. (1952) established that the production of β-galactosidase in the cell constituted *de novo* protein synthesis, rather than the conversion of a pre-existing molecule. They named the associated gene "z". By 1956, Monod and his collaborators also established that a second enzyme, which they named "galactoside permease", was also necessary for lactose to be metabolized (Monod, 1956; Rickenberg et al., 1956). As the name indicates, this enzymes was hypothesized to help the passage of lactose through the cell membrane. It was assigned a gene "y".

Meanwhile Lederberg had found mutant strains of *E. coli* that produced β-galactosidase even in the absence of an inducer: these were called "constitutive" mutants. Monod and his collaborators had shown that the production of that enzyme could be inhibited by molecules similar to the inducer. They interpreted these findings to mean that the formation of an active enzyme required the presence of an inducer, that the inducer was naturally present in constitutive mutants, and inhibition resulted because molecules similar to the inducer acted as a surrogate for them but led to the formation of a variant, less effective enzyme. According to this scheme, part of the specification of an active enzyme was determined by the inducer. Lederberg (1956), however, provided a strikingly different interpretation of these findings. "Dr. Monod asked whether the inducer carries information needed for the specification of the enzyme", he observed.

> One permissible view holds that the enzyme, on its critical surface, is directly molded on the inducing substrate. The alternative, which I prefer, is that all the specifications are already inherent in the genetic constitution of the cell: the inducer signals a regulatory system to accelerate the synthesis of the corresponding enzyme protein. On this notion, substrate-induced or, better, substrate-regulated enzyme formation is an evolved adaptation to relieve the organism from always having to produce a full quota of its genetic potential of enzymes regardless of their immediate utility. (p. 161)

With the idea of a genetic regulatory system, connected with associated concepts of signals and control, a new conceptual apparatus for the discussion of gene expression began to be formed.

However, progress towards a model of the lactose system was relatively slow. Monod discovered that constitutive mutants were altered only at one locus which was labeled "i". In what became a famous experiment, Pardee et al. (1959) conjugated wild-type inducible male bacteria (denoted "i^+z^+") with constitutive females (denoted "i^-z^-") to produce partially diploid (or merozygotic) i^-z^-/i^+z^+ bacteria. Two results were obtained: (i) β-galactosidase started to be produced immediately but further production stopped after about two hours in the absence of an inducer; (ii) however, the new merozygotes became inducible after that period, indicating that i^+ was dominant over i^-. The kinetics of the expression suggested that the i and z genes interacted through a cytoplasmic molecule ultimately produced by the i gene. The interpretation of the experiment which Pardee et al. preferred was that the i^+ gene produced a "repressor" molecule that prevented the z^+ gene from being expressed. The external inducer interacted with the repressor to disable it, thus permitting expression of the z gene. The constitutive mutants (i^-) simply did not produce the correct version of the repressor.

Jacob and Monod (1959) then distinguished between genes such as i which were "regulator genes" and ordinary genes such as z which were "structural genes". What was striking about the i locus, however, was that mutations in it affected not only the z locus, but also the y, and in the same fashion in which they affected the z. Jacob et al. ([1960] 1965) showed that the repressor bound to yet another locus, termed the "operator" locus and labeled "o", which was tightly linked to y and z and controlled their expression. They called the o-y-z system an "operon": "The hypothesis of the operator implies that between the classical gene . . . and the entire chromosome, there exists an intermediate genetic organization. The latter would include the *units of coordinated expression* (*operons*), comprising an operator and the group of genes for structure which it coordinates" (1965, p. 200). In common usage, however, "operon" came to signify the entire system, including the regulatory gene which need not be linked to the others.

"Microscopic Cybernetics"

"Feedback", "regulation", "control", "inhibition", etc. – the new concepts that had been introduced in the 1950s and were incorporated into the operon model seem to be direct applications of the conceptual apparatus of cybernetics. Nevertheless, explicit references to "cybernetics" are rare in the molecular biology literature from that period. In 1961, Chance (1961) even argued that it would be inappropriate to use "cybernetics" in biochemistry. Chance distinguished between feedback systems that were cybernetic and those that were not. The former, according to him, necessarily involved the idea of "steering" (1961, p. 289). Biochemical control, he argued, "involve[d] basically different concepts". Chance's objection does not now appear to be particularly important, or even interesting, but it remains a curious fact that none of the proponents of cybernetics ever proposed a detailed and clear definition of "cybernetic regulation".[15] In fact, in spite of Chance's strictures, "cybernetics" largely seemed to have been taken to mean little more than feedback regulation. However, even "feedback regulation" did not turn out to be a concept that was easy to characterize (Wimsatt, 1971). The original attempt, inside cybernetics, by Rosenblueth et al. (1943) was unduly behavioristic, in keeping with the temper of those times. The same behaviors that they invoked were, under some formally (though not biologically) plausible assumptions, shown by Kleene (1956) not to require feedback in the sense of cyclical patterns of state determination. More recent attempts to revive a non-trivial definition of cybernetic systems attribute internal states to these systems and give them mechanisms of self-regulation and make all of these externally accessible (Shimony, 1995). Such attempts trivially satisfy the strictures of behaviorism. To the extent that they show that such feedback behavior is trivially possible from the known relations between the internal states and mechanisms of a cybernetic system, they "demystify" cybernetics. However, these analyses are really about metaphysical issues of what "determines" or "causes" what; the more interesting – epistemological – question is whether cybernetic thinking affords potentially interesting explanations, either by making certain systems more intelligible, or by affording strategies for the study of new systems.

A rather remarkable discussion by Monod (1971) seemed to suggest that even the latter option is possible. For Monod, a "cybernetic system" amounted to no more than one governed by feedback regulation. In his view both allosteric enzymes and operons were examples of cybernetic systems since both were regulated through feedback.[16] For Monod, cybernetic systems were not necessarily hierarchical in organization, or even complex because even systems as simple as allosteric enzymes could exhibit cybernetic, that is feedback-regulated behavior. He conceived of the cell as a complex "cybernetic network [that] guarantees the functional coherence of the intracellular chemical machinery" (1971, p. 63). Even for allosteric interactions, Monod argued, cybernetic concepts became more important when regulation was effected by more than one metabolite and reactions could be branched.

Gene regulation, for Monod, was one level of organization higher than allostery since the latter regulated the metabolites whereas the former regulated the production of the allosteric enzymes themselves. Thus, what begins to emerge from this picture is a view of the organism as a hierarchically organized, cybernetic system, with successive levels of parts also being cybernetic systems. Perhaps expectedly, Monod paid particular attention to the lactose operon. His description of the system was terse:

1. The regulator gene directs the synthesis, at a constant and very slow rate, of the repressor protein.
2. The repressor specifically recognizes the operator segment to which it binds, with it forming a very stable complex . . .
3. In this state, synthesis of messenger . . . is blocked, presumably by simple steric hindrance, the beginning of this synthesis having to occur on the level of the promoter.
4. The repressor also recognizes β-galactosides, but binds them firmly only when in a free state: hence in the presence of the β-galactosides the operator-repressor complex is dissociated, thus permitting the synthesis of messenger and consequently or protein. (Monod, 1971, p. 75)

The "logic" of the system, he argued, was simple and similar to that of a computer. As he put it:

the repressor inactivates transcription; it is inactivated in turn by the inducer. From this double negation results a positive effect, an 'affirmation'. The logic of this negation of the negation, we may add, is not dialectical: it does not result in a new statement but in the reiteration of the original one, written within the structure of DNA in accordance with the genetic code. The logic of biological regulatory systems abides not by Hegelian laws, but, like the workings of computers, by the propositional algebra of George Boole. (Monod, 1971, p. 76)

He drew three conclusions. First, the "repressor, having no activity of its own, is purely a transducer – a mediator – of chemical signals (p. 76)". Second, the role of the β-galactoside in enzyme formation was "indirect, due exclusively to the repressor's recognition properties and to the fact that two states, each exclusive of the other, are accessible to it" (p. 76). What seems to have impressed Monod is that the β-galactoside almost inadvertently initiated a chain of reactions that looped back to affect it. Third, and most important, "[t]here is no *chemically necessary* relationship between the fact that β-galactosidase hydrolyzes β-galactosides, and the fact that its biosynthesis is induced by the same compounds Physiologically useful or 'rational', this relationship is chemically arbitrary – 'gratuitous', one might say" (pp. 76–77).

The concept of "gratuity", that is, "the independence, chemically speaking, between the function itself and the nature of the chemical signals controlling it" (p. 77), is critical to Monod's interpretation of the operon (and also of allostery) as cybernetic systems. Though, no doubt, the chemical interactions determine the behavior of the operon, these chemical interactions do not explain the behavior as well as the description of the system using feedback and signals that are responsible for control and have no "chemical requirements to answer to" (p. 78). Ultimately, the reason why they exist is to be given an evolutionary explanation: such controls must have been

> selected for the extent to which they confer heightened coherence and efficiency upon the cell or organism. In a word, the very gratuitousness of these systems, giving molecular evolution a practically limitless field for exploration and experiment, enabled it to elaborate the huge network of cybernetic interconnections which makes each organism an autonomous functional unit, whose performances appear to transcend the laws of chemistry if not to ignore them altogether. (p. 78)

Monod's position, that the cybernetic account is of more explanatory value than a purely physicalist alternative is persuasive but does it truly satisfy the adequacy condition on explanations, that is, does it help codify gene regulation in general? After all, his discussion was limited to the operon model which, in 1971, was known to hold for only a few sets of bacterial genes. The superiority of cybernetic explanation over a purely chemical one would be demonstrated only if two conditions could be satisfied: (i) that other cases of gene regulation could also be easily described as cybernetic systems; and (ii) the claim of gratuity would continue to hold for these systems. Satisfaction of (i) is required for the cybernetic accounts to be explanations and not merely interesting descriptions of some special cases. However, satisfaction of (i) alone would not preclude the possibility that there are equally attractive chemical accounts that even are, perhaps, of greater explanatory value. However (ii) would make such attractive chemical explanations implausible.

Eukaryotic Gene Regulation

I will, in fact, argue later (in section 'Reduction and the Physicalist Alternative') that, far from vindicating Monod's introduction of the concept of gratuity, the operon model represents one of the most significant triumphs of ordinary chemical (or reductionist) explanations in molecular biology. For the time being, however, I wish to turn to the other question, whether the cybernetic account can even be extended beyond bacterial gene regulation so that we can attribute some explanatory role to it, whether or not it ultimately fares better than its chemical alternatives. The obvious arena to investigate is eukaryotic gene regulation. Unfortunately, the situation there is complex and poorly understood at present. Nevertheless, what is clear is that attempts to generalize the operon model to eukaryotic gene regulation have so far shown no trace of success.

The most significant attempt in that direction was that of Britten and Davidson (1969). Their model was a straightforward generalization of the operon. They postulated the existence of four classes of eukaryotic DNA: producer genes (which are the direct analog of structural genes in bacteria), receptor genes linked to them; integrator genes (the analog of regulatory genes in bacteria) and sensor genes linked to them. Each linked pair, receptor–producer or sensor–integrator, is contiguous on a chromosome but the two pairs may be on different chromosomes altogether. The integrator gene can produce an activator RNA transcript which binds to a receptor and activates its linked producer gene. The transcription of activator RNA is controlled by an inducing agent that may or may not be bound to a sensor. Note that altering the activator molecule from RNA to protein would not change the formal properties of the model. As with operons, both positive and negative feedback control is possible.

In more complicated versions of this model, each sensor can activate a series of integrator genes next to it; an activator molecule (from an integrator gene) can bind to more than one receptor; several activators can bind to the same receptor; and several receptors can control the same producer gene. The set of producer genes ultimately under the control of a single sensor is called a "battery". Batteries can overlap creating a complex network, and there is considerable redundancy in the interactions between integrators and receptors (mediated by the activators). Britten and Davidson suggested that repetetive DNA sequences (which Britten and Kohne (1968) had discovered) played the role of integrators and receptors ensuring the redundancy invoked in the model.

Besides being a straightforward extension of familiar ideas, note that this model explicitly incorporated most of what was then known about eukaryotic gene expression: (i) that functionally related genes are not necessarily physically clustered in eukaryotes; (ii) that there were a large number of repeated sequences throughout the genome; (iii) that there appeared to be genomic sequences that were transcribed in the nucleus but, nevertheless, absent in the cytoplasm; and (iv) that cell differentiation, based on differential gene expression could be induced by "simple external signals". The Britten-Davidson model was greeted with a fair degree of enthusiasm in the early 1970s (see, e.g., Lewin, 1974, pp. 366–374). Its success, however, was short-lived. Regulatory mutations in eukaryotes have shown no definite pattern. No definite correlation between repetitive DNA sequences and regulatory functions has yet been proven. It is possible that further work will revive some cybernetic model, like the Britten–Davidson model, as a candidate for eukaryotic gene regulation. For the time being, however, the only reasonable conclusion is that, Monod's enthusiasm notwithstanding, cybernetics is of little value for molecular biology.

The failure of cybernetic models to capture eukaryotic gene regulation no doubt dooms the prospects for a cybernetic notion of "information" to explicate information in molecular biology. Nevertheless, it should not go unnoticed that even if some model such as the Britten–Davidson model been successful, the notion of "information" it would have incorporated would have been radically different from that which is invoked in the three precepts that I started out with. To the extent that there is a clear notion of "information" in cybernetics, it must be what provides "feedback" for regulation. In fact, it is hard to see what "information", in this context, is other than simply feedback. But this is far from the "information" that allegedly is contained in the DNA, gets transferred from DNA to RNA to protein and never from protein to nucleic acid. A similar negative assessment will be made for a third potential candidate for "biological information", namely, the notion of "information" that has formed part of information theory since the late 1940s.

INFORMATION THEORY

Molecular biology came of age around the same time as the mathematical theory of communication, also known as information theory (Shannon, 1948). The temptation to apply it to molecular biology was, therefore, probably irresistible. The first attempts to do so, however, were far from promising. According to Shannon's information theory, the amount of "information" is (roughly) measured by the logarithm of the number of choices available during a communication process. This notion of information connotes uncertainty; its numerical value is determined by an "entropy" function that is formally similar to the usual entropy of statistical mechanics. Shannon considered a message consisting of a linear sequence of symbols obtained from a symbol set. Let p_i be the probability of occurrence of the i-th symbol. Then, the entropy (of that occurrence in a message), **H**, is defined by $-p_i \log_2 p_i$. Shannon (1948) justified the choice of this measure by showing that $-Kp_i \log p_i$, where K is a positive constant, is the only function that satisfied three assumptions: (i) that **H** be continuous in the p_i; (ii) if all the p_i are equal to $1/n$, **H** should be a monotonic increasing function of n; and (iii) if a choice (such as that for the choice of a symbol in the message) can be analyzed into two successive choices, **H** should be the weighted sum of what it is for each of these choices. K can be fixed by a choice of units. In particular it is equal to 1 if the base of the logarithm is chosen to be 2. This definition of information as entropy was incorporated into a rigorous theory of communication systems.

Attempts to apply Shannon's information theory to molecular biology were most systematically promoted by Quastler (e.g., 1953a,b) who attempted to give an information-theoretic definitions of the specificity of enzyme action and other biological interactions. The series of definitions he proposed proved to be of little use. Branson (1953) calculated the information content of a polypeptide sequences using the empirical frequencies of the various residues to calculate the probabilities of their occurrence at each position. Linschitz (1953) attempted to estimate the information content of a bacterial cell. By 1956 even proponents of information theory in molecular biology had begun to question how valuable this approach was. When Quastler (1958) summarized a round table discussion at the end of a "Symposium on Information Theory in Biology", he felt compelled to begin:

> Information theory is very strong on the negative side, i.e. in demonstrating what cannot be done; on the positive side its application to the study of living things has not produced many results so far; it has not led to the discovery of new facts, nor has its application to known facts been tested in critical experiments. To date, a definitive judgment of the value of information theory in biology is not possible. (p. 399)

though he concluded, rather optimistically "that information theory is here to stay in biology (p. 402)". The absence of any insight from these early attempts precluded systematic developments along these lines though sporadic efforts to apply information theory to molecular biology continued.[17]

Information and Natural Selection

Though somewhat outside what is usually considered to be molecular biology, the most intriguing of attempts to apply information theory in biology was Kimura's (1961) attempt to use it to estimate the amount of information that accumulated through natural selection. Consider an infinite population of haploid organisms. Let p be the frequency of an advantageous allele at some locus. The probability of fixation of this allele in the population by random genetic drift alone is p; by natural selection, it is 1. Thus, the uncertainty that natural selection removes is $1/p$. Therefore, Kimura argued, the information that accumulates through natural selection, $\mathbf{H} = \log_2(1/p) = -\log_2 p = L/\ln 2$ where L is the substitutional load of the population. Kimura proposed $\mathbf{H} = L/\ln 2$ as a definition of the information gained through natural selection in general, that is, for all organisms. The consequences of this proposal for evolution theory were never worked out. However, Kimura went on to estimate 10^8 bits as the amount of information that had accumulated since the Cambrian epoch. From the estimated size of chromosomes, he also estimated 10^{10} bits as the amount of information that could be stored in a typical diploid chromosome set. Redundancy in the form of repeated DNA sequences could explain this discrepancy but another possibility was that "that amount of genetic information which has been accumulated [though natural selection] is a small fraction of what can be stored in the chromosome set" (p. 135).

This was the first inkling of what Kimura (1968, 1983) would eventually advocate as the "neutral theory of molecular evolution", that is, at the molecular level, the principal mechanism of change is random genetic drift rather than natural selection. Nevertheless, when that theory came to be formulated and systematically developed, from about 1968 (see Kimura 1983), information theory played no role, though considerations of the substitutional load were critical. "Information" thus provided little more than a redescription of "load" and information theory played no explanatory role whatsoever.

Within evolutionary biology, however, in the 1960s, a systematic analysis of Kimura's calculation was given by Williams (1966) in his influential *Adaptation and Natural Selection*. Williams explicitly defined the gene to be "that which segregates and recombines with appreciable frequency" (p. 24), implicitly assumed – with no argument whatsoever – that this definition was equivalent to the gene being "any hereditary information for which there is a favorable or unfavorable selection bias equal to several or many times its rate of endogenous change [mutation]" (p. 25), and concluded, therefore, that the gene was a "cybernetic abstraction" (p. 33). Kimura's calculation was taken to be the "most notable contribution on progress as accumulation of information" (p. 35). Williams accepted Kimura's account of information accumulation but denied that information accumulation could be identified with evolution. This formed part of his general arguments against the notion of evolutionary progress. Williams' book was important because it was critical in bringing the "units of selection" problem to the forefront of biology and, eventually, philosophy. Nevertheless, Kimura's calculation seems to have received no further attention.[18]

It is impossible to predict whether Kimura's proposal may yet prove to be fruitful. For our purposes here, however, it is irrelevant: it simply does not use "information" in the way that it is invoked in the three precepts with which we started. Kimura's "genetic information" is, in the final analysis, at best a measure of he improbability of our current DNA sequences and, what is worse, if this "information" is to have any interesting applications, any two sequences (of the same length) would contain the same amount of information! That it can be simply related to the substitutional load is no doubt an interesting result, if it can be fully generalized. Nevertheless, that "information" is not the "information" of the central dogma. In fact, all three precepts that we started out with are conceptually independent of this notion of "information" and, since as far as these precepts are concerned, all DNA sequences potentially carry information, and the same amount of information if they are of the same length, it is not even clear that these percepts are consistent with Kimura's result" that "information" is what gets accumulated through evolution.

Information Content of Sequences

Meanwhile, with the accumulation of protein and DNA sequences, information theory has begun to be used to compare sequences. Some useful techniques were developed as early as 1972 by Gatlin (1972), but the most promising results so far have come from methods developed and popularized by Schneider and his collaborators (starting with Schneider et al., 1986). For simplicity, restrict attention to DNA sequences. suppose that several sequences, corresponding to some functional region of DNA such as a binding site for a given protein, are given. Let **B** be the set of nucleotide bases, A, C, G and

T. For $b \in \mathbf{B}$, let $p(b)$ be the probability that a base is a b (as obtained from the frequencies of each of the bases in the set of sequences). Then the uncertainty of a base chosen at random is defined by $\mathbf{H}_g = -\sum_{b \in \mathbf{B}} p(b)\log_2 p(b)$. This is the usual Shannon measure for this situation. Now, let l be a particular position along the sequences and let $p(l, b)$ be the probability of finding base b at position l. Then the uncertainty at that position, $\mathbf{H}_s(l) = -\sum_{b \in \mathbf{B}} p(l, b)\log_2 p(l, b)$. The information content at that position is defined as $\mathbf{R}_s(l) = \mathbf{H}_g - \mathbf{H}_s(l)$ and the total information content for the sequence is defined by $\mathbf{R}_s = \sum_l \mathbf{R}_s(l)$. Schneider et al. (1986), Herman and Schneider (1992), Stephens and Schneider (1992) computed these measures for a variety of sequences from *E. coli*, phage and humans. The measure $\mathbf{R}_s(l)$ behaved as expected: it was highest where the different sequences in each set had the same bases, and least where they varied most. Meanwhile, $\mathbf{R}_s$ permitted the information content of different sets to be compared.

Schneider et al. (1986) also introduced a different measure, $\mathbf{R}_f = -\log_2 f$, where f is the frequency of a given site in a genome. $\mathbf{R}_f$ is therefore, roughly, the information required to locate that site in the genome. For most sets, they found that $\mathbf{R}_s = \mathbf{R}_f$. Drift, they argued, probably kept $\mathbf{R}_s$ from becoming higher than $\mathbf{R}_f$, which is all that is required for recognition. Similarly, Schneider (1988) has argued that natural selection prevented $\mathbf{R}_s$ from becoming smaller than $\mathbf{R}_f$ because such sites would not be found, thus leading to a decreased fitness of the organism. These interpretations are speculative but, nevertheless, suggestive. In two cases, however, $\mathbf{R}_s$ was different than $\mathbf{R}_f$. For a promoter site in the genome of the phage T7, $\mathbf{R}_s$ is almost exactly twice $\mathbf{R}_f$. For the 12 *incD* repeated segment in the F plasmid, $\mathbf{R}_s$ is almost exactly thrice $\mathbf{R}_f$ (Herman and Schneider, 1992). Schneider and his collaborators have interpreted these intriguing results to mean that these sites are recognized by two or three different proteins respectively. This is a prediction from their analysis: information theory has actually yielded a testable result in this context.

It is certainly too early to judge whether Schneider's program will live up to its initial promise. Nevertheless, once again, for our purposes here, even if it is successful, it is irrelevant. Schneider's "information content" is a property of sets of sequences. It is highest at those regions where the sequences are most constant, and lowest where they are most variable. The results obtained from those analyses, at best, begin to provide a method for quantifying the intuition that functionally important sequences are more likely to be conserved through evolution. If Schneider's measure continues to capture that insight, it will no doubt be a valuable innovation, especially for evolutionary theory. Nevertheless, it is not a measure of the information content of a single sequence, the "information" that is invoked in our three precepts (and which, for example, is what gets "transferred" through the genetic code). As far as capturing the notion of "information" that is customary in molecular biology, we are back where we started.

REDUCTION AND THE PHYSICALIST ALTERNATIVE

Where does all this leave us? Well, it is clear that cybernetics and whatever concept of information that might come with it are irrelevant in molecular biology. It is also clear that information theory, no matter to what use it might eventually be successfully put in molecular biology, does not involve a concept of information similar to the customary one. This only leaves "information" as sequence, as is used in the context of coding. There we saw that the concepts of "information" and "coding" no longer play any significant explanatory role, primarily due to the complexity of eukaryotic genetics, which they do not help organize in any significant way. However, had that account succeeded, and especially if the comma-free code which, as we saw, did make significant predictions, had turned out to be correct, what would have been philosophically most significant about such a development is that it would have constituted a very significant success of the non-reductionist explanation in biology.

In order to formulate this point more precisely, let me clarify exactly what I mean by "reduction". First, I am concerned here only with epistemological aspects of reduction. When I am suggesting that a particular pattern of explanation or reasoning is not reductionist, I am making no ontological claim about whether some of the underlying processes are physical or novel or whatever.[19] Second, I am leaving open the possibility that there can be several explanations of the same phenomena, some of which might be reductionist and others not. Third, as before (see the 'Introduction'), I do not wish to get involved in disputes about reductionist explanation that are ultimately disputes about the nature of explanation.[20] Fourth, the account that I will give here is geared towards understanding reductionist explanation in molecular biology, though with some relatively trivial modifications, which are indicated in the footnotes, it can potentially be used in many other contexts. With these disclaimers, I will call an explanation – something that already satisfies whatever strictures that are put on explanations – reductionist if and only if it satisfies the following three criteria:

(i) *Physicalism*: The explanatory factors invoked in the explanation have "physical warrants", that is, they are either obtained (perhaps informally) from physical theory, or recognized as mechanisms from physical experiments alone;

(ii) *Hierarchical structure*: The complex entity whose behavior is being explained must be characterized as having a hierarchical organization in which the properties of entities at lower levels alone are used to explain its behavior,[21]

(iii) *Spatial instantiation*: The hierarchical structure of the entity must be realized in physical space, that is, entities at lower levels of the hierarchy must be spatial parts of entities at higher levels of organization.

This is, no doubt, a very stringent sense of "reduction". For instance, it has long been known that there is a fully legitimate type of "reduction" in which only criterion (i), but not the other two, is satisfied.[22] Examples include the reduction of Newtonian gravitation to general relativity, Newtonian kinematics to special relativity, geometrical optics to physical optics and, arguably, the putative reduction of psychology to neurophysiology.[23] It is also possible that criterion (ii), or criteria (ii) and (iii), are satisfied without satisfying criterion (i). All that would be required is the admission of explanatory factors that do not yet have physical warrants.

What is more interesting is that criterion (ii) can be satisfied without satisfying criterion (iii). A hierarchical organization can exist without being realized as such in physical space. Consider genetic explanations of various features of diploid organisms before the 1940s (when exactly how a gene and its effects are to be physically characterized began to be clarified). Some features were explained by one allele at one locus, others by both alleles at that locus and yet others by a set of alleles at several loci. There was a clear hierarchical structure but, in these explanations, this hierarchy was not *necessarily* a physical hierarchy. While it was known from linkage studies (that is, the study of the probabilities of different loci to be inherited together) that there was a linear order of loci on chromosomes, and it was strongly suspected that this corresponded to an underlying physical order of genes, that suspicion played no part in these explanations. Indeed, the complications of eukaryotic genetics that I have already mentioned, such as the existence of overlapping loci, show that the relation between the two orders is not one of trivial one-to-one correspondence. In fact, in 1939, hoping that ordinary mechanistic explanation would break down in genetics, Haldane (1939) predicted that the two orders would turn out to be inconsistent. Of course, that prediction has turned out to be largely false – the extent of overlap between genes that is known today does not yet significantly challenge the order established by linkage studies. By 1951, linkage studies had even led Lederberg et al. (1951) to suggest a branched genetic map which, as they explicitly emphasized, was formal and did not refer to a branched chromosome.[24] It is clear that in genetics at least, the possibility that criterion (ii) be satisfied while criterion (iii) was violated was historically a real possibility.

Moreover, during the 1930s and 1940s, the mode of action of genes – what chemicals were involved, and what mechanisms were necessary – was not known. Therefore, in these "genetic explanations", criterion (i) was also violated. Nevertheless. I think that is quite reasonable to say that features at the organismic (or, in this context, phenotypic) level were being "reduced" to genetics (that is, the genotypic level). There are thus at least two other senses of "reduction" than the one in which the satisfaction of all three of the criteria is required: one in which only criterion (i) is satisfied and one in which only criterion (ii) is satisfied. Moreover, it should be clear that (i) and (ii) can both be satisfied without satisfying (iii). Just imagine the situation when the action of genes receives a full physical characterization, but there is so much overlap and other complicating features discovered at the DNA sequence level that the genetic hierarchy does not, even at a coarse level, correspond to any physical hierarchy of nucleotide bases and groups of bases on chromosomes. Such a situation generates yet another sense of "reduction". Finally, it is possible that both criteria (ii) and (iii) are satisfied but not (i): an explanation involves a hierarchical structure, and this structure is realized in ordinary physical space, but the interactions between the parts that are invoked in the explanation do not have physical warrants.[25] I will argue below, that explanations involving information as sequence fit this pattern. In order to avoid misunderstanding, even at the expense of reiteration, let me emphasize here that such an explanation, which is not physicalist because criterion (i) is violated, nevertheless does not involve any ontological commitment to non-physical interactions. The concern, here, is purely epistemological, with what the factors are that offer good explanations in some context.

There are thus at least five senses of "reduction" in biological explanation that can be distinguished: (i) when all three criteria are satisfied; (ii) when only criterion (i) is satisfied; (iii) when only criterion (ii) is satisfied; (iv) when only criteria (ii) and (iii) are satisfied; and (v) when only criteria (i) and (ii) is satisfied.[26] The first sense, then, is the most stringent of all of these (and, elsewhere, I have called it "strong reduction"). Nevertheless, that is the sense in which many of the most successful explanations in molecular biology have turned out to be reductionist.

However, before turning to some characteristic examples, note that the satisfaction of any or all of these criteria does not guarantee that a reductionist explanation is of much value as an instance of reduction. Such explanations routinely involve the use of approximations and limiting procedures such as assuming that a phenotypic trait is governed by an infinite number of loci. An important biological example, though one that has not received the kind of philosophical scrutiny that it deserves, is Fisher's (1918) celebrated demonstration that, if there are that many loci involved, and each had an infinitely small effect, but all acted additively, a phenotypic trait (such as body weight or height) would vary continuously and follow a normal (Gaussian) distribution in a population. That famous "reduction" established the relation between biometry and ordinary Mendelian genetics. Nevertheless, such a reduction generates some unease because of the contrary-to-plausibility assumptions incorporated in the limiting procedure: an infinite number of loci, additive action, and infinitely small effects. This example suggests that there are at least some other conditions that can be used to gauge the significance of a reduction: (i) whether the approximations or limiting procedures involved are conceptually (e.g., mathematically) well-defined; (ii) whether they are known to be at least approximately empirically plausible; and (iii) whether they truly are "independent" of the properties of the features to be "reduced" in the sense they have not been expressly invoked to carry out a particular reduction. It is an open problem as to whether these conditions can be made more precise in a context-independent fashion. Suffice it here merely to note that even these conditions are usually satisfied in the most of the reductionist explanations in molecular biology.[27]

If what I have said about reduction in molecular biology is true, it affords examples of the most stringent and significant instances of reductionist explanations known. Let me attempt to demonstrate the truth of this claim by just noting two striking examples. The first is the molecular explanation of allostery: the "cooperative" behavior of proteins, such as hemoglobin, which are composed of several sub-units. The composite hemoglobin structure has greater affinity for oxygen or carbon dioxide, depending on the acidity (pH) of its fluid environment, than the total affinity of each of the units taken separately. The reason why this example is important is because such cooperative behavior is usually regarded as the special province of traditional anti-reductionists: a "whole" is said to be more than the "sum of the parts". The various contemporary models of allostery (which differ primarily in detail) each explains this phenomenon by showing how the structure of the sub-units changes in the composite whole, but changes entirely due to localized ordinary physical interactions (see Perutz (1990) for a review). All of these interactions, except the slightly stronger ones involving the heme group, are entirely explained on the basis of molecular shape underscoring how the "lock-and-key" fit generally reigns over all other physical interactions between biological macromolecules. The changed structure explains the non-linear increase in affinity. Thus, the explanation involves a hierarchical organization in ordinary space, and the usual interactions of physics and chemistry. All three criteria for reduction listed above are satisfied.

The second example is one we have already encountered: the operon. This example is important because the cell exhibits "goal-directed" behavior, also grist to the mill of anti-reductionist. As noted before (section 'Cybernetics and Gene Regulation'), the operon model explains this phenomenon: the various genetic units are hierarchically organized in chains along DNA segments, a physical interaction between the substrate, sometimes other molecules, and one of the genetic units initiates a sequence of interactions that leads to the production of the enzyme in question. When the substrate is no longer present, the initiating step of these reactions ceases, and enzyme production stops. Moreover, molecules that have similar chemical properties to the substrate, but which do not react with the enzyme, can also induce this process.[28] To the extent that we understand the details of this interaction at present, molecular shape is of paramount importance. Once again, this explanation is based on a spatially realized hierarchical structure and interactions with physical warrants. All three criteria for reduction that I gave above are trivially satisfied.

Though, in order to return soon to considerations about information, I will not pursue this point in detail here, it is also true that these explanations neither involve implausible or false approximations and limiting procedures, nor those that are designed only for biological applications. The characterization of the chemical interactions underlying allostery or operon-based gene regulation, such as van der Waals "repulsion" or hydrogen bonding, is no different from what physics or chemistry admits in a non-biological context. The van der Waals radii of the atoms, which determine molecular shape, are not obtained from analytic solutions of Schrödinger equations. Hydrophobic bonds are only approximately explained by the statistical mechanics of water molecules. Nevertheless, the same approximations are made in ordinary inorganic chemistry and molecular physics. Perhaps what is most striking, since it is quite rare in reductions, is that the biological context, in such explanations in molecular biology, does *not* require the introduction of specially designed approximations. Reductionist explanations of this kind abound in molecular biology. In fact, at least one of the fairly well-known precepts of the field, that the "structure determines the function" of biological macromolecules, where "function" only means behavior, can be taken to be a succinct statement of the reasoning behind reductionist explanation.

This long digression about reduction has not been without a purpose. I am now in a position to defend my claim that the assumptions about coding and information, especially the rather strong ones that led to the "comma-free" code, do not satisfy all of the criteria for the strong sense of reduction that I have been using above. Thus they run counter to one of the most striking patterns of explanation in molecular biology. Consider, first, the comma-free code model of Crick et al. (1957). Criteria (ii) and (iii) for reduction are trivially satisfied. There is a hierarchy of triplets simply because there is a contiguous set of triplets. This is obviously not an interesting hierarchy in any sense but, nevertheless, it is a hierarchy. Moreover, the units, that is, the triplets are physical objects – three bonded nucleotide bases – and their contiguity is in ordinary physical space. However, criterion (i) is violated. The restrictions on what sequences were permissible, which define the interactions between the units, were based on implicit assumptions about simplicity and efficiency of information storage and transfer, and these are clearly not explained on the basis of physical principles.[29] Explanations based on this model are not physicalist, in the epistemological sense in which that criterion was formulated above.[30] If anything, if the assumptions about information that were implicit in the comma-free code were made explicit and incorporated into a systematic account of information, there could have been a kind of reduction satisfying criteria (ii), (iii), and (i)' where (i)' is some requirement about information that replaces physicalism. However, this reduction would not be the type of physicalist reductionist explanation that, as was argued above, is characteristic of molecular biology.

What happened with the comma-free code is perhaps not very important since that code is little more than a historical curiosity any more. Nevertheless, what was just said about it can be said about our usual notion of "coding" and its associated notion of "information" as sequence. Remember that all that can be said about this code is linear, symbolic and local. Unlike the case of the comma-free code, there is no further underlying explanation of these properties from deeper assumptions about information. Consequently, even to the limited extent that it can be used for explanatory purposes, it satisfies criteria (ii) and (iii), does not satisfy (i), and there is little hope of even finding some criterion (i)' from some general account of information. Should such an account be found, then the conclusions of this paper will have to be modified. The philosophical point being made here would be lost but that is not much of a price to pay for the scientific advance that would be achieved instead.

The failure of explanations involving codes and information, and the success of the usual reductionist explanations in molecular biology, together suggest a rather striking possibility: abandon the notions of codes and information altogether and pursue a thoroughly physicalist reductionist account of the interactions between DNA, RNA and protein (and explain away the conceptual framework from the 1950s as an artifact of the coincidental co-linearity of DNA and protein). In principle, this does not present any real difficulty. We would treat the DNA-RNA-protein system as a network of chemical reactions, and write down a system of linear differential equations to describe the process.[31]

The main difficulty with such an account is that this model would have a rather large number of variables. These variables would have to keep track of the concentrations of each different type of DNA, RNA and protein that could potentially arise in the cell, not just the ones that emerge during normal gene expression, but also those that could arise through errors. Nevertheless, in most contexts, this level of complexity is not beyond what can be quite easily numerically analyzed. The coefficients describing the formation of a particular RNA type from a given DNA type and for forming a protein type from an RNA type would incorporate the specificity of the code. Moreover, these coefficients could quite naturally incorporate all that is known to happen to an RNA segment in a particular environment including editing, as well as incorporating non-standard translation. Moreover, such a model would be dynamic and actually allow the exact description of the concentration changes of the various components over time.

Why not abandon "codes" and "information" and pursue this possibility? There is no fully convincing reason not to do so. One possible reason why it has so far not been pursued is that only what might be called the "statics" of gene expression has usually been studied in any context. Molecular biology, especially, molecular genetics has not yet gone very far even in characterizing what genes are expressed in what part of an organism, let alone the finer details of the temporal regulation of genes within a cell except, of course, for *Escherichia coli*. In this static context, the coding account still serves some organizing function and, in spite of its increasing failures, it is being retained while few *explanations* are actually being pursued in this kind of work in genetics. Thus, the inability of the code to provide *significant* explanations is no major handicap. Another possible reason is that there is a useful sense in which "coding", "information", "translation" and related notions make the relations between DNA, RNA and protein transparent which no dynamical account involving reaction coefficients can. The intuition, here, is that the code is "natural" in a sense that these coefficients are not. Perhaps lurking behind this usefulness there is some insight to be grabbed, which the conventional information-based account of molecular biology has grasped even if, so far, very shakily.

Nevertheless, what does seem obvious is that a dynamical account, whether it is physicalist, informational, or whatever, will eventually be necessary if even approximate accounts of gene expression, interaction and cellular behavior, let alone the development of complex organisms, are to be pursued at the molecular level.[33] There is, moreover, one development whose importance even the most conventional molecular biologists have recognized, which implicitly relies on such a dynamical model and has no concern whatsoever for coding and information even though it deals with DNA. This is the polymerase chain reaction (PCR), which permits the rapid amplification of a given (double-stranded) DNA segment. The process begins by creating complementary single strands of DNA by heating the double-stranded segments to temperatures close to boiling. Using primers, which are a few nucleotides long to start the reaction, entire double stranded segments are created from each of the single stranded segments. This results in two copies of the original double-stranded segment. After n cycles of this process, there are (on average, that is, ignoring stochastic effects) 2^n such copies and they can be used for any purpose. It has been claimed with some justice that the PCR technique has "revolutionized molecular genetics" (Watson et al., 1992, p. 79). There is no concern for coding or information in this process. What is at stake is that a particular single-stranded DNA sequence catalyzes the formation of a specific other such sequence, namely, the one that is complementary to it. The dynamical equations to model this process are formally those for DNA replication in general. These can be solved to calculate the rate of DNA formation. This is reductionist reasoning, through and through, and this is what molecular biologists using PCR are relying on, even if only implicitly, when they predict the time required to produce a certain amount of DNA. "Codes" and "information" are irrelevant in this context.

CONCLUSIONS

It appears, therefore, that we are faced with a quandary. The conventional account of information as sequence is of little explanatory value in the novel contexts we are beginning to encounter. The alternative physicalist account of rate equations is, at present, of use in only very few contexts. I wish to suggest that we are faced with a situation that will require a rather radical departure from the past. For a coherent account of the relations between DNA, genes and biological behavior that has significant explanatory value, we will have to avoid both these accounts. Not adopting the physicalist account is neither controversial nor difficult since it has so far never been invoked in this context except, implicitly, in the development of PCR technology. However, because of the extent to which the idea of coding has been central to how molecular biology has been conceptualized, abandoning the idea of information as sequence will have important consequences. Let me list some of these and argue that they are, in fact, desirable:

(i) If biological "information" is not DNA sequence alone then, trivially, other features of an organism can also characterize information. And this is precisely what recent developments in molecular biology indicate. In particular, the developmental fate of a cell, might be largely a result of features such as methylation patterns of DNA that are not, as far as we know today, even ultimately determined only by DNA sequences. These "epigenetic" patterns can be inherited for several cell generations. Different cells in the same organism, presumably with identical DNA sequences, can be epigenetically different. Because of these differences, cell specialization and differentiation, the usual prelude to developmental changes, can take place. Epigenetic specifications are also critical in generating those differences in offspring (of sexually reproducing organisms) which depend on whether an allele is inherited from the mother or the father. Epigenetic specifications can sometimes be transmitted across organismic generations. It would be highly unintuitive not to regard these determinations as "transfers of information" if "information" is to have any plausible biological significance;

(ii) The central dogma of molecular biology, that information only flows from nucleic acid to protein, and never in the reverse direction (precept (iii) of the 'Introduction'), is false. However, a less grandiose claim, that protein sequences do not directly determine nucleic acid sequences in the way in which the latter determine the former, remains true as far as we currently know. No doubt this humbler claim does not have the majestic power of the central dogma but the question that I wish to raise is whether this retreat really undermines some putative insight that was enshrined in that dogma. The usual defence of the general importance of the central dogma, that is, its importance for biology in general (and not just molecular biology) has been that it is a statement, at the molecular level, of the non-inheritance of acquired characteristics (precept (iv) of the 'Introduction'). However, this is nothing but egregious misinterpretation. Acquired characteristics are sometimes inherited, though usually not.[33] What ensures that even those acquired characteristics that involve changes in the DNA component of genomes of cells are not inherited in higher animals is the segregation of the germ-line from the soma. But plants have no germ-line, and the extent of its segregation in animals greatly varies across the phyla. Nevertheless, whatever the relationship between nucleic acid and protein may be, it is the same across the organic world: *ipso facto* the central dogma, even if it were true, could not be either an explanation or an alternative synonymous statement of the alleged non-inheritance of acquired characteristics. I do not mean to deny that there is something peculiar, and extremely interesting, about how DNA resists easy change across the phyla. But this is something to be studied and understood, not something to be "explained" away on the basis of some alleged law about some incoherent notion of "information";

(iii) Many of the more influential contemporary discussions of the origin of life have concentrated on the origin of information, where information is construed simply to be nucleic acid sequences (e.g., Eigen, 1992). Implicit in these discussions is the assumption that these sequences ultimately encode all that is necessary for the genesis of living forms and, therefore, a solution to the problem of the initial generation of these sequences is all that is required to solve the problem of the origin of life. The move away from sequences would put these efforts in proper perspective: to explain the possible origin of persistent segments of DNA is a far cry from explaining the origin of living cells. However, I do not wish to harp on this point since, quite justifiably, most molecular biologists think that such discussions of the origin of life are little other than idle speculation;

(iv) The position that I am advocating certainly suggests that we remove the amount of emphasis on the DNA sequence that we see today. In turn, this suggests that the sorts of arguments that were mustered to initiate the Human Genome Project, a crash program to sequence DNA blindly, that is, without first determining the functional (or behavioral) roles of the segments to be sequenced, does not have much scientific rationale. This is not a new point. It has previously been made, on the basis of other considerations, by many critics of the Human Genome Project as it was initially conceived (Sarkar and Tauber, 1991; Davis, 1992; Tauber and Sarkar, 1992; Lederberg, 1993). If my arguments here are sound, then the Human Genome Project would best be limited to its first stage, that is, the mapping of all known genetic loci to specific positions on chromosomes, and then proceed to sequencing segments as and when they come to be known to have some functional interest.

Each of these consequences is desirable. Nevertheless, before the notion of information as sequence and the picture of molecular biology that comes with it are abandoned, some alternative is necessary. Let me end by noting two possibilities. the first would be a return to the old concept of specificity and develop it further, beyond the stereochemical theory whose limitations are, in any case, gradually becoming apparent. For instance, in the immunological context, it has already become clear that not all residues that are in contact between antibodies and antigens contribute equally to the free energy of the interaction (see Sarkar, 1996). Of about twenty residues that are in contact at the interaction site, only four or less dominate the interaction. This is not a total failure of the stereochemical model. Rather, it is a modification of that part of it which asserts that only molecular shape matters.

Perhaps a similar account can be given that will explain the specific interaction between complementary base paris during DNA replication and transcription, and all this will be incorporated in a general systematic account of specificity. Coding will be retained only as a short-hand description of the usual triplet specification of amino acid residues, but it will not be assumed to have any explanatory value. Nevertheless, the special role played by DNA triplets would be incorporated in the account of specificity that would be developed. Finally, the differential equations that I mentioned above (in the section 'Reduction and the Physicalist Alternative') would be incorporated to provide a dynamical account of the entire DNA-RNA-protein and other interactions in the cell. However, the new account of specificity would remove the taint of artificiality that the reaction coefficients had. Perhaps the most interesting feature of such an account would be that it would be purely physicalist and, consequently, reductionist in the strongest sense, in sharp contrast to the informational account that is currently prevalent. So far little, if any, effort has been expended towards the elaboration of such a picture, probably because there is, as yet, no plausible candidate for such a generalized theory of specificity.

The second possibility is the elaboration of a new informational account in which "information" is construed to be broader than just DNA sequence. For instance, Shapiro (1991, 1992) has argued that the entire genome should be viewed as "a dynamic information storage system that is subject to rapid modification" (1992, p. 99). From Shapiro's point of view, there exists not only the usual genetic code relating amino acid residues to DNA base triplets, but an additional coding relation for sequences that serve regulatory and other roles. The latter code is clearly not triplet, and is not even symbolic in any conventional sense since it would be "interpreted" as a process rather than some other entity. The hierarchical organization of the entire genome determines the behavioral repertoire of a cell. "Information" is no longer determined by local sequence alone, nor is it linear, since these hierarchies can show considerable complexity. Though Shapiro is not explicit on this point, "information" in such a picture need not be constrained to DNA patterns – patterns of methylation and other heritable features of the genome could well carry information.

There is little doubt that such a picture is intriguing because it automatically retains those insights that the conventional view has, such as the existence of a peculiar triplet relationship between DNA and amino acid residues, while extending this view to incorporate recent discoveries in molecular biology. However, as Shapiro (1991) has acknowledged, no testable claim has yet emerged from this picture. Whether, eventually, any will, I cannot say. I am willing to acknowledge that "testability", as it is usually understood – that is, demanding new predictions – might well be too strong a criterion for most biological contexts. However, for a theory to be worth admission into serious discourse, it should at least allow the systematic and clear organization of known facts. Shapiro's approach has not yet even been developed to that extent. However, should an admissible theory emerge, it will, as in the case of the comma-free code, be a theory based on assumptions about information storage and utilization. It would be manifestly non-physicalist. It would also give a new lease of life to information-oriented thinking in biology and, perhaps, even finally begin to explain what "biological information" actually happens to be.

McGill University,
Montréal, Québec, Canada

NOTES

[1] I have benefited from extensive discussions with Angela Creager, Lily Kay, Evelyn Fox Keller, Joshua Lederberg, Richard Lewontin and William C. Wimsatt. Sections of this analysis were presented to the conference on 'Methods in Philosophy and the Sciences', New York (January 16, 1993), the Boston colloquium for the Philosophy of Science (April 13, 1993) and the Department of Philosophy, McGill University (April 23, 1993). Remarks by members of the audiences, especially David Thaler and Abner Shimony, the official commentators on the first two occasions, were particularly helpful. I am grateful to all of these individuals for their comments and criticism; none of them, however, should be presumed to agree with what I say. The work reported here was partially supported by a Senior Research Fellowship from the Sidney M. Edeltstein Centre for the History and Philosophy of Science, Technology and Medicine, Hebrew University, Jerusalem, a Resident Fellowship from the Dibner Institute for the History of Science and Technology, MIT, and NIH Grant No. HG 00912-01/2. I would like to thank all these organizations for their support.

[2] The distinctions between the three senses of "information" that will be used throughout this paper are apparently being *explicitly* made for the first time. As should be evident from what follows, these distinctions significantly help the philosophical analysis being attempted here. I am not suggesting that these three construals of "information" were always historically distinguished by the molecular biologists, especially in the confusing conceptual landscape of the 1950s when "information" began to infiltrate molecular biology. However, even in that rhetoric, explicit conflation of these senses is rare, and I know of no example where the different senses were

conflated in any *technical* argument or inference. For an analysis of the complexities of this process which begins from a rather different methodological perspective than the one used here but, generally, reaches the same sort of conclusion, see Keller (1995). For one that reaches strikingly different conclusions by denying these distinctions, see Kay (1994).

[3] Note that this is a purely epistemological construal of "physicalism" and does not invoke any of the usual ontological connotations of that doctrine that have been advocated by the later logical empiricists (e.g., Neurath or Carnap after 1931) and their followers (such as Quine). "Physicalism" will only be used in this sense in this paper.

[4] Part of such a sketch, at least, has been attempted by Keller (1995). See also, Judson (1979) for a journalistic account.

[5] Even before Crick's (1958) explicit identification of information as specificity, the same idea was implicitly being used quite systematically. In 1955, for instance, during a symposium on enzymes in Detroit, Mazia (1956, p. 262) argued that the role of RNA was to carry "information" from the nuclear DNA to the cytoplasm for the synthesis of proteins. At the same conference, Spiegelman (1956, p. 77) argued that the required "informational complexity" made RNA and DNA the only two plausible candidates for being templates for protein formation and Lederberg (1956, p. 167) noted that "information" was what "specificity" was "called nowadays". See also, Gamow et al. (1955) which will be discussed in the text.

[6] This section is based largely on the more detailed account in Sarkar (1989).

[7] There were 8 diamonds in which the top and bottom bases were the same, A, T, C or G-for each of the 4 possibilities at top and bottom, either a C-G or an A-T pair could be in the middle. Different bases at top and bottom can be chosen in $\binom{4}{2}$, that is, 6 ways. These give rise to 12 diamonds, once again because both C-G and A-T can occur in the middle. Together, there are 20.

[8] If it is interpreted as a triplet code, this means the code had an overlap of two. The terminology used here is that of the 1950s: "overlapping" means an overlap of two; "partially overlapping" means an overlap of one, where only the last position of one triplet is shared with the next.

[9] Any meaningful attempt at prediction in the reverse direction, that is, from polypeptide to DNA sequence is, of course, hopeless because of the degeneracy of the code, even independent of the other problems that will emerge from the discussion in the text.

[10] For example, the virus $Q\beta$, which preys on *Escherichia coli* has a coat protein that is usually produced by having UGA read as a termination codon. However, 2% of the time, it is ignored, resulting in a longer coat protein whose presence turns out to be necessary for the normal behavior of the phage (Fox, 1987).

[11] Of course, if one had available all the DNA sequences in the world today, and all that have ever existed in organisms, and all the polypeptides these ever coded for, then it is possible to assert that, subject to an important qualification, one would be able to identify all intron-exon boundaries on basis of sequence information alone. The qualification is that novel sequences could arise through mutation, recombination, etc. So, all this information would still be of no avail for the purpose at hand. In any case, if prediction is the desideratum as it is in the discussion in the text, even without worrying about the insurmountable obstacles presented by the possibilities of mutation and recombination, the sparse information available at present about the exact relationships between known DNA and protein sequences makes attempts at prediction perilous in exactly those new contexts where they would be valuable.

[12] For instance, in the intestines of mammals, the mRNA apolipoprotein undergoes a deamination of a C, converting it to a U in such a way that a stop codon is created. This behavior is tissue-specific. The same kind of deamination, and the reverse U $\to$ C amination process, have also been observed in several plant mitochondrial mRNA transcripts.

[13] In the most extreme case known to date, in the human parasite *Trypanasoma brucei*, in the mRNA transcript leading to the formation of NADH dehydrogenase subunit 7 (a protein), as

many as 551 U's are inserted throughout the transcript while 88 are deleted (Koslowski et al. 1990). In such a case it is hard to see why the DNA segment encoding such a transcript should be called the "gene for NADH dehydrogenase subunit 7" – by looking at the DNA sequence it would be impossible to predict beforehand that this was the protein that would eventually be produced

[14] The University of Chicago Press plans to publish this material, along with Haldane's unpublished lectures on Darwinism.

[15] Consequently, it is hard to judge whether Chance's position, as well as his failure to cite Wiener as one of the originators of cybernetics, was quite as idiosyncratic as it appears to be.

[16] The importance of allostery in the development of Monod's thinking has been emphasized by Angela Creager and Jean-Paul Gaudillière (personal communication, and in several forthcoming publications; see also Creager, 1994).

[17] See Yockey (1992) for a recent, though unilluminating, review.

[18] Though they do not refer either to Kimura or to Williams, Waddington and Lewontin (1968) presented an alternative argument for the irrelevance of this notion of information as a measure of evolutionary progress. In fact, they argued that selection would act to prevent the continued accumulation of this kind of information. This argument, too, has received no further attention, underscoring the general perception that formal information theory is irrelevant to discussions of evolution.

[19] I am not, of course, denying either that epistemological success often leads to ontological commitments or that the latter generate epistemological programs. However, since, with the sole exception of some religious fanatics, there appears to be consensus that there is nothing non-physical going on in biological systems, ontological aspects of reduction are of little more than formal interest in this context.

[20] One consequence of this move is that, though the account given below generally makes no reference to "theories", it does not explicitly preclude the construal of reduction as a relation between theories. I have three reasons for no longer being willing to get involved in the dispute between "theory reduction" and "explanatory reduction" (contrary to Sarkar, 1992): (i) it has become less and less clear that there is any single notion of "theory" applicable to all of science, which philosophers should try to explicate; (ii) it has also become clear that there are few, if any, universal theories. (I am even skeptical that the usual examples drawn from physics – quantum mechanics or general relativity – truly are as universal as philosophers generally believe. This will be taken up in a future paper.) Moreover, there is no harm to accept as "theories" even those claims that are acknowledged to have small domains of applicability; (iii) finally, and this is the most important of these three reasons, the philosophical disputes about the role of theories in explanation have largely degenerated into disputes about *formal* issues and have led to little attention being paid to the *substantive* – and, I think, much more philosophically and scientifically interesting – problems with reduction. The criteria suggested below are substantive, not formal. This point is elaborated in Sarkar (1996).

[21] Note that, if both criteria (i) and (ii) are satisfied, the properties referred to in (ii) must be the physical warrants invoked in (i). (We are looking at the same putative explanation). Obviously, certain kinds of explanations satisfying (i) would automatically preclude the satisfaction of (ii), for instance, if the explanation manifestly invoked properties that can only be defined by reference to higher levels of organization.

[22] That there is an important distinction between reductions involving only criterion (i), and those also involving the other criteria, seems to have been first suggested by Nickles (1973) though, in his account, reduction occurs in the opposite direction than the one suggested here which follows the treatments of Wimsatt (1976) and Sarkar (1989, 1992). The relations between most of the formal accounts of reduction offered so far are reviewed in Sarkar (1989, 1992). Note that, in those previous accounts criteria (ii) and (iii) were not distinguished.

[23] Note that, in the context of reductionist explanation outside the natural sciences, the major

modification of this account that would be required is the replacement of criterion (i) by an assumption that the warrants for the explanatory factors that are invoked are provided by the specified theory that is supposed to play the explanatory role in that context. If criterion (ii) is also satisfied, this would mean that the warrants come from the lower level. However, criterion (i) can be satisfied without the satisfaction of criterion (ii) — in fact, the explanation might involve higher levels of organization (as noted before). In biology this is endemic in selectionist explanations. In physics explanatorily privileged scaling laws can operate at the same level as what is being explained.

[24] I am particularly indebted to David Thaler for drawing my attention to this episode. See Thaler (1996) for a more extended discussion.

[25] Beyond the biological context, this last sense of reduction can be particularly important. For instance, if social behavior is to be explained on the basis of individual behavior, criteria (ii) and (iii) would be satisfied but not criterion (i) since, presumably, physical warrants would not have to be invoked to justify whatever rules are said to govern the interactions between individuals. However, in these cases, presumably, the requirement of physicalism would itself be replaced by some other kind of "fundamentalist" assumption. For instance, it might be required that the rules in question should refer only to individuals and their properties.

[26] Note that since criterion (iii) is formulated in terms of a hierarchical organization, it is not possible to satisfy (iii) without satisfying (ii). This precludes the possibilities of having sense of "reduction" involving the satisfaction only of criterion (iii) or only criteria (i) and (iii). There are no other logical possibilities than the ones that have so far been considered.

[27] Somewhat ironically, the satisfaction of (i) is probably much more problematic in physics than in biology (see Primas (1991) and the references therein). What is at stake here, is the status of the approximations routinely required in physics to establish connections between levels of organization. Leggett (1987) has called these "physical approximations". Strangely, besides Shimony (1987), no philosopher of physics seems to have acknowledged their significance.

[28] It is perhaps ironic that both the processes discussed here — allostery and operon-based gene regulation — are the ones that led Monod (1971) to suggest the concept of gratuity and a role for cybernetics in molecular biology!

[29] Whether or not these requirements can truly be given a selectionist interpretation is irrelevant: as noted before, selectionist arguments refer to a higher (rather than a lower) level of organization. (That they can be given a selectionist interpretation is argued in Sarkar (1989), though — I now think — not very successfully.)

[30] Once again, let me emphasize that this does not mean that anything other than ordinary physical and chemical interactions are occurring in DNA in the comma-free code model. All that is at stake here is what enters into an explanation, not the ontological concern for what is "ultimately causing" some phenomena, even if that notion can be made sensible.

[31] The techniques necessary for this sort of development are not particularly profound or difficult. See Sarkar (1988) for a trivial example, and Küppers (1983) for a more systematic treatment.

[32] It is, of course, open to dispute whether such a pursuit will yield anything except a morass of inchoate detail. That, in turn, will show the extent of the success, and the limitations, of the type of reductionist explanations that are characteristic of molecular biology.

[33] See Landman (1992), as well as Jablonka et al. (1992) for recent reviews.

REFERENCES

Atkins, J.F., Weiss, R.B., Thompson, S. and Gesteland, R.F. (1991). 'Towards a Genetic Dissection of the Basis of Triplet Decoding, and its Natural Subversion: Programmed Reading Frame Shifts and Hops'. *Annual Review of Genetics* **25**: 201–28.

Avery, O.T., Macleod, C.M. and McCarthy, M. (1994). 'Studies of the Chemical Nature of the

Substance Inducing Transformation of Pneumococcal Types: Induction of Transformation by a Deoxyribonucleic Acid Fraction Isolated from Pneumococcus III'. *Journal of Experimental Medicine* **79**: 137–57.
Branson, H.R. (1953). 'A Definition of Information from the Thermodynamics of Irreversible Processes'. In: Quastler, H. (ed.), *Essays on the Use of Information Theory in Biology*. Urbana: University of Illinois Press, pp. 25–40.
Brenner, S. (1957). 'On the Impossibility of All Overlapping Triplet Codes in Information Transfer from Nucleic Acids to Proteins'. *Proceedings of the National Academy of Sciences* **43**: 687–93.
Britten, R.J. and Davidson, E.H. (1969). 'Gene Regulation for Higher Cells: A Theory'. *Science* **165**: 349–57.
Britten, R.J. and Kohne, D.E. (1968). 'Repetitive Sequences in DNA'. *Science* **161**: 529–40.
Cattaneo, R. (1991). 'Different Types of Messenger RNA Editing'. *Annual Review of Genetics* **25**: 71–88.
Chance, B. (1961). "Control Characteristics of Enzyme Systems'. *Cold Spring Harbor Symposia on Quantitative Biology* **26**: 289–99.
Chargaff, E. (1963). *Essays on Nucleic Acids*. Amsterdam:, Elsevier.
Creager, A. (1994). 'Reconciling Experimental Systems and Institutions: The Invention of "Allostery" in Paris and Berkeley, 1959–1968'. Paper presented at the Fourth Mellon Workshop, Program in Science, Technology and Society, Massachusetts Institute of Technology, April 30, 1994.
Crick, F.H.C. (1958). 'On Protein Synthesis'. *Symposium of the Society for Experimental Biology* **12**: 138–63.
Crick, F.H.C., Barnett, L., Brenner, S., and Watts-Tobin, R.J. (1961). 'General Nature of the Genetic Code for Proteins'. *Nature* **192**: 1227–32.
Crick, F.H.C., Griffith, J.S. and Orgel, L.E. (1957). 'Codes Without Commas'. *Proceedings of the National Academy of Sciences (USA)* **43**: 416–21.
Davis, B.D. (1992). 'Sequencing the Human Genome: A Faded Goal'. *Bulletin of the New York Academy of Medicine* **68**: 115–25.
Eigen, M. (1992). *Steps Towards Life*. Oxford: Oxford University Press.
Ephrussi, B., Leopold, U., Watson, J.D. and Weigle, J.J. (1953). 'Terminology in Bacterial Genetics'. *Nature* **171**: 701.
Fisher, R.A. (1918). 'The Correlation Between Relatives on the Supposition of Mendelian Inheritance'. *Transactions of the Royal Society of Edinburgh* **52**: 399–433.
Fox, T.D. (1987). 'Natural Variation in the Genetic Code'. *Annual Review of Genetics* **21**: 67–91.
Gamow, G. (1954a). 'Possible Relation Between Deoxyribonucleic Acid and Protein Structures'. *Nature* **173**: 316.
Gamow, G. (1954b). 'Possible Mathematical Relation Between Deoxyribonucliec Acid and Proteins'. *Biologiske Meddelelser udviket af Det Kongelige Danske Videnskabernes Selskab* **22**(3): 1–11.
Gamow, G. and Metropolis, N. (1954). 'Numerology of Polypeptide Chains'. *Science* **120**: 779–80.
Gamow, G., Rich, A., and Ycas, M. (1955). 'The Problem of Information Transfer from the Nucleic Acids to Proteins'. *Advances in Biological and Medical Physics* **4**: 23–68.
Gamow, G. and Ycas, M. (1955). "Statistical Correlation of Protein and Ribonucleic Acid Composition'. *Proceedings of the National Academy of Sciences* **41**: 1011–19.
Gatlin, L. (1972). *Information Theory and the Living System*. New York: Columbia University Press.
Golomb, S.W. (1962). 'Efficient Coding for the Desoxyribonucleic Acid Channel'. *Proceedings of the Symposium for Applied Mathematics* **14**: 87–100.
Golomb, S.W. Welch, L.R. and Delbrück, M. (1958). 'Construction and Properties of Comma-

Free Codes'. *Biologiske Meddelelser udviket af Det Kongelige Danske Videnskabernes Selskab* **23**(9): 1–34.
Haldane, J.B.S. (1939). *The Marxist Philosophy and the Sciences*. New York: Random House.
Herman, N.D. and Schneider, T.D. (1992). 'High Information Conservation Implies That at Least Three Proteins Bind Independently to F Plasmid *incD* Repeats'. *Journal of Bacteriology* **174**: 3558–60.
Jablonka, E., Lachmann, L. and Lamb, M.J. (1992). 'Evidence, Mechanisms and Models for the Inheritance of Acquired Characteristics'. *Journal of Theoretical Biology* **158**: 245–68.
Jacob, F. and Monod, J. (1959). 'Gènes de structure et gènes de régulation dans la biosynthèse des protéines'. *Comptes Rendus des Séances de l'Academie des Sciences* **249**: 1282–84.
Jacob, F., Perrin, D., Sanchez, E. and Monod, J. [1960] (1965). 'The Operon: A Group of Genes Whose Expression is Coordinated by an Operator'. In Adelberg, E.A. (ed.), *Papers on Bacterial Genetics*. Boston: Little, Brown and Co., pp. 198–200.
Judson, H.F. (1979). *The Eighth Day of Creation*. New York: Simon and Schuster.
Kalmus, H. (1950). 'A Cybernetical Aspect of Genetics'. *Journal of Heredity* **41**: 19–22.
Kay, L.E. (1994). 'Who Wrote the Book of Life? Information and the Transformation of Molecular Biology, 1945–1955'. In: Hagner, M. and Rheinberger, H.-J. (eds.), *Experimentalsysteme in den Biologischen-Medizinischen Wissenschaften: Objekt, Differenzen, Konjunkturen*. Berlin: Akademie Verlag.
Keller, E.F. (1995). *Refiguring Life: Metaphors of Twentieth Century Biology*. New York: Columbia University Press.
Kimura, M. (1961). 'Natural Selection as the Process of Accumulating Genetic Information in Adaptive Evolution'. *Genetical Research* **2**: 127–40.
Kimura, M. (1968). 'Evolutionary Rate at the Molecular Level'. *Nature* **217**: 624–6.
Kimura, M. (1983). *The Neutral Theory of Molecular Evolution*. Cambridge: Cambridge University Press.
Kleene, S.C. (1956). 'Representation of Events in Nerve Nets and Finite Automata'. In: Shannon, C.E. and McCarthy, J. (eds.), *Automata Studies*. Princeton: Princeton University Press, pp. 3–41.
Koslowski, D.J., Bhat, G.J., Perollaz, A.L., Feagin, J.E. and Stuart, K. (1990). 'The MURF3 Gene of *T. brucei* Contains Multiple Domains of Extensive Editing and is Homologous to a Subunit of NADH Dehydrogenase'. *Cell* **62**: 901–11.
Küppers, B.-O. (1983). *The Molecular Theory of Evolution*. Berlin: Springer-Verlag.
Landman, O. E. (1991). 'The Inheritance of Acquired Characteristics'. *Annual Review of Genetics* **25**: 1–20.
Landsteiner, K. (1936). *The Specificity of Serological Reactions*. Springfield, IL: C.C. Thomas.
Lanni, F. (1964). 'The Biological Coding Problem'. *Advances in Genetics* **12**: 1–141.
Lederberg, J. (1951). 'Genetic Studies with Bacteria'. In Dunn, L.C. (ed.), *Genetics in the 20th Century*. New York: Macmillan, pp. 263–89.
Lederberg, J. (1956). 'Comments on the Gene-Enzyme Relationship'. in Gaebler, O.H. (ed.), *Enzymes: Units of Biological Structure and Function*. New York: Academic Press, pp. 161–9.
Lederberg, J. (1993). 'What the Double Helix (1953) Has Meant for Basic Biomedical Science'. *Journal of the American Medical Association* **269**: 1981–5.
Lederberg, J., Lederberg, E.M., Zinder, N.D. and Lively, E.R. (1951). 'Recombination Analysis of Bacterial Heredity'. *Cold Spring Harbor Symposia on Quantitative Biology* **16**: 413–43.
Lederberg, J. and Tatum, E.L. (1946a). 'Gene Expression in *Escherichia coli*'. *Nature* **158**: 588.
Lederberg, J. and Tatum, E.L. (1946b). 'Novel Genotypes in Mixed Cultures of Biochemical Mutants of Bacteria'. *Cold Spring Harbor Symposia on Quantitative Biology* **11**: 113–4.
Leggett, A.J. (1987). *The Problems of Physics*. Oxford: Oxford University Press.
Lewin, B. (1974). *Gene Expression-2: Eucaryotic Chromosomes*. New York: John Wiley.

Linschitz, H. (1953). 'The Information Content of a Bacterial Cell'. In Quastler, H. (ed.), *Essays on the Use of Information Theory in Biology*. Urbana, University of Illinois Press, pp. 251–62.
Luria, S.E. and Delbrück, M. (1943). 'Mutations of Bacteria from Virus Sensitivity to Virus Resistance'. *Genetics* **28**: 491–511.
Matthaei, J.H. and Nirenberg, M.W. (1961a). 'Characteristics and Stabilization of DNA Sensitive Protein Synthesis in *E. coli* Extracts'. *Proceedings of the National Academy of Sciences* **47**: 1580–8.
Matthaei, J.H. and Nirenberg, M.W. (1961b). 'The Dependence of Cell-Free Protein Synthesis in *E. coli* upon Naturally Occurring or Synthetic Polyribonucleotides'. *Proceedings of the National Academy of Sciences* **47**: 1588–94.
Mazia, D. (1956). 'Nuclear Products and Nuclear Reproduction'. In: Gaebler, O.H. (ed.), *Enzymes: Units of Biological Structure and Function*. New York: Academic Press, pp. 261–78.
Monod, J. (1956). 'Remarks on the Mechanism of Enzyme Induction'. In: Gaebler, O.H. (ed.), *Enzymes: Units of Biological Structure and Function*. New York: Academic Press, pp. 7–28.
Monod, J. (1971). *Chance and Necessity*. New York: Knopf.
Monod, J., Cohen-Bazire, G. and Cohn, M. (1951). 'Sur la biosynthese de la β-galactosidase (lactase) chez *Esherichia coli* la specificite de l'induction'. *Biochimica et Biophysica Acta* **7**: 585–99.
Monod, J., Pappenheimer, A., and Cohen-Bazire, G. (1952). 'La cinétique de la biosynthèse de la β-galactosidase chez *E. coli* considérée comme fonction de la croissance'. *Biochimica et Biophysica Acta* **9**: 648–60.
Neveln, B. (1990). 'Comma-Free and Synchronizable Codes'. *Journal of Theoretical Biology* **144**: 209–12.
Nickles, T. (1973). 'Two Concepts of Inter-Theoretic Reduction'. *Journal of Philosophy* **70**: 181–201.
Olby, R.C. (1971). 'Schrödinger's Problem: What is Life?'. *Journal of the History of Biology* **4**: 119–48.
Pardee, A.B., Jacob, F. and Monod, J. (1959). 'The Genetic Control and Cytoplasmic Expression of 'Inducibility' in the Synthesis of β-galactosidase by *E. coli*'. *Journal of Molecular Biology* **1**: 165–78.
Pauling, L. (1940). 'A Theory of the Structure and Process of Formation of Antibodies'. *Journal of the American Chemical Society* **62**: 2643–57.
Perutz, M. (1990). *Mechanisms of Cooperativity and Allosteric Regulation in Proteins*. Cambridge: Cambridge University Press.
Pierce, J. R. (1962). *Symbols, Signals and Noise*. New York: Harper and Brothers.
Primas, H. (1991). 'Reductionism: Palaver without Precedent'. In: Agazzi, E. (ed.), *The Problem of Reductionism in Science*. Dordrecht: Kluwer Academic Publishers, pp. 161–172.
Quastler, H. (1953a). "The Measure of Specificity'. In: Quastler, H. (ed.), *Essays on the Use of Information Theory in Biology*. Urbana, University of Illinois Press, pp. 41–71.
Quastler, H. (1953b). 'The Specificity of Elementary Biological Functions'. In: Quastler, H. (ed.), *Essays on the Use of Information Theory in Biology*. Urbana, University of Illinois Press, pp. 170–88.
Quastler, H. 1958. 'The Status of Information Theory in Biology: A Round-Table Discussion'. In: Yockey, H. P. (ed.), *Symposium on Information Theory in Biology*. New York: Pergamon Press, pp. 399–402.
Rickenberg, H.V., Cohen, G.N., Buttin, G. and Monod, J. (1956). 'La galactoside-permease d'*Escherichia coli*'. *Annales de l'Institut Pasteur* **91**: 829–57.
Sanger, F. and Thompson, O.P. (1953a). 'The Amino Acid Sequence in the Glycine Chain of Insulin. 1. The Identification of Lower Peptides from Partial Hydrolysates'. *Biochemical Journal* **53**: 353–66.

Sanger, F. and Thompson, O.P. (1953b). 'The Amino Acid Sequence in the Glycine Chain of Insulin. 2. The Investigation of Peptides from Enzymic Hydrolysates'. *Biochemical Journal* **53**: 366–74.

Sanger, F. and Tuppy, H. (1951a). 'The Amino-Acid Sequence in the Phenylalanyl Chain of Insulin. 1. The Identification of Lower Peptides from Partial Hydrolysates'. *Biochemical Journal* **49**: 473–81.

Sanger, F. and Tuppy, H. (1951b). 'The Amino-Acid Sequence in the Phenylalanyl Chain of Insulin. 2. The Investigation of Peptides from Enzymic Hydrolysates'. *Biochemical Journal* **49**: 481–90.

Sarkar, S. (1988). 'Natural Selection, Hypercycles and the Origin of Life'. In: Fine, A. and Leplin, J. (eds), *PSA 1988: Proceedings of the 1988 Biennial Meeting of the Philosophy of Science Association*, vol. 2. East Lansing: Philosophy of Science Association, pp. 197–206.

Sarkar, S. (1989). 'Reductionism and Molecular Biology: A Reappraisal'. Ph.D. Dissertation. Department of Philosophy, University of Chicago.

Sarkar, S. (1992). 'Models of Reduction and Categories of Reductionism'. *Synthese* **91**: 167–94.

Sarkar, S. (1996). *Reductionism and Genetics: A Primer*. Draft manuscript.

Sarkar, S. and Tauber, A.I. (1991). 'Fallacious Claims for HGP'. *Nature* **353**: 691.

Schaffner, K. (1974a). 'Logic of Discovery and Justification in Regulatory Genetics'. *Studies in the History and Philosophy of Science* **4**: 349–385.

Schaffner, K. (1974b). 'The Unity of Science and Theory Construction in Molecular Biology'. *Boston Studies in the Philosophy of Science* **58**: 497–533.

Schneider, T.D. (1988). 'Information and Entropy of Patterns in Genetic Switches'. In: Erickson, G.J. and Smith, C.R. (eds.), *Maximum-Entropy and Bayesian Methods in Science and Engineering*, vol. 2. Dordrecht: Kluwer Academic Publishers, pp. 147–54.

Schneider, T.D., Stormo, G.D., Gold, L. and Ehrenfeucht, A. (1986). 'Information Content of Binding Sites of Nucleotide Sequences'. *Journal of Molecular Biology* **188**: 415–31.

Schrödinger, E. (1944). *What is Life?* Cambridge: Cambridge University Press.

Shannon, C. E. (1948). 'A Mathematical Theory of Communication'. *Bell System Technical Journal* **27**: 379–423, 623–56.

Shapiro, J.A. (1991). 'Genomes as Smart Systems'. *Genetica* **84**: 3–4.

Shapiro, J.A. (1992). 'Natural Genetic Engineering in Evolution'. *Genetica* **86**: 99–111.

Shimony, A. (1987). 'The Methodology of Synthesis: Parts and Wholes in Low-Energy Physics'. In: Kargon, R. and Achinstein, P. (eds.), *Kelvin's Baltimore Lectures and Modern Theoretical Physics*. Cambridge: MIT Press, pp. 399–423.

Shimony, A. (1995). 'Cybernetics and Social Entities'. *Boston Studies in the Philosophy of Science* **164**: 181–96.

Smith, C.W., Patton, J.G. and Nadal-Ginard, B. (1989). 'Alternative Splicing in the Control of Gene Expression'. *Annual Review of Genetics* **23**: 527–77.

Spiegelman, S. (1956). 'On the Nature of the Enzyme-Forming System'. In: Gaebler, O.H. (ed.), *Enzymes: Units of Biological Structure and Function*. New York: Academic Press, pp. 67–89.

Stephens, R.M. and Schneider, T.D. (1992). 'Features of Spliceosome Evolution and Function Inferred from an Analysis of the Information at Human Splice Sites'. *Journal of Molecular Biology* **228**: 1124–36.

Tauber, A.I. and Sarkar, S. (1992). 'The Human Genome Project: Has Blind Reductionism Gone Too Far?'. *Perspectives in Biology and Medicine* **35**: 220–35.

Thaler, D.S. (1996). 'Paradox as Path: Pattern as Map-Classical Genetics as a Source of Non-Reductionism in Molecular Biology'. In: Sarkar, S. (ed.), *The Philosophy and History of Molecular Biology*. Dordrecht: Kluwer Academic Publishers.

Timoféeff-Ressovsky, H.A. and Timoféeff-Ressovsky, N.W. (1926). 'Über das Phänotypische Manifestieren des Genotyps. II. Über Idio-Somatische Variationsgruppen bei Drosophila funebiris'. *Roux Archiv für Entwicklungsmechanik der Organismen* **108**: 146–70.

Umbarger, H.E. (1956). 'Evidence for a Negative-Feedback Mechanism in the Biosynthesis of Isoleucine'. *Science* **123**: 848.
Waddington, C.H. and Lewontin, R.C. (1968). 'A Note on Evolution and Changes in the Quantity of Genetic Information'. In: Waddington, C.H. (ed.), *Towards a Theoretical Biology*, vol. 1. Chicago: Aldine Publishing Company, pp. 109–10.
Watson, J.D. and Crick, F.H.C. (1953a). 'Molecular Structure of Nucleic Acids: A Structure for Deoxyribose Nucleic Acid'. *Nature* **171**: 737–8.
Watson, J.D. and Crick, F.H.C. (1953b). 'Genetical Implications of the Structure of Deoxyribose Nucleic Acid'. *Nature* **171**: 964–7.
Wiener, N. (1948). *Cybernetics*. Cambridge, MA: MIT Press.
Williams, G.C. (1966). *Adaptation and Natural Selection: A Critique of Some Current Evolutionary Thought*. Princeton: Princeton University Press.
Wimsatt, W.C. (1971). 'Some Problems with the Concept of 'Feedback'. *Boston Studies in the Philosophy of Science* **8**: 241–56.
Wimsatt, W.C. (1976). 'Reductive Explanation: A Functional Account'. *Boston Studies in the Philosophy of Science* **32**: 671–710.
Woese, C. (1967). *The Genetic Code*. New York: Harper and Row.
Yanofsky, C., Carlton, B.C., Guest, J.R., Helinski, D.R. and Henning, U. (1964). 'On the Colinearity of Gene Structure and Protein Structure'. *Proceedings of the National Academy of Sciences* **51**: 266–72.
Yates, R.A. and Pardee, A.B. (1956). 'Control of Pyrimidine Biosynthesis in *Escherichia coli* by a Feed-Back Mechanism'. *Journal of Biological Chemistry* **221**: 757–70.
Ycas, M. (1969). *The Biological Code*. New York: American Elsevier.
Yockey, H.P. (1992). *Information Theory and Molecular Biology*. Cambridge: Cambridge University Press.
Yoxen, E.J. (1979). 'Where Does Schroedinger's 'What is Life?' Belong in the History of Molecular Biology?'. *History of Science* **17**: 17–52.

DAVID S. THALER

PARADOX AS PATH: PATTERN AS MAP

Classical Genetics as a Source of Non-Reductionism in Molecular Biology

Each discipline cultivates a particular culture, set of attitudes and methods of seeking truth. These characteristics of a discipline are inoculated into the participants, the disciples of that field. This lineage of style is as distinct and notable as that visible between those who have been schooled in Shao Lin kung fu and those schooled in Shotokan karate. Differences may even be obvious in body build, posture and likely mental attitudes towards other topics. This essay will outline some of the characteristics of the genetics that contributed to the development of molecular genetics and, through molecular genetics, to molecular biology.

Molecular biology was woven from several intellectual and cultural strands, among them classical genetics and quantum physics. In his chapter, Sarkar emphasizes the proposition that computer science and cybernetics were important sources of non-reductionist styles for explanation and research in molecular biology. In the case of the genetic code, the

> immediate sources were ideas about information, coding and cybernetics that came into vogue after world war II as the computer age began. . . . On occasion there were systematic attempts to apply information theory or cybernetics to biology. However, more often than not, these ideas filtered into biology in a fairly unsystematic way, though no doubt helped by the explicit biological concerns of many early cyberneticists.

This essay argues that classical genetics is the key source of the non-reductionist style of explanation in molecular biology. The intellectual threads elsewhere attributed to cybernetics and information theory are herein attributed to classical genetics. After all, classical genetics was part of the intellectual milieu of the molecular biologists. The facts of genetics constituted the facts of life to be explained, and the workers were certainly aware of the intellectual approach used to gather those facts.

Before considering the source of non-reductionist ideas in molecular genetics it is worthwhile to review reductionism. This summary draws on the work of Sarkar which is now another chapter in this volume. Sarkar's criteria for a reductionist explanation follow:
i. The mechanisms or properties used in an explanation are ultimately physical and chemical.
ii. The structure must be hierarchical, properties at any level are explained in terms of properties at a lower level.
iii. The hierarchy must exist in physical space.

Explanations must help codify a body of experimental results such that they may be viewed as conforming to some general principle. The explanation must be of use in predicting the results of new experiments.

Therefore, for an explanation to be reductionist, it must codify and predict with two characteristics:

i. separability: the explanation is in terms of "independently specifiable spatial parts of the system". The precise interaction of these parts automatically creates properties seen at higher levels.
ii. physicalism: the interactions invoked in the explanation must have physical warranty, that is, the governing interaction must be derived, even if only informally, from known physical or chemical laws, or be determined from purely physical and chemical measurements (the last part is questionable because all sensory input and observation might be here subsumed).

Sarkar defines non-reductionist explanations by negative criteria. Any explanation must codify in a way which indicates a general explanation and must have predictive value. A non-reductive explanation must violate, or ignore, either the principle of separability or the principle of physicalism. Since separability is itself based on a definite location in space, these two criteria overlap.

Non-reductionist explanations or analyses do not deny a physical basis to what is being described. The assumption of physicalism is not at stake but is not important at the moment. The physical basis is in the background and somehow, someday a physical-reductionist model may be coupled to a contemporary non-reductionist model. However, such coupling of reductionist and non-reductionist approaches is not mandatory and may never happen. Non-reductionist explanation goes on its own terms, at its own level, with only arm-waving reference to the physical basis as a courtesy to fellow academics with different styles. A non-reductionist explanation is one that would be perfectly satisfied if any of several underlying physical mechanisms are at work. A non-reductive explanation need make only minimal reference to the smaller parts whose behavior may be underneath the system being explained.

Sarkar goes on to discuss the coding problem in genetics and non-reductionist approaches which were used to address it. The translation problem of coding is defined as defining the sequence of bases that determine a particular amino acid. The problem is stated: "How does the linear sequence of bases in DNA code for the linear sequence of amino acids in a protein?" The solution to this problem is known as the "Genetic Code". The solution is said to be encompassed in code tables now available in most textbooks of genetics. The problem was addressed within the context of the double helix structure and base pair complementarity, down to the atomic level. The possibility of adaptor molecules was one way to suggest a future link between a contemporary non-reductive explanation and a possible future reductive one.

CLASSICAL GENETICS ORIGINS OF NON-REDUCTIONIST APPROACHES

A case is presented here that the intellectual tradition of genetic analysis dating from Mendel fits the above criteria for non-reductionist approaches. Furthermore a case is argued that these aspects of classical genetics were appreciated and assimilated by the phage and bacterial geneticists who founded molecular genetics. Biochemical traditions are another key input to molecular biology, one with a much more reductionist bent than classical genetics. This author suspects that molecular geneticists were, and are, more influenced by the non-reductionist traditions of classical and molecular genetics than by the reductionist roots of biochemistry. Some tension between the disciplines follows from their different attitudes toward reductionism as method and goal. Sadly, since they are often in the same environs, practitioners of one school sometimes fail to appreciate those of the other.

Aspects of classical genetic analysis that merit discussion here include the work of Mendel himself with some reference to the explanatory style contemporary to it: plant breeding in 1865. Also considered are the resurrection of Mendel in the form of William Bateson's work as published in his 1909 book *Mendel's Principles of Heredity*; the extension of linkage groups into linkage maps as carried out by Morgan and colleagues; the use of genetic analysis in phage and bacterial crosses of the Delbrück school in phage and Lederberg and colleagues. Contemporary applications of non-reductionist genetics in framing problems of mutation, mapping and chiasmata interference are also related to their non-reductionist context. Pattern formation is discussed as a positive example of a non-reductionist style, and one which is important in genetics.

Mendel's unique framing of the problems of inheritance was pointed out by Bateson and later by Jacob (1973). Bateson footnoted his translation of Mendel's classic paper Mendel ([1865] 1909) to emphasize Mendel's relation to previous work in the field. What follows is first Bateson's translation of Mendel and then Bateson's footnote for the paragraph:

Those who survey the work done in this department will arrive at the conviction that among all the numerous experiments made, not one has been carried out to such an extent and in such a way as to make it possible to determine the number of different forms under which the offspring of hybrids appear, or to arrange these forms with certainty according to their separate generations, or definitely to ascertain their statistical relations. (Mendel [1865] 1909, p. 318)

Bateson's footnote for this paragraph:

It is to the clear conception of these three primary necessities that the whole success of Mendel's work is due. So far as I know this conception was absolutely new in his day. (Bateson, 1909, p. 318)

Some years later Jacob quoted the same paragraph of Mendel's paper and commented on it as follows:

There were three entirely novel elements in Mendel's approach: the way of envisaging experiments and choosing appropriate material; the introduction of discontinuity and the use of large populations, which meant that results could be expressed numerically and treated mathematically; the use of simple symbolism, which permitted a continuous interchange between experiment and theory. (Jacob [1973] 1982, p. 203)

The traits Mendel worked with, flower color, plant height and wrinkled or smooth seeds, are described primarily as "characters" and nothing further. The paper is mainly an algebraic treatment of hereditary characters and a demonstration that they segregate independently in subsequent generations.

To what extent was Mendel non-reductionist? The formal breeding analysis itself in Mendel stands on its own, in my opinion, as a non-reductionist piece of work and insight. As Olby (1985) says in summarizing the primary contribution of Mendel's work and comparing Mendel to his rediscoverers:

Mendel's analysis of hybridisation in terms of the independent character-pairs and their conformity with a combination series sets him apart from his contemporaries and his rediscoverers. *The treatment of hybrid variability in terms of the character-pair gave rise to the conception of two-unit hereditary determination, whereas cytological theory stimulated speculation in terms of multiple-unit hereditary determination.* (Italics in original; Olby, 1985, p. 133)

On the other hand, Mendel was aware of the cell theory and proposed that the segregation he saw in hybrids was due to the formation of pollen and egg cells of different types. Thus Mendel grounds his breeding results in a physical hypothesis which would explain the breeding results in terms of the behavior of actions at a lower level. Mendel, therefore, partly qualifies as a reductionist.

My own reading of Mendel is that he was aware of the urge in science to tie findings to a physical entity and to explain the actions at one level in terms of those at a lower level. He fulfilled his obligations to those urges by invoking the cell theory. Mendel refers to the idea that his results are the inevitable outcome of movements and relations of smaller entities, the pollen and egg cell. However, Mendel's novel insight flows from the way he deals with hereditary characters on their own terms. Mendel entertained physicalist and reductionist ideas but was not imprisoned by them.

The alternatives to Mendel's approach included the ideas of mixing or blending inheritance and ideas of gemmules. In the latter case, it was supposed that each organ contributed tiny versions of itself to the construction of a homunculus which was the gamete. These ideas are reductionist since they invoke explanations among smaller and physical entities. Mendel made courteous attempts to form a synthesis with the physical entities of pollen and egg, but his analysis is based only on the formal relationships among hereditary characters. The power in Mendel's approach is in his use of a non-reductionist style. This style has remained with genetics as a discipline. In an age in which interdisciplinary studies are almost deified, it is worthwhile to articulate and appreciate as precious the diverse and distinct culture invoked by each discipline.

A few quotes will be used to indicate the continuation of a non-reductionist thread in genetics which is continuous from Bateson through the origins of molecular biology and in the contemporary intellectual flavor of the field.

William Bateson (1909) championed Mendel:

> Purity of type thus acquires a precise meaning. It is dependent on gametic segregation, and has nothing to do with a prolonged course of selection, natural or artificial.
>
> All this of course is consonant with the visible facts that have been discovered by cytologists, in so far as the nucleus of each somatic cell is a double structure, while the nucleus of each gametic cell is a single structure. *It is, in my judgement, impossible as yet to form definite views as to the relations of the various parts of the cell to the function of heredity.* (1909, p. 16)

(The italics are not in the original. The emphasis here is to point out the independence of genetic analysis from the precise nature of, or necessary for, a physical-chemical mechanism).

Bateson is very aware that the formal laws of heredity are not depending on being linked to chromosomes, i.e. on being tied to any known physical entity. William Bateson liked Mendel so much he named his son after him. Gregory Bateson became concerned with the cross-fertilization of cybernetics and genetics a generation later (Bateson, 1971, 1988; Bateson and Bateson, 1988).

By 1922 William Bateson came to accept the chromosomes as the bearers of many Mendelian traits:

> We have turned still another bend in the track and behind the gametes we see the chromosomes. For the doubts – which I trust may be pardoned in one who had never seen cytology, save through a glass darkly – can not, as regards the main thesis of the *Drosophila* workers, be any longer maintained. (Bateson ([1922] 1928, p. 392)

Tschermak, one of the three rediscoverers of Mendel, is quoted by Olby (1985) as follows:

> The rules of inheritance, quite intentionally, I expressed at first purely descriptively or phenomenologically, in order not at once to anchor the newly-beginning experimental phase of the doctrine of heredity – as had happened inexpediently with Darwinism – to define theoretical terms. (Olby, 1985, p. 123)

To me it appears that Tschermak is defending the disciplinary method of genetics as the analysis of the results of breeding. I take "theoretical entities" to mean terms which are offered in the name of reduction. Tschermak appears to say that he has no use for even giving reductionism a courtesy call. In quoting the above passage Olby has another purpose in mind. Olby develops an argument that Tschermak was underestimating the degree of influence that Mendel's paper had on his (Tschermak's) primary publications on heredity. Whether or not Tschermak was ingenuous in his (non) attribution of Mendel, my point remains. The fact that Tschermak invokes a non-reductionist style as the reason for his manner of presentation presupposes that the value of such a style was widely appreciated among the geneticists of the era.

The conceptual independence of genetics from a physical basis continues into the work of Morgan and colleagues who created the first genetic maps out of linkage groups, *circa* 1915. A linkage map is an association of genes which do not show independent assortment. It is based purely on the results of breeding experiments and does not rely on assumptions about the physical nature of the gene.

From Morgan et al. (1915), *The Mechanism of Mendelian Heredity*:

Exception may perhaps be taken to the emphasis we have laid on the chromosomes as the material basis of inheritance. Whether we are right here, the future – probably a very near future – will decide. *But it should not pass unnoticed that even if the chromosome theory be denied, there is no result dealt with in the following pages that may not be treated independently of chromosomes; for, we have made no assumption concerning heredity that cannot also be made abstractly without chromosomes as bearers of the postulated markers.* (Morgan et al., 1915, p. viii)

Again the emphasis is not found in the original. I included more so that the context can be seen. It might be argued that Morgan is simply being careful, that he fully believes in a physical basis for heredity, he simply doesn't want to overstate the evidence. I agree with this view. However, it is not Morgan's beliefs which are at issue here. The point is that the genetic method of analysis is independent of those beliefs. The authors' approach to the relation of Mendelian segregation and chromosomes is considerably altered in the revised 1926 edition of this book.

The non-reductionist nature of genetics was appreciated by Max Delbrück from the early stage of his inspiration of the phage group. Delbrück dwells on the failure of genetics to get at the physical nature of the gene. The key finding being the definition of DNA as the genetic material (Avery and Macleod, 1944) which might be considered in large part a contribution from the biochemical-reductionist school. I quote now from a lecture Delbrück delivered in 1949:

All of natural history operates with a system of concepts which has very little contact with the physical and chemical sciences. The habits of animals and plants, their reproduction and development, their relations to their symbionts and enemies, can all be described and analyzed with very little reference to the concepts of physics and chemistry. Perhaps the most notable of these independent branches of biology is genetics, which in its pure form operates with "hereditary factors" and "phenotypic characters" in a perfectly logical system, as an exact science without ever having to refer to the processes by which the characters originate from the factors. The root of this science lies in the existence of natural units of observation, the individual living organisms, which in genetics play somewhat the same role as the atoms and molecules in chemistry. (Delbrück [1949] 1992, p. 17)

Formal genetic analysis flourished in the emerging disciplines of prokaryotic genetics. To illustrate that it was alive and well in the golden age of phage and bacterial genetics I will discuss a study of bacterial conjugation and one of phage crosses. Lederberg et al. (1951) presented the results of bacterial conjugation crosses as a branched genetic map. They were explicit that they did not assert this map to have a physical correspondence, i.e., they did not expect a branched chromosome. A branched map was the best representation of the genetic data and therefore was fully justified within the tradition of classical genetics. Figure 1 of their paper is a branched genetic map. The last sentence of the figure legend states: "This diagram is purely formal and does not imply a true branched chromosome." The issue of a branched genetic map is also considered in the text of the paper. The character of the discussion along with associated citations shows that such considerations were well within the nature of classical genetics discourse:

In a purely formalistic way, these data could be represented in terms of a 4-armed linkage group, Figure 1a, without supposing for a moment that this must represent the physical situation. This recalls the branched chromosome representation . . . of translocation heterozygotes in *Drosophila* before the cytogenetics of the situation was well understood. Newcomb and Nyholm . . . have, however, interpreted or described the deviations in linkage behavior as due to "negative interference. (p. 417)"

Another example of continuity in conceptual style between classical and molecular genetics is evident when William Bateson's diagrammatic representation of Mendelian segregation (Bateson, 1909, p. 12) is compared with the schematic representation put forth by Meselson and Stahl (1958) for the semiconservative replication of DNA. These two figures are juxtaposed as Figure 1 of the present paper. The conceptual similarity is especially poignant

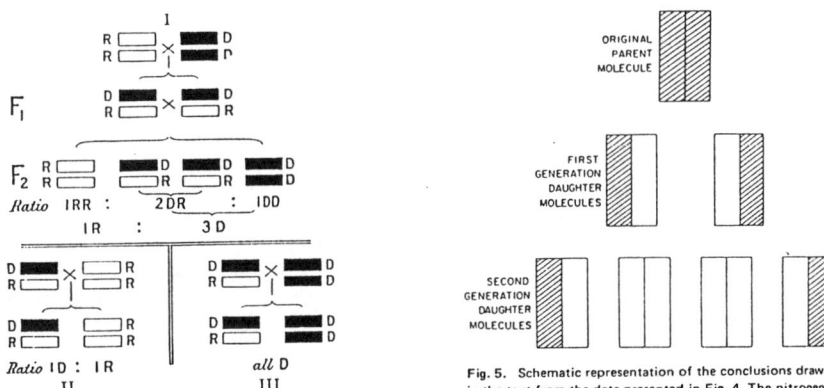

Figure 1. (Reprinted with permission of Cambridge University Press from Bateson (1909) (Fig. 2); reprinted with permission of Stahl from Meselson and Stahl (1958) (Fig. 5).)

because virtually every introductory genetics student goes through a phase of confusing these two concepts.

In 1965 a major conference was held to celebrate the hundredth anniversary of Mendel's paper. The period was also just after the genetic code was solved. In his contribution to the Mendel symposium, Max Delbrück assessed the role of classical genetics in the elucidation of the genetic code. Note that the aspects which Dr. Sarkar attributes to the influence of cybernetics, such as information transmission and error processes, seem similar to those attributed by Delbrück to classical genetics (Delbrück, 1967).

> The two genetic error processes, mutation and recombination, have a longer, richer, and more complex research history than do normal functions. For decades mutation and recombination of the genes were the only genetic properties that might lead to an unraveling of the mystery of the genetic material, and an enormous amount of effort has been spent refining the tools of research in these areas. However, the historical development of molecular genetics has bypassed these lines of research. Our understanding of the basic chemistry of the genes did not come from research on mutation and recombination but from the new approaches opened up by the study of the genetics of microorganisms. We are now reaching the point where we may be able to interpret mutation and recombination in molecular terms and to apply these insights toward analysis of the higher organization of the chromosomes. (1967, p. 66)

In 1967, F. Stahl presented a discussion of how circular genetic maps may arise. Maps which are circular in genetic terms may correspond to either linear or circular chromosomes (Stahl, 1967). This article reviewed extant cases, such as that of phage T4, in which a physically linear chromosome gave rise to a circular genetic map whereas phage lambda whose chromosome is physically circular during intracellular growth, yields a linear genetic map. Prokaryotic workers in the thick of molecular biology both at its origins and subsequent development were well aware of the independence of genetic analysis from assumptions about its physical underpinnings.

Stahl's non-reductionist approach was continued ten years later as evidenced in his summary statement at the 1978 Cold Spring Harbor symposium (Stahl, 1978):

> At this meeting, we all noticed a disparity in the state of the art between the studies in replication and those in recombination. The former has become serious biochemistry with clean substrates and clean enzymes, whereas the latter is still propelled largely by colored chalk on smudgy blackboards. I predict, however, that this 1978 Symposium will someday be recognized as the one at which successful in vitro analyses marked the beginning of the end of recombination as a geneticist's playground. I can already hear the biochemists circling in the night! (1967, p. 1356)

In 1971 Gregory Bateson published a collection of essays many of which had first appeared decades before. In these essays Bateson is explicit about connections between genetics and cybernetics. He also makes clear his debt to his father and the primacy of genetic analysis in his thought style. I quote here from one of Bateson's essays which was originally published in 1960:

> My father was a geneticist, and he used to say "It's all vibrations," and to illustrate this he would point out that the striping of the common zebra is an octave higher than that of Grevy's zebra. While it is true that in this particular case the "frequency" is doubled, I don't think that it is entirely a matter of vibrations as he endeavored to explain it. Rather, he was trying to say that it is all a matter of the sort of modifications which could be expected among systems whose determinants are not a matter of physics in the crude sense, but a matter of messages and modulated systems of messages. (Bateson, 1971, p. 232)

Gregory Bateson's influence may have been largely felt − at least so far − in anthropology and psychology rather than in molecular biology. However, there is little doubt that Gregory's father, William Bateson, was one of the most influential geneticists of this century. The value of Gregory's quote then, is that it came from a father who certainly predisposed his son to his lifelong quest for a synthesis of biology and consciousness, often via the conceptual approaches of genetics.

Genetics has been accused of being a reductionist science conditioned ultimately by a social context of eugenics and bio-social control (Hubbard, 1982). Certainly genetics has played that role in the justification of, and some geneticists have profited from, racist and social-Darwinian contexts (Mueller-Hill, 1988). A contention by the author of the present essay is that there are non-reductionist threads in the science of genetics. A hope of the author is that cultivation of these non-reductionist threads may be, to a certain extent, prophylactic with regard to the corruption of genetics into eugenics. Prophylactic effects could follow from two aspects of a non-reductionist paradox-seeking approach to truth seeking: (i) when reductionism is treated as an ultimate goal of truth seeking, it is easy to view people as things and every thing as an object to be manipulated. Proof is considered in terms of predictable manipulation; (ii) the method of paradox-seeking tends to sustain humor and irony on the part of the participants since they are constantly working to prove themselves and each other wrong. One should value a path for what it does to those on it as well as assessing value for their creations considered alone.

A non-reductionist style of truth seeking tends to use concepts of hierarchy in different ways than does a reductionist style. First of all there is a tendency to use a top-down versus down-up order of consideration and even determination. Upon characterizing a phenomenon (and in the process of characterization) a reductionist asks: "What are the smaller pieces which are acting in concert to produce the effect, and how can I separate these smaller pieces and set about recreating the interaction in a defined and simplified way?" In biochemical terms this is approach of purifying the components with the goal of reconstituting the phenomenon in a defined in vitro system, what has become known as "doing a Kornberg" in honor of Arthur Kornberg who first purified and reproduced *in vitro* an enzymatic system for replication of DNA (Kornberg, 1989).

A non-reductionist approach to understanding, especially relevant to a living system, tends to ask "What intelligence or evolved system might manifest such behavior, and what other behaviors might such intelligence or system manifest?" That is, the hypothesis is first that there is such an intelligence or evolved system. From the hypothesized existence follows the corollary that an interesting research program of questions is to seek the goals and methods that are being used. The second corollary is that the experimental approach involves forming and testing hypotheses about the input or sensory data acquired and how it is processed. Different questions and experiments follow from the non-reductionist approach, but experiments do indeed follow!

PATTERN AS MAP

So far we've defined non-reductionist in largely negative terms; in this section one non-reductionist approach will be described in a positive way; pattern apprehension.

The study of pattern in biology is ancient. D'Arcy Thompson (1917) showed pattern in many aspects of biology relating overall form in terms of either – and ideally both – its function and the mechanism by which it arose. The study of embryogenesis in terms of cell lineage followed closely on the origins of cell theory. Developmental biology and genetics found each other through Drosophila (Lewis, 1978) and corn (McClintock, 1983). In each of these cases mutants were identified as altered in pattern formation of the organism. Study of these mutants proceeded along three lines: (i) their meiotic behavior was studied as Mendelian markers; (ii) their effects on pattern formation via cell lineage studies was undertaken; and (iii) cytogenetic analysis was undertaken. The coming together of these three lines of work led to a major reductionist goal: the proof that Mendelian characters reside on chromosomes. It has also led to a tradition of coupling pattern and genetic analyses in a manner that set the stage on which the reductionist production was played out.

The contemporary work of James Shapiro is an example of non-reductionist approaches in microbiology. The work involves a genetic and morphologic analysis of pattern formation in colony morphology. Shapiro is explicit that he believes that the reductionist approach – in this case sequencing the products of genome rearrangement – is not the appropriate route for analysis: ". . . We cannot expect to answer questions like those listed above simply by defining the nature of DNA changes that have taken place" (Shapiro, 1984). He is also clear that this line of work is in direct lineage with the analysis by pattern formation in corn kernels (Shapiro, 1992).

Systems may be related by metaphor or by cause and effect; the latter alone is of interest in a reductionist hierarchy. Metaphor may apply by coincidence or it may apply because patterns of relationships are the consequential parameters rather than the reductionist mechanism by which parts are connected.

In gestalt formulations the whole is perceived and understood first, then the parts are apprehended in terms only of their interaction with the whole (Polanyi, 1969). Consider a system in which there are 1,000 or better yet, an infinite number of qualitatively exclusive gestalts. Only by knowing the whole does it become possible to cogently assign roles to the parts. (On the other hand it may be that the once some of the parts are specified then the relationships of the other parts become certain and the whole is thereby derived through a reductionist approach.)

What kind of systems are especially appropriate to try to understand with a reductionist or non-reductionist approach? To some extent this is a matter of taste and individual researchers are often more or less versed in both approaches, although perhaps not by name. One way is to ask for cases in which the overall behavior is not easy to derive from the action of parts at a lower level. If a system is sensitive to the singularities of quantum or thermodynamic fluctuations then cause and effect prediction will be difficult (Schrödinger, 1944).

Balancing on a quantum high wire is not the only way to be highly sensitive. Exquisite sensitivity an also be obtained through iterative processes (Kauffman, 1993). Small effects can be amplified, *especially when the mode of variation itself is part of what changes.* Biological evolution may well be such a process.

The generation of living variation in evolution is in large measure the generation of new DNA sequences. DNA sequence alterations, like most other processes inside the cell, are mediated via enzymes and these enzymes are themselves at least partially consequential to particular genes. The specifics of DNA enzymes and the changes they engender are thus partially a property of the alleles of genes for DNA metabolism. Because of cellular packaging, particular alleles and sets of alleles for the genes of DNA metabolism will themselves be co-inherited with any other alleles that are more directly selected. Thus the "creators" of a particular subset of all possible genetic diversity will tend to be co-inherited with their "creations". This process seems to possess the necessary characteristics to evolve the way in which change is generated (Thaler, 1994).

PARADOX AS A PATH

In addition to classical genetics, quantum mechanics and the theory of relativity were important intellectual influences on those who created molecular biology. The synthesis of the questions of classical genetics melded with the intellectual approaches of both genetics and of physics has been summed up in the name of one individual, Max Delbrück. Delbrück was himself deeply influenced by Bohr and the Copenhagen school. A tenet of the Copenhagen school also expressed by Delbrück was that the search for paradoxes constituted a high and profound style of scientific investigation.

The crucial point in this abbreviated account of an historical episode is the appearance of a conflict between separate areas of experience, which gradually sharpens into a paradox and must then be resolved by a radically new approach. (Delbrück [1949] 1992, p. 18).

It has been argued that the search for paradoxes in genetics was a failure (Stent, 1992). This is true only insofar as the paradoxes found so far are not as permanent as those encountered in quantum physics. On the other hand, the method of searching for paradoxes has been, and will continue to be, a vital path, or method, in the discipline of molecular genetics. A paradox resolved may not be a paradox any longer, but the historical (or evolutionary) route to the new synthesis was through the thesis and antithesis.

The interface of physics and genetics is of great interest not because of the juxtaposition facts with doctrines or the possibility of analogies between different levels (Capra, 1977). More interesting is the question of what style of truth-seeking are used and what do these styles do for the people involved in them. The thread of the Copenhagen bent for seeking paradox, for expecting, craving, and loving paradox interwove in the most beautiful way with the non-reductionist thread of classical genetics. Together they interwove with the pure reductionist style of biochemistry to create some of the prettiest tapestries in the patchwork quilt of modern science. Two examples, mismatch correction and chiasmata interference, are mentioned below.

The thread of paradox-seeking brought from the Copenhagen school and cultivated in union with the abstraction of classical genetics has something in common with certain forms of buddhism. I am not referring to the products of science or mystical literature describing the Bardo planes. Rather I am referring to the method of seeking truth in which doubt and the seeking of paradox are emphasized as path. In the Tibetan Buddhist tradition the cultivation of doubt is encouraged: "Doubt is the incentive to research, and research is the path which leads to true knowledge" (David-Neel, 1967). Further on in the same lineage is the Zen tradition in which "dai-didan" is defined as: "Great doubt. One of the three essentials of Zen practice" (Wood, 1957). A topic deserving further exploration is the comparative study of the meaning and uses of doubt as articulated in the Cartesian tradition with those in Buddhism (M. Cook, 1994, personal communication). The concept of paradox and its use in reasoning was not alien to early geneticists. William Bateson alludes to the "spirit of paradox" (Bateson [1922] 1928, p. 393) in a discussion concerning problems and prospects for unifying Mendel and Darwin in evolutionary theory.

One case in which the role of paradox and its resolution was particularly clear occurs in the work which eventually led to the postulation and discovery of mismatch correction. The gist of this work is that mapping studies indicated paradoxes between physical, i.e. reductionist and genetic, i.e., non-reductionist, distances or mapping studies contained internal paradoxes (non-reductionist paradoxes). These only become paradoxes if one does not allow branched maps, thus, they too imply a paradox between levels, if one allowed the constraint of a linear or circular genome (Lieb, 1991). Paradox occurred during the pursuit of reduction.

The subject of chiasma interference is an example of a contemporary problem to which both the non-reductionist thread of molecular genetics and the method of paradox-framing are being applied. A definition of chiasma interference: when one crossover occurs there is not likely to be another one nearby. Thus close double crossovers occur much less often than they should be chance alone. The phenomenon does not depend on the number of base pairs, that is the physical distance, between sites but rather on the genetic distance between sites which varies with the organism (Foss et al., 1993). Foss et al. specifically reject any extant physical models: "Most models for interference have assumed, at least implicitly, that the intensity of interference depends inversely on the physical distance separating the intervals" (Foss et al., 1993). No physical model is presented for their purely genetic analysis.

CONCLUSIONS

1. A working definition of reductionism has been abstracted from the work of Sarkar. The character of non-reductionism has been similarly indicated through negative criteria.
2. A case has been outlined that classical genetics contains a strong non-reductionist aspect.
3. A case is made that the non-reductionist aspect of classical genetics had a profound influence on the founders of molecular genetics.
4. The point is argued that the intellectual influence on molecular biology attributed by Sarkar to cybernetics and information theory is instead due to the influence of classical genetics.
5. A case is outlined that the influence of the Copenhagen school (searching for a paradox, for quantum complementarity) on molecular genetics included a wonderful synergy of the Copenhagen school and the non-reductionist aspect of classical genetics.
6. The method of Pattern Apprehension is presented as one example of a non-reductionist approach in contemporary biology. Two other examples of non-reductionist analysis are mentioned: mismatch correction and chiasmata interference.
7. Metaphor, hierarchy and social implications of scientific styles are alluded to.
8. The analogy is implicitly developed between the three-fold path in

Buddhism (Buddha, Dharma, Sangha) and modern molecular genetics. In this formulation the role of Buddha is that of the individual seeker of paradox; the dharma is represented by pattern apprehension, and finally the community of truth seekers is related to the culture and set of attitudes within each discipline.
9. Interesting paradoxes can and should be sought at the intersection of reductionist and non-reductionist disciplines. Purity of discipline must be maintained to give the maximum possibility for this creative conflict. At the same time, maximum opportunity occurs when the conflicting disciplines are well represented in the minds of the same individuals.
10. In the formation of molecular biology, genetics and quantum-influenced physics were the non-reductionist seeds. Biochemistry provided the reductionist foil against which the non-reductionist tendencies in genetics were played by Copenhagen physicists intent on searching out a strong paradox.

ACKNOWLEDGMENTS

Thanks to Marx Wartofsky for inviting me to the conference on Methods in Philosophy and the Sciences as a commentator on Sahotra Sarkar's paper *Codes, Cybernetics and Networks: Aspects of non-reductionist Research in Molecular Biology*. Further thanks to both for encouraging me to write up my comments from the conference. Thanks to Joshua Lederberg for support and encouragement in this and many other projects. Mike Cook, Fiona Doetsch, Shumo Liu, Robert Olby, John Robbins, Sahotra Sarkar, Trudee Tarkowski and Greg Tombline improved the draft manuscript. R.O. especially helped me clarify the section on what Mendel may have thought, Mike Cook suggested a comparison of Cartisian and Buddhist meanings of doubt, and Trudee Tarkowski helped eliminate nonsense from the interference section. Lindley Darden gave many comments in a reasonably timely fashion but because of the insane brinkspersonship with which I have conducted this, as well as other writing projects, I was not able to incorporate her insights into the submitted manuscript. Thanks to Frank Stahl for a cup of tea. Support from the Markey Foundation to the Raymond and Beverly Sackler Laboratory of Molecular Genetics and Informatics is acknowledged.

Rockefeller University,
New York, USA

REFERENCES

Avery, O., Macleod, C. and McCarty, M. (1944). 'Studies on the Chemical Nature of the Substance Inducing Transformation of Pneumococcal Types'. *J. Expt. Med.* **79**: 137–58.

Bateson, G. (1971). *Steps to an Ecology of Mind*. New York: Ballantine Books.
Bateson, G. (1988). *Mind & Nature: A Necessary Unity*. San Francisco: Harper.
Bateson, G. and Bateson, M.C. (1988). *Angel's Fear: Towards an Epistemology of the Sacred*. New York: Bantam Books.
Bateson, W. (1909). *Mendel's Principles of Heredity*. London: Cambridge University Press.
Bateson, W. (1922). 'Evolutionary Faith and Modern Doubts'. In: Bateson, B. (1928) *William Batson*. Cambridge: Cambridge University Press.
Capra, F. (1977). *The Tao of Physics*. Toronto: Bantam Books.
David-Neel, A. (1967). *Secret Oral Teachings of Tibetan Buddhist Sects*. San Francisco: City Lights Books.
Delbrück, M. ([1949] 1992). 'A Physicist Looks at Biology'. In: Cairns, J., Stent, G.S. and Watson, J.D. (eds.), *Phage and the Origins of Molecular Biology*. Cold Spring Harbor: Cold Spring Harbor Laboratory Press, pp. 9–22.
Delbrück, M. (1968). 'Molecular Aspects of Genetics'. In: Brink, R.A. (ed.), *Heritage from Mendel*. Madison, WI: University of Wisconsin Press, pp. 65–6.
Foss, E., Lande, R., Stahl, F.W. and Steinberg, C.M. (1993). 'Chiasma Interference as a Function of Genetic Distance'. *Genetics* 3: 681–91.
Hubbard, R. (1982). 'The Theory and Practice of Genetic Reductionism – From Mendel's Laws to Genetic Engineering'. In: Rose, S. (ed.), *Towards a Liberatory Biology. The Dialectics of Biology Group*. London: New York: Allison & Busby, pp. 62–78.
Jacob, F. (1973). *The Logic of Life. A History of Heredity*. 1982 edition. New York: Pantheon Books.
Kauffman, S.A. (1993). *The Origins of Order*. New York: Oxford University Press.
Kornberg, A. (1989). *For the Love of Enzymes: The Odyssey of a Biochemist*. Cambridge, MA: Harvard University Press.
Lederberg, J., Lederberg, E.M., Zinder, N.D., Lively, E.R. (1951). 'Recombination Analysis of Bacterial Heredity'. *Cold Spring Harbor Symposium of Quantitative Biology* 16: 413–43.
Lewis, E. (1978). 'A Gene Complex Controlling Segmentation in Drosophila'. *Nature* 276: 565–70.
Lieb, M. (1991). 'Spontaneous Mutation at a 5-Methylcytosine Hotspot is Prevented by Very Short Patch (VSP) Mismatch Repair'. *Genetics* 128: 23–7.
McClintock, B. (1983). 'The Significance of Responses of the Genome to Challenge'. In: Odelberg, W. (ed.), *Les Prix Nobel*. Stockholm: Almquist & Wiksell Int., pp. 174–93.
Meselson, M. and Stahl, F.W. (1958). 'The Replication of DNA in *Escherichia coli*'. *Proceedings of the National Academy of Science USA* 44: 671–82.
Morgan, T., Sturtevant, A.H., Muller, H.J. and Bridges, C.B. (1915). *The Mechanism of Mendelian Heredity*. New York: Henry Holt and Company.
Mueller-Hill, B. (1988). *Murderous Science. Elimination by Scientific Selection of Jews, Gypsies, and Others, Germany, 1933–1945*. Oxford: Oxford University Press.
Olby, R. (1985). *Origins of Mendelism*. Chicago: University of Chicago Press.
Polanyi, M. (1969). *Knowing and Being*. Chicago: University of Chicago Press.
Schrödinger, E. (1944). *What is Life? and Mind and Matter*. Cambridge: Cambridge University Press.
Shapiro, J.A. (1984). 'Observations on the Formation of Clones Containing *araB-lacZ* Cistron Fusions'. *Molecular and General Genetics* 194: 79–90.
Shapiro, J.A. (1992). 'Kernels and Colonies: The Challenge of Pattern'. In: Federoff, N. and Botstein, D. (eds.), *The Dynamic Genome*. Cold Spring Harbor, NY: Cold Spring Harbor Laboratory Press, pp. 213–21.
Stahl, F.W. (1967). 'Circular Genetic Maps'. *Journal of Cell and Comparative Physiology* 70: sup. 1–12.
Stahl, F.W. (1978). 'Summary'. *Cold Spring Harbor Symposium of Quantitative Biology* 43: 1353–6.

Stent, G. (1992). 'Introduction: Waiting for the Paradox'. In: Cairns, J., Stent, G.S. and Watson, J.D. (eds.), *Phage and the Origins of Molecular Biology*. Cold Spring Harbor, NY: Cold Spring Harbor Laboratory Press, pp. 3–8.
Thaler, D.S. (1994). 'The Evolution of Genetic Intelligence'. *Science* **264**: 224–5.
Thompson, D.W. (1917). *On Growth and Form*. Cambridge: Cambridge University Press.
Wood, E. (1957). *Zen Dictionary*. Rutland: Charles E. Tuttle Co.

NAME INDEX

Abir-Am, P. 6, 7, 103
Alexander, J. 137
Allfrey, V. 115
Altman, S. 16
Arnold, W. 54
Arnon, D. 53
Arrhenius, S. 129
Astbury, W.T. 48, 56
Atkins, J.F. 198, 200
Auerbach 21
Avery, O.T. 17, 190, 238

Bagg, H.J. 141
Bail, O. 137
Baltimore, D. 48
Bareiss, R. 36
Bassham, J.A. 51
Bateson, G. 235, 237
Bateson, M.C. 237
Bateson, W. 105, 235, 237, 239–241, 244
Baumgarten, P. 128
Baur, E. 174
Bawden 17
Beadle, G. 19, 69, 70, 91, 108
Beatty, J. 27
Beckner, M. 10
Beckwith, J. 31
Beerman, W. 114
Benacerref, B. 144
Bennett, J.C. 139
Benson, A.A. 51
Benzer, S. 69
Bergson, H. 131
Bernal, J.D. 48
Bernstein, F. 142
Berrill, N.J. 103, 104, 119

Billingham, R. 144
Bohr, N. 138, 186
Boole, G. 206
Boveri, T. 102
Brachet, A. 72
Brachet, J. 67, 69, 71–79, 110, 118
Bragg, L. 48
Bragg, W. 48
Brandon, R. 10
Branson, H.R. 209
Breinl, F. 137
Brenner, S. 194, 196
Brent, L. 144
Briggs, R. 114
Britten, R.J. 207, 208
Brooks, W.K. 102
Buchner, H. 136
Burian, R. 6, 8–10
Burnet, MacF. 138, 139, 141, 151, 153, 156
Buss, L. 161

Callebaut, W. 10
Calvin, M. 51, 58
Cannon, W. 97, 133
Capra, F. 244
Carlson, E.A. 11
Carnap, R. 1, 224
Caskey, C.T. 16
Caspersson, T. 77, 110
Castle, W. E. 141
Cattaneo, R. 200
Cech, T. 16
Chance, B. 205
Chargaff, E. 18, 68, 199
Chase 18
Claude, A. 73

NAME INDEX

Cohen, I. 155, 156
Cohen, R.S. 10
Cohen, S. 68
Cohn, M. 113, 114, 155, 246
Coutinho, A. 154
Creager, A. 6, 223
Crick, F.H.C. 15, 48, 49, 60, 69, 72, 75, 77, 78, 135, 191, 192, 194, 197, 198, 224
Culp, S. 27

Darden, L. 246
Darwin, C.R. 127, 241, 244
Davidson, E.H. 207, 208
Davis, B.D. 221
De Robertis, E. 119
Delbrück, M. 8, 18, 48, 55, 78, 91, 138, 162, 186, 190, 191, 197, 235, 238, 240, 243
Dobzhansky, Th. 174
Doetsch, F. 246
Dreyer, W. 139
Driesch, H. 1, 107, 131

Ehrlich, P. 4, 129, 134–137, 139, 141, 142, 151, 162
Eigen, M. 221
Emerson, R. 51, 54, 57
Ephrussi, B. 70, 71, 191, 202

Fanck, J. 55
Feulgen 18
Fischer, E. 174
Fischer, H. 60
Fisher, E. 136, 138, 215
Flexner, S. 90
Fluegge, C. 129
Fosdick, R.D. 56
Foss, E. 245
Fox, T.D. 199, 200, 224
Franklin, R. 18

Gall, J. 116
Gamow, G. 191, 192, 194, 196, 197, 224
Garrod, A. 19
Gatlin, L. 211
Gaudillière, J.-P. 6, 69
Gayon, J. 70
Gilbert, S. 6, 8, 69
Gilbert, W. 7, 177
Glass, B. 114
Goldschmidt, R. 105–107
Goldschmidt, R.B. 120
Golomb, S.W. 195, 197

Goodwin, B. 119
Gorer, P.A. 143
Gorini, L. 114, 115
Goudge, T.A. 10
Grüneberg, H. 108
Greenstein, J.P. 108
Grene, M. 10
Griffith 17
Grossman, Z. 155, 156

Hämmerling, J. 76, 112
Hacking, I. 79
Hadorn, E. 108
Haldane, J.B.S. 1, 120, 142, 143, 174, 202, 214
Harrison, R.G. 106, 113
Haurowitz, F. 137
Helmholtz, H. 131
Henderson, L. 133
Herman, N.D. 212
Hershey, A.D. 18, 48, 71
Hill, R. 51
Hirschfeld, L. 142
Hogben, L. 1, 2
Hogness, T.R. 54
Holtfreter, J. 108, 118
Hubbard, R. 241
Hull, D. 2, 8, 9
Huxley, J.S. 174

Jacob, F. 69, 77, 116, 117, 204, 235, 236
Janeway Jr., C. 152, 155
Jensen, A. 175
Jensen, C.O. 141
Jerne, N.K. 134, 135, 138, 139, 153, 154, 160, 162
Jordon, P. 162
Judson, H. 4, 72
Judson, J.F. 8
Just, E.E. 101, 107

Kabat, E. 129
Kalmus, H. 202
Kan, Y.W. 16
Kauffman, S.A. 243
Kay, L. 6, 223, 224
Kazatchkine, M. 154
Keller, E.F. 5, 6, 11, 89, 223, 224
Kendrew, J. 4, 8, 58, 68
Kevles, D. 5
Khorana 19
Kimura, M. 210, 211

King, T.J. 14
Kitasato, S. 128
Kitcher, P. 8, 10, 27
Kleene, S.C. 205
Kluyver, A.J. 50, 54
Koch, R. 135
Kohne, D.E. 208
Koltzoff, N. 73
Konner, M. 179
Kornberg, A. 16, 19, 241
Koshland, D. 177, 179
Kostellow, A. 115
Krimsky, S. 4
Kuhn, T. 28

Løvtrop, S. 113
Landsteiner, K. 94, 137, 138, 142, 190
Lanni, F. 197
Lederberg, J. 7, 71, 139, 156, 190, 203, 214, 221, 223, 224, 235, 239, 246
Lederman, M. 6
Leggett, A.J. 10
Lehman, H. 10
Lenz, F. 174
Levan 20
Levene, P. 16
Lewin, B. 208
Lewis, E. 242
Lewontin, R. 3, 223
Lieb, M. 245
Lillie, F.R. 106
Lindegren, C.C. 71
Linschitz, H. 209
Little, C.C. 141, 142
Liu, S. 246
Lloyd, E.A. 3, 10
Loeb, J. 1, 133
Loeb, L. 142
Lubas, T. 10
Luria, S.E. 48, 71, 190
Lwoff, A. 70, 77

Maas, W.K. 114, 115
MacLeod, C. 17, 238
Magasanik, B. 114, 115
Mangold, H. 113
Markert, C. 114, 116
Mason, M. 91
Mather, K. 112
Matthaei, J.H. 197
Matzinger, P. 155, 156
Maxwell 27

Maynard Smith, J. 3
Mayr, E. 3, 34, 132
Mazia, D. 224
McCarty, M. 17
McClintock, B. 20, 242
McCoy, E. 5
McDevitt, H. 144
Medawar, P. 143, 144
Mendel, G.J. 19, 102, 172, 235–237, 239, 240, 242, 244
Merton, R.K. 28
Meschelke, F. 117
Meselson, M. 7, 239
Metchnikoff, E. 127–131, 133–135, 162
Metropolis, N. 193
Miescher 16
Mirsky, A.E. 115
Mitchell, P. 52, 71
Mitchison, A. 144
Monod, J. 6, 69–71, 77, 104, 108, 113, 116, 117, 203–206, 208
Moore, J. 118
Morgan, T.H. 57, 89, 102, 104, 107, 117, 118, 141, 235, 238
Morgenroth, J. 142
Morowitz, H. 30
Mudd, S. 137
Mueller-Hill, B. 241
Muller, H.J. 21, 174

Nagel, E. 1, 2, 9, 10, 31
Needham, J. 103–105, 118
Neel, J.V. 96
Neurath, O. 224
Neveln, B. 198
Newcomb, H.B. 239
Newton, I. 27
Nijhout, H.F. 103
Nirenberg, M.W. 197
Northrop, J. 17
Nossal, G. 139
Nuttal, G. 129
Nyholm, M.H. 239

Olby, R.C. 4, 5, 7, 8, 190, 237, 246
Opitz, J. 119
Ornstein, L. 54
Oyama, S. 103

Pardee, A.B. 203, 204
Pasteur, L. 128

NAME INDEX

Pauling, L. 18, 48, 87, 137, 138, 162, 190, 191
Pavan, C. 114
Pei, D. 22
Perutz, H. 8
Perutz, M. 216
Pfeiffer, R. 128
Pierce, J.R. 202
Pirie 17
Polanyi, M. 243
Pollock, M.R. 113
Pontecorvo, G. 71
Popper, K. 3
Popper, P. 10

Quastler, H. 209
Quine, W.V.O. 3, 224

Rabinowitch, E. 53
Rich, A. 193, 224
Rickenberg, H.V. 203
Robbins, J. 246
Rosenberg, A. 5, 27
Rosenblueth 205
Ruse, N. 2, 27
Ryan, F.J. 117

Sager, R. 117
Salmon, W. 2
Sandri, G. 10
Sanger, F. 193
Sarkar, S. 6, 7, 9–11, 31, 185, 186, 221, 224, 233, 234, 240, 245, 246
Saxén, L. 103
Schaffner, K.F. 2, 5, 8, 9, 80, 203
Schneider, T.D. 211, 212
Schrödinger, E. 189, 191, 198, 243
Schultz, J. 77
Sela, M. 144
Shannon, C.E. 209
Shapiro, J.A. 222, 223, 242
Shimony, A. 2, 10, 185, 205, 223
Shrader-Frechette, K. 5
Smith, C.W. 200
Snell, G. 143
Sober, E. 3
Spemann, H. 113
Spiegelman, S. 71, 108–110, 112, 115, 120, 224
Stahl, F.W. 239, 240, 246
Stanier, R.Y. 113
Stanley, W. 92

Stearns, R. 115
Stein, H. 2
Steinberg, A.G. 174
Stent 68
Stephens, R.M. 212
Stern, C. 56, 107, 113
Strong, L.C. 141
Sturtevant, A. 105
Summers, W. 6
Sumner 17

Talmage, D. 139, 156
Tamarin, R. 10
Tarkowski, T. 246
Tarski, A. 1
Tatum, E.L. 19, 69, 70, 108, 190
Tauber, A.I. 8, 221
Thaler, D.S. 7, 9, 10, 185, 186, 223, 233, 243
Thompson, D'A. 242
Thompson, O.P. 193
Timoféeff-Ressovsky, H.A. 190
Timoféeff-Ressovsky, N.N. 190
Tiselius, A. 129
Tjio 20
Tolin, S.A. 6
Tonegawa, S. 139, 140
Topley, W. 137
Tschermak, E. 237, 238
Tuppy, H. 193
Tyzzer, E. 142

Uchida, H. 6
Umbarger, H.E. 203

van der Steen, W.J. 27
Van Niel, C.B. 50
Vogel, H.J. 114
Von Behring, E. 128, 129, 135
Von Dungern, E. 142

Waddington, C.H. 27, 106, 110–112, 117
Waddington, J. 118
Warburg, O. 50, 53, 60
Wartofsky, M. 10, 246
Waters, C.K. 8
Watson, J.D. 15, 32, 48, 49, 60, 69, 77, 78, 152, 177, 191, 192, 219
Weaver, W. 56, 57, 90
Weigert, C. 128
Weismann, A. 102
Weiss, P. 97, 103, 111

NAME INDEX

Wendell Stanley 17
Whitehead, A.N. 132
Wiener, N. 202
Wilkes, D. 10
Wilkins, M. 18
Williams, D.A. 210, 211
Wilson, E.B. 16, 103
Wilson, E.O. 175
Wimsatt, W.C. 2, 3, 9, 10, 205, 223
Woese, C. 198
Wood, E. 244
Woodger, J.H. 1, 2, 9
Woodward, R.B. 60
Wright, S. 1
Wurmser, R. 50

Yanofsky, C. 19, 198
Yates, R.A. 203
Ycas, M. 193, 194, 198, 224
Yockey, H.P. 198
Yoxen, E. 4

Zallen, D. 6, 8, 67, 70,
Zamecnik, P. 74
Ziegler, E. 128
Zscheile Jr., F.P. 54

Boston Studies in the Philosophy of Science

123. P. Duhem: *The Origins of Statics*. Translated from French by G.F. Leneaux, V.N. Vagliente and G.H. Wagner. With an Introduction by S.L. Jaki. 1991
 ISBN 0-7923-0898-0
124. H. Kamerlingh Onnes: *Through Measurement to Knowledge*. The Selected Papers, 1853-1926. Edited and with an Introduction by K. Gavroglu and Y. Goudaroulis. 1991 ISBN 0-7923-0825-5
125. M. Čapek: *The New Aspects of Time: Its Continuity and Novelties*. Selected Papers in the Philosophy of Science. 1991 ISBN 0-7923-0911-1
126. S. Unguru (ed.): *Physics, Cosmology and Astronomy, 1300-1700*. Tension and Accommodation. 1991 ISBN 0-7923-1022-5
127. Z. Bechler: *Newton's Physics on the Conceptual Structure of the Scientific Revolution*. 1991 ISBN 0-7923-1054-3
128. É. Meyerson: *Explanation in the Sciences*. Translated from French by M-A. Siple and D.A. Siple. 1991 ISBN 0-7923-1129-9
129. A.I. Tauber (ed.): *Organism and the Origins of Self.* 1991
 ISBN 0-7923-1185-X
130. F.J. Varela and J-P. Dupuy (eds.): *Understanding Origins*. Contemporary Views on the Origin of Life, Mind and Society. 1992 ISBN 0-7923-1251-1
131. G.L. Pandit: *Methodological Variance*. Essays in Epistemological Ontology and the Methodology of Science. 1991 ISBN 0-7923-1263-5
132. G. Munévar (ed.): *Beyond Reason*. Essays on the Philosophy of Paul Feyerabend. 1991 ISBN 0-7923-1272-4
133. T.E. Uebel (ed.): *Rediscovering the Forgotten Vienna Circle*. Austrian Studies on Otto Neurath and the Vienna Circle. Partly translated from German. 1991
 ISBN 0-7923-1276-7
134. W.R. Woodward and R.S. Cohen (eds.): *World Views and Scientific Discipline Formation*. Science Studies in the [former] German Democratic Republic. Partly translated from German by W.R. Woodward. 1991
 ISBN 0-7923-1286-4
135. P. Zambelli: *The Speculum Astronomiae and Its Enigma*. Astrology, Theology and Science in Albertus Magnus and His Contemporaries. 1992
 ISBN 0-7923-1380-1
136. P. Petitjean, C. Jami and A.M. Moulin (eds.): *Science and Empires*. Historical Studies about Scientific Development and European Expansion.
 ISBN 0-7923-1518-9
137. W.A. Wallace: *Galileo's Logic of Discovery and Proof.* The Background, Content, and Use of His Appropriated Treatises on Aristotle's *Posterior Analytics*. 1992 ISBN 0-7923-1577-4
138. W.A. Wallace: *Galileo's Logical Treatises*. A Translation, with Notes and Commentary, of His Appropriated Latin Questions on Aristotle's *Posterior Analytics*. 1992 ISBN 0-7923-1578-2
 Set (137 + 138) ISBN 0-7923-1579-0

Boston Studies in the Philosophy of Science

139. M.J. Nye, J.L. Richards and R.H. Stuewer (eds.): *The Invention of Physical Science.* Intersections of Mathematics, Theology and Natural Philosophy since the Seventeenth Century. Essays in Honor of Erwin N. Hiebert. 1992
ISBN 0-7923-1753-X
140. G. Corsi, M.L. dalla Chiara and G.C. Ghirardi (eds.): *Bridging the Gap: Philosophy, Mathematics and Physics.* Lectures on the Foundations of Science. 1992
ISBN 0-7923-1761-0
141. C.-H. Lin and D. Fu (eds.): *Philosophy and Conceptual History of Science in Taiwan.* 1992
ISBN 0-7923-1766-1
142. S. Sarkar (ed.): *The Founders of Evolutionary Genetics.* A Centenary Reappraisal. 1992
ISBN 0-7923-1777-7
143. J. Blackmore (ed.): *Ernst Mach – A Deeper Look.* Documents and New Perspectives. 1992
ISBN 0-7923-1853-6
144. P. Kroes and M. Bakker (eds.): *Technological Development and Science in the Industrial Age.* New Perspectives on the Science–Technology Relationship. 1992
ISBN 0-7923-1898-6
145. S. Amsterdamski: *Between History and Method.* Disputes about the Rationality of Science. 1992
ISBN 0-7923-1941-9
146. E. Ullmann-Margalit (ed.): *The Scientific Enterprise.* The Bar-Hillel Colloquium: Studies in History, Philosophy, and Sociology of Science, Volume 4. 1992
ISBN 0-7923-1992-3
147. L. Embree (ed.): *Metaarchaeology.* Reflections by Archaeologists and Philosophers. 1992
ISBN 0-7923-2023-9
148. S. French and H. Kamminga (eds.): *Correspondence, Invariance and Heuristics.* Essays in Honour of Heinz Post. 1993
ISBN 0-7923-2085-9
149. M. Bunzl: *The Context of Explanation.* 1993
ISBN 0-7923-2153-7
150. I.B. Cohen (ed.): *The Natural Sciences and the Social Sciences.* Some Critical and Historical Perspectives. 1994
ISBN 0-7923-2223-1
151. K. Gavroglu, Y. Christianidis and E. Nicolaidis (eds.): *Trends in the Historiography of Science.* 1994
ISBN 0-7923-2255-X
152. S. Poggi and M. Bossi (eds.): *Romanticism in Science.* Science in Europe, 1790–1840. 1994
ISBN 0-7923-2336-X
153. J. Faye and H.J. Folse (eds.): *Niels Bohr and Contemporary Philosophy.* 1994
ISBN 0-7923-2378-5
154. C.C. Gould and R.S. Cohen (eds.): *Artifacts, Representations, and Social Practice.* Essays for Marx W. Wartofsky. 1994
ISBN 0-7923-2481-1
155. R.E. Butts: *Historical Pragmatics.* Philosophical Essays. 1993
ISBN 0-7923-2498-6
156. R. Rashed: *The Development of Arabic Mathematics: Between Arithmetic and Algebra.* Translated from French by A.F.W. Armstrong. 1994
ISBN 0-7923-2565-6

Boston Studies in the Philosophy of Science

157. I. Szumilewicz-Lachman (ed.): *Zygmunt Zawirski: His Life and Work.* With Selected Writings on Time, Logic and the Methodology of Science. Translations by Feliks Lachman. Ed. by R.S. Cohen, with the assistance of B. Bergo. 1994 ISBN 0-7923-2566-4
158. S.N. Haq: *Names, Natures and Things.* The Alchemist Jābir ibn Ḥayyān and His *Kitāb al-Aḥjār* (Book of Stones). 1994 ISBN 0-7923-2587-7
159. P. Plaass: *Kant's Theory of Natural Science.* Translation, Analytic Introduction and Commentary by Alfred E. and Maria G. Miller. 1994
 ISBN 0-7923-2750-0
160. J. Misiek (ed.): *The Problem of Rationality in Science and its Philosophy.* On Popper vs. Polanyi. The Polish Conferences 1988–89. 1995
 ISBN 0-7923-2925-2
161. I.C. Jarvie and N. Laor (eds.): *Critical Rationalism, Metaphysics and Science.* Essays for Joseph Agassi, Volume I. 1995 ISBN 0-7923-2960-0
162. I.C. Jarvie and N. Laor (eds.): *Critical Rationalism, the Social Sciences and the Humanities.* Essays for Joseph Agassi, Volume II. 1995 ISBN 0-7923-2961-9
 Set (161–162) ISBN 0-7923-2962-7
163. K. Gavroglu, J. Stachel and M.W. Wartofsky (eds.): *Physics, Philosophy, and the Scientific Community.* Essays in the Philosophy and History of the Natural Sciences and Mathematics. In Honor of Robert S. Cohen. 1995
 ISBN 0-7923-2988-0
164. K. Gavroglu, J. Stachel and M.W. Wartofsky (eds.): *Science, Politics and Social Practice.* Essays on Marxism and Science, Philosophy of Culture and the Social Sciences. In Honor of Robert S. Cohen. 1995 ISBN 0-7923-2989-9
165. K. Gavroglu, J. Stachel and M.W. Wartofsky (eds.): *Science, Mind and Art.* Essays on Science and the Humanistic Understanding in Art, Epistemology, Religion and Ethics. Essays in Honor of Robert S. Cohen. 1995
 ISBN 0-7923-2990-2
 Set (163–165) ISBN 0-7923-2991-0
166. K.H. Wolff: *Transformation in the Writing.* A Case of Surrender-and-Catch. 1995 ISBN 0-7923-3178-8
167. A.J. Kox and D.M. Siegel (eds.): *No Truth Except in the Details.* Essays in Honor of Martin J. Klein. 1995 ISBN 0-7923-3195-8
168. J. Blackmore: *Ludwig Boltzmann, His Later Life and Philosophy, 1900–1906.* Book One: A Documentary History. 1995 ISBN 0-7923-3231-8
169. R.S. Cohen, R. Hilpinen and Q. Renzong (eds.): *Realism and Anti-Realism in the Philosophy of Science.* Beijing International Conference, 1992. 1995
 ISBN 0-7923-3233-4
170. I. Kuçuradi and R.S. Cohen (eds.): *The Concept of Knowledge.* The Ankara Seminar. 1995 ISBN 0-7923-3241-5

Boston Studies in the Philosophy of Science

171. M.A. Grodin (ed.): *Meta Medical Ethics*: The Philosophical Foundations of Bioethics. 1995 ISBN 0-7923-3344-6
172. S. Ramirez and R.S. Cohen (eds.): *Mexican Studies in the History and Philosophy of Science*. 1995 ISBN 0-7923-3462-0
173. C. Dilworth: *The Metaphysics of Science*. An Account of Modern Science in Terms of Principles, Laws and Theories. 1995 ISBN 0-7923-3693-3
174. J. Blackmore: *Ludwig Boltzmann, His Later Life and Philosophy, 1900–1906* Book Two: The Philosopher. 1995 ISBN 0-7923-3464-7
175. P. Damerow: *Abstraction and Representation*. Essays on the Cultural Evolution of Thinking. 1996 ISBN 0-7923-3816-2
176. G. Tarozzi (ed.): *Karl Popper, Philosopher of Science*. (in prep.)
177. M. Marion and R.S. Cohen (eds.): *Québec Studies in the Philosophy of Science*. Part I: Logic, Mathematics, Physics and History of Science. Essays in Honor of Hugues Leblanc. 1995 ISBN 0-7923-3559-7
178. M. Marion and R.S. Cohen (eds.): *Québec Studies in the Philosophy of Science*. Part II: Biology, Psychology, Cognitive Science and Economics. Essays in Honor of Hugues Leblanc. 1996 ISBN 0-7923-3560-0
    Set (177–178) ISBN 0-7923-3561-9
179. Fan Dainian and R.S. Cohen (eds.): *Chinese Studies in the History and Philosophy of Science and Technology*. 1995 ISBN 0-7923-3463-9
180. P. Forman and J.M. Sánchez-Ron (eds.): *National Military Establishments and the Advancement of Science and Technology*. Studies in 20th Century History. 1996 ISBN 0-7923-3541-4
181. E.J. Post: *Quantum Reprogramming*. Ensembles and Single Systems: A Two-Tier Approach to Quantum Mechanics. 1995 ISBN 0-7923-3565-1
182. A.I. Tauber (ed.): *The Elusive Synthesis: Aesthetics and Science*. 1996 ISBN 0-7923-3904-5
183. S. Sarkar (ed.): *The Philosophy and History of Molecular Biology: New Perspectives*. 1996 ISBN 0-7923-3947-9
184. J.T. Cushing, A. Fine and S. Goldstein (eds.): *Bohemian Mechanics and Quantum Theory: An Appraisal*. 1996 ISBN 0-7923-4028-0
185. K. Michalski: *Logic and Time*. An Essay on Husserl's Theory of Meaning. 1996 ISBN 0-7923-4082-5
186. G. Munévar (ed.): *Spanish Studies in the Philosophy of Science*. 1996 ISBN 0-7923-4147-3

Also of interest:
R.S. Cohen and M.W. Wartofsky (eds.): *A Portrait of Twenty-Five Years Boston Colloquia for the Philosophy of Science, 1960-1985*. 1985 ISBN Pb 90-277-1971-3
Previous volumes are still available.

KLUWER ACADEMIC PUBLISHERS – DORDRECHT / BOSTON / LONDON